Neuroexistentialism

Neuroexistentialism

Meaning, Morals, and Purpose in the Age of Neuroscience

Edited by
Gregg D. Caruso *and* Owen Flanagan

Oxford University Press is a department of the University of Oxford. It furthers
the University's objective of excellence in research, scholarship, and education
by publishing worldwide. Oxford is a registered trade mark of Oxford University
Press in the UK and certain other countries.

Published in the United States of America by Oxford University Press
198 Madison Avenue, New York, NY 10016, United States of America.

© Oxford University Press 2018

All rights reserved. No part of this publication may be reproduced, stored in
a retrieval system, or transmitted, in any form or by any means, without the
prior permission in writing of Oxford University Press, or as expressly permitted
by law, by license, or under terms agreed with the appropriate reproduction
rights organization. Inquiries concerning reproduction outside the scope of the
above should be sent to the Rights Department, Oxford University Press, at the
address above.

You must not circulate this work in any other form
and you must impose this same condition on any acquirer.

CIP data is on file at the Library of Congress
ISBN 978-0-19-046072-3 (hbk.)
ISBN 978-0-19-046073-0 (pbk.)

9 8 7 6 5 4 3

Paperback printed by Webcom, Inc., Canada
Hardback printed by Bridgeport National Bindery, Inc., United States of America

In memory of
Louis J. Caruso

CONTENTS

Preface ix
List of Contributors xi

1. Neuroexistentialism: Third-Wave Existentialism   1
   Owen Flanagan and Gregg D. Caruso

PART I: Morality, Love, and Emotion   23

2. The Impact of Social Neuroscience on Moral Philosophy   25
   Patricia Smith Churchland
3. All You Need Is Love(s): Exploring the Biological Platform of Morality   38
   Maureen Sie
4. Does Neuroscience Undermine Morality?   54
   Paul Henne and Walter Sinnott-Armstrong
5. The Neuroscience of Purpose, Meaning, and Morals   68
   Edmund T. Rolls
6. Moral Sedimentation   87
   Jesse Prinz

PART II: Autonomy, Consciousness, and the Self   109

7. Choices Without Choosers: Toward a Neuropsychologically Plausible Existentialism   111
   Neil Levy
8. Relational Authenticity   126
   Shaun Gallagher, Ben Morgan, and Naomi Rokotnitz
9. Behavior Control, Meaning, and Neuroscience   146
   Walter Glannon
10. Two Types of Libertarian Free Will Are Realized in the Human Brain   162
    Peter U. Tse

PART III: Free Will, Moral Responsibility, and Meaning in Life   191

11. Hard-Incompatibilist Existentialism: Neuroscience, Punishment, and Meaning in Life   193
    Derk Pereboom and Gregg D. Caruso
12. On Determinism and Human Responsibility   223
    Michael S. Gazzaniga
13. Free Will Skepticism, Freedom, and Criminal Behavior   235
    Farah Focquaert, Andrea L. Glenn, and Adrian Raine
14. Your Brain as the Source of Free Will Worth Wanting: Understanding Free Will in the Age of Neuroscience   251
    Eddy Nahmias
15. Humility, Free Will Beliefs, and Existential Angst: How We Got from a Preliminary Investigation to a Cautionary Tale   269
    Thomas Nadelhoffer and Jennifer Cole Wright
16. Purpose, Freedom, and the Laws of Nature   298
    Sean M. Carroll

PART IV: Neuroscience and the Law   309

17. The Neuroscience of Criminality and Our Sense of Justice: An Analysis of Recent Appellate Decisions in Criminal Cases   311
    Valerie Hardcastle
18. The Neuroscientific Non-Challenge to Meaning, Morals, and Purpose   333
    Stephen J. Morse

Index   359

PREFACE

An aim of perennial philosophy is to locate deep, satisfying answers that make sense of the human predicament, that explain what makes human life meaningful, and what grounds and makes sense of the quest to live morally. Existentialism is a philosophical expression of the anxiety that there are no deep, satisfying answers to these questions, and thus that there are no secure foundations for meaning and morality, no deep reasons that make sense of the human predicament. Existentialism says that the quest of perennial philosophy to locate firm foundations for meaning and morals is quixotic, largely a matter of tilting at windmills.

There are three kinds of existentialism that respond to three different kinds of grounding projects—grounding in God's nature, in a shared vision of the collective good, or in science. The first-wave existentialism of Kierkegaard, Dostoevsky, and Nietzsche expressed anxiety about the idea that meaning and morals are made secure because of God's omniscience and good will. The second-wave existentialism of Sartre, Camus, and de Beauvoir was a post-Holocaust response to the idea that some uplifting secular vision of the common good might serve as a foundation. Today, there is a third-wave existentialism, neuroexistentialism, which expresses the anxiety that, even as science yields the truth about human nature, it also disenchants. The theory of evolution together with advances in neuroscience remove the last vestiges of an immaterial soul or self that can know the nature of what is really true, good, and beautiful. We are gregarious social animals who evolved by descent from other animals and who are possessed of all sort of utterly contingent dispositions and features that result from having evolved as such an animal. Our fate is the fate of other animals.

This collection explores the anxiety caused by this third-wave existentialism and some responses to it. It brings together some of the world's leading philosophers, neuroscientists, cognitive scientists, and legal scholars to tackle our neuroexistentialist predicament and explore what the mind sciences can tell us about morality, love, emotion, autonomy, consciousness, selfhood, free will, moral responsibility, law, the nature of criminal punishment, meaning in life, and purpose.

The collection begins with an introduction to neuroexistentialism by Owen Flanagan and Gregg D. Caruso. This chapter sets the stage for the chapters to follow. It explains what neuroexistentialism is and how it is related to, but differs from, the first two waves of existentialism. Eighteen original chapters divided into four parts follow the introduction. There are contributions by Sean M. Carroll, Gregg D. Caruso, Patricia Smith Churchland, Farah Focquaert, Shaun Gallagher, Michael S. Gazzaniga, Walter Glannon, Andrea L. Glenn, Valerie Hardcastle, Paul Henne, Neil Levy, Ben Morgan, Stephen J. Morse, Thomas Nadelhoffer, Eddy Nahmias, Derk Pereboom, Jesse Prinz, Adrian Raine, Naomi Rokotnitz, Edmund Rolls, Maureen Sie, Walter Sinnott-Armstrong, Peter U. Tse, and Jennifer Cole Wright.

Readers will find that each chapter explores a different component of neuroexistentialism, and many draw on different traditions and disciplines. There are several chapters, for instance, that combine insights from the European traditions of existentialism and phenomenology with recent empirical work in the behavioral, cognitive, and neurosciences. There are others that draw on legal scholarship to explore the implications of neuroscience for criminal punishment and the law. Others still take an empirical approach and report here for the first time their findings. The result is a diverse collection of essays that sheds new light on the human predicament and suggests new and potentially fruitful areas of research. We hope you enjoy.

LIST OF CONTRIBUTORS

Sean M. Carroll is a Research Professor of Theoretical Physics at the California Institute of Technology. He received his PhD in 1993 from Harvard University. His research focuses on fundamental physics and cosmology, quantum gravity and spacetime, and the evolution of entropy and complexity. He is the author of *The Big Picture: On the Origins of Life, Meaning, and the Universe Itself*; *The Particle at the End of the Universe: How the Hunt for the Higgs Boson Leads Us to the Edge of a New World*; *From Eternity to Here: The Quest for the Ultimate Theory of Time*; and the textbook *Spacetime and Geometry: An Introduction to General Relativity*.

Gregg D. Caruso is Associate Professor of Philosophy at SUNY Corning and Co-Director of the Justice Without Retribution Network (JWRN) housed at the University of Aberdeen School of Law. His research focuses on free will and moral responsibility, philosophy of mind, cognitive science, and criminal punishment. He is the author of *Free Will and Consciousness: A Determinist Account of the Illusion of Free Will* (2012) and the editor of *Exploring the Illusion of Free Will and Moral Responsibility* (2013) and *Science and Religion: 5 Questions* (2014).

Patricia Smith Churchland is a Professor Emerita of Philosophy at the University of California, San Diego, and an adjunct Professor at the Salk Institute. Her research focuses on the interface between neuroscience and philosophy. She is author of the pioneering book, *Neurophilosophy* (1986), and co-author with T. J. Sejnowski of *The Computational Brain* (1992). Her current work focuses on morality and the social brain—e.g., *Braintrust: What Neuroscience Tells Us About Morality* (2011). *Touching a Nerve* (2013) portrays how to get comfortable with this fact: I am what I am because my brain is as it is. She has been president of the American Philosophical Association and the Society for Philosophy and Psychology, and she won a MacArthur Prize in 1991, the Rossi Prize for neuroscience in 2008, and the Prose Prize for science for the book *Braintrust*. She was chair of the Philosophy Department at the University of California San Diego from 2000 to 2007.

Owen Flanagan is the James B. Duke Professor of Philosophy, Duke University, Durham North Carolina. He is the author of several books, including *The Science of the Mind* (1984/1991), *Varieties of Moral Personality* (1991), *Consciousness Reconsidered* (1992), *Self Expressions: Minds, Morals and Meaning of Life* (1996), *The Problem of the Soul: Two Visions of Mind and How to Reconcile Them* (2003), *The Really Hard Problem: Meaning in a Material World* (2007), *The Bodhisattva's Brain: Buddhism Naturalized* (2011), and most recently *The Geography of Morals: Varieties of Moral Possibility* (2017).

Farah Focquaert obtained her PhD in 2007 from Ghent University in Belgium. She is a Research Fellow at the Bioethics Institute Ghent, Ghent University. Her work is predominantly situated in the domain of neuroethics and the philosophy of free will. Her current research focuses on the ethics of moral enhancement, the ethics of neuromodulation for treatment and enhancement, and the implications of the philosophy and neuroscience of free will for the criminal justice system. In the United States, she was a Visiting Research Fellow at the Center for Cognitive Neuroscience, Dartmouth College, a visiting scholar at the Department of Bioethics at the National Institutes of Health, and a visiting scholar at the Department of Criminology and the Center for Neuroscience and Society at the University of Pennsylvania. She is a Research Fellow of the Research Foundation Flanders, a member of the Belgian College of Neurological and Biological Psychiatry, and a member of the ethics committee of the Forensic Psychiatric Center Ghent.

Shaun Gallagher is the Lillian and Morrie Moss Professor of Excellence in Philosophy at the University of Memphis. His areas of research include phenomenology and the cognitive sciences, especially topics related to embodiment, self, agency and intersubjectivity, hermeneutics, and the philosophy of time. Dr. Gallagher has a secondary research appointment at the University of Wollongong, Australia, and is Honorary Professor at Durham University (UK), and the University of Tromsø, Norway. He has held visiting positions at the Cognition and Brain Sciences Unit, Cambridge University; the Center for Subjectivity Research, University of Copenhagen; the Centre de Recherche en Epistémelogie Appliquée (CREA), Paris; the Ecole Normale Supériure, Lyon; the Humboldt University in Berlin; and, most recently, at Keble College, University of Oxford. Professor Gallagher holds the Humboldt Foundation's Anneliese Maier Research Award (2012–2018). Gallagher is a founding editor and a co-editor-in-chief of the journal *Phenomenology and the Cognitive Sciences*.

Michael S. Gazzaniga is the Director of the SAGE Center for the Study of Mind at the University of California, Santa Barbara. In 1964, he received a PhD from the California Institute of Technology, where he worked under the guidance of Roger Sperry, with primary responsibility for initiating human split-brain research. In his subsequent work, he has made important advances in our understanding of functional lateralization in the brain and how the cerebral hemispheres communicate with one another. He has published many books accessible to a lay audience, such as *The Social Brain*, *Mind Matters*, *Nature's Mind*, *The Ethical Brain*, and *Who's in Charge? Free Will and the Science of the Brain*. Dr. Gazzaniga's teaching and mentoring career has included beginning and developing Centers for Cognitive Neuroscience at Cornell University Medical Center, University of California-Davis, and Dartmouth College. He founded the Cognitive Neuroscience Institute and the *Journal of Cognitive Neuroscience*, of which he is the Editor-in-Chief Emeritus. Dr. Gazzaniga is also prominent as an advisor to various institutes involved in brain research and was a member of the President's Council on Bioethics from 2001 to 2009. He is a member of the American Academy of Arts and Science, the Institute of Medicine, and the National Academy of Sciences. His new book is *Tales from Both Sides of the Brain*.

Walter Glannon is Professor of Philosophy at the University of Calgary. He is the author of *Bioethics and the Brain* (2007), *Brain, Body and Mind: Neuroethics with a Human Face* (2011), and editor of *Free Will and the Brain: Neuroscientific, Philosophical, and Legal Perspectives* (2015).

Andrea L. Glenn, PhD, is Assistant Professor in the Center for the Prevention of Youth Behavior Problems and the Department of Psychology at the University of Alabama. Her research focuses on understanding the biological correlates of psychopathy and using biological information in the development of interventions for youth with conduct problems. She is also interested in the ethical implications of this research.

Valerie Gray Hardcastle is Professor of Philosophy, Psychology, and Psychiatry and Behavioral Neuroscience at the University of Cincinnati. She co-directs the Weaver Institute for Law and Psychiatry there, as well as directs the Program in Medicine, Health, and Society. An internationally recognized scholar, she is the author of five books and more than 150 essays. She studies the nature and structure of interdisciplinary theories in the cognitive sciences and has focused primarily on developing a philosophical framework for understanding conscious phenomena responsive to neuroscientific, psychiatric, and psychological data. Currently, she is investigating the implications neuroscience should have on the criminal justice system. Hardcastle received a bachelor's degree with a double major in philosophy and political science from the University of California, Berkeley, a master's degree in philosophy from the University of Houston, and an interdisciplinary PhD in cognitive science and philosophy from the University of California, San Diego.

Paul Henne is a PhD student at Duke University. He received his bachelor's degree from Lake Forest College in 2011 and his master's degree from Arizona

State University in 2013. He primarily works on causation and causal cognition, but he also has interests in moral psychology and metaphilosophy.

Neil Levy is Professor of Philosophy at Macquarie University, Sydney, and Director of Research at the Oxford Centre for Neuroethics. He is the author, most recently, of *Consciousness and Moral Responsibility*. .

Ben Morgan is Fellow and Tutor in German at Worcester College, Associate Professor of German, and Co-Convenor of the Oxford Comparative Criticism and Translation Programme at the University of Oxford. He is author of *On Becoming God: Late Medieval Mysticism and the Modern Western Self* (2013) and articles on modernist literature, film, and philosophy in the German-speaking world (Trakl, Kafka and Kierkegaard, Benjamin and Heidegger, Fritz Lang, Leni Riefenstahl, the Frankfurt School). He is also editor with Carolin Duttlinger and Anthony Phelan of *Walter Benjamins anthropologisches Denken* (2012), and with Sowon Park and Ellen Spolsky of a Special Issue of *Poetics Today* on "Situated Cognition and the Study of Culture" (2017).

Stephen J. Morse is Ferdinand Wakeman Hubbell Professor of Law, Professor of Psychology and Law in Psychiatry, and Associate Director of the Center for Neuroscience and Society at the University of Pennsylvania. He was previously Co-Director of the MacArthur Foundation Law and Neuroscience Project and is a member of the MacArthur Foundation Law and Neuroscience Research Network. Trained in both law and psychology at Harvard, Professor Morse's work focuses on individual responsibility in criminal and civil law and social control. Dr. Morse has published *A Primer on Neuroscience and Criminal Law* (with Adina Roskies), and he has written widely about the relation of neuroscience to law and social policy. Professor Morse is a Diplomate in Forensic Psychology of the American Board of Professional Psychology, a recipient of the American Academy of Forensic Psychology's Distinguished Contribution Award, and a recipient of the American Psychiatric Association's Isaac Ray Award for distinguished contributions to forensic psychiatry and the psychiatric aspects of jurisprudence.

Thomas Nadelhoffer is an Associate Professor of Philosophy, an Affiliate Member of the Psychology Department, and a Roster Faculty in the Neuroscience Program at the College of Charleston. Before teaching at the College of Charleston, Professor Nadelhoffer spent two years as a postdoctoral fellow with the MacArthur Foundation Law and Neuroscience Project (2009–2011) under the direction of Michael Gazzaniga (UCSB) and Walter Sinnott-Armstrong (Duke University). Professor Nadelhoffer's main areas of research include free will, moral psychology, neuroethics, and punishment theory. He is particularly interested in research at the crossroads of moral and political philosophy and the sciences of the mind. Professor Nadelhoffer co-edited *Moral*

Psychology (2010) with Eddy Nahmias and Shaun Nichols. More recently, he edited *The Future of Punishment* (2013). Professor Nadelhoffer has also published nearly fifty articles and chapters in top philosophy and psychology journals and with top academic presses.

Eddy Nahmias is a Professor in the Philosophy Department and the Neuroscience Institute at Georgia State University. He specializes in philosophy of mind and cognitive science, free will, moral psychology, and experimental philosophy, focusing on questions about human agency: what it is, how it is possible, and how it accords with scientific accounts of human nature. He has published numerous chapters and articles in these areas and is coeditor of *Moral Psychology: Historical and Contemporary Readings* (2010).

Derk Pereboom is Stanford H. Taylor '50 Chair and Susan Linn Sage Professor in the Philosophy Department at Cornell University. He is the author of *Living Without Free Will* (2001), *Consciousness and the Prospects of Physicalism* (2011), *Free Will, Agency, and Meaning in Life* (2014), and of articles on free will, philosophy of mind, and in the history of modern philosophy. In his work on free will, he defends the position that we lack free will defined as the control in action required for attributions of desert, and he explores the implications of this view for rational deliberation, ethics, personal relationships, and treatment of criminals.

Jesse Prinz is a Distinguished Professor of Philosophy and Director of Interdisciplinary Science Studies at the City University of New York, Graduate Center. His research focuses on the perceptual, emotional, and cultural foundations of human psychology. He is author of *Furnishing the Mind*, *Gut Reactions*, *The Emotional Construction of Morals*, *Beyond Human Nature*, and *The Conscious Brain*. Two other books are forthcoming: *The Moral Self* and *Works of Wonder*.

Adrian Raine is the Richard Perry University Professor of Criminology, Psychiatry, and Psychology at the University of Pennsylvania. He gained his undergraduate degree in Experimental Psychology at the University of Oxford and his PhD in Psychology from the University of York. His interdisciplinary research focuses on the etiology and prevention of antisocial, violent, and psychopathic behavior in children and adults. He has published 375 journal articles and book chapters, 7 books, and given 335 invited presentations in 26 countries. His latest book, *The Anatomy of Violence* (2013), reviews the brain basis to violence and draws future implications for the punishment, prediction, and prevention of offending, as well as the neuroethical concerns surrounding this work. He is past-President of the Academy of Experimental Criminology and received an honorary degree from the University of York (UK) in 2015.

Naomi Rokotnitz is the Massada Fellowships Programme Coordinator at Worcester College, University of Oxford. Her work explores the intersections of literature, philosophy, and science. She is author of *Trusting Performance: A Cognitive Approach to Embodiment in Drama* (2011) and numerous journal articles that investigate how literature affects behavior and beliefs and influences conceptions of subjectivity, agency, and authenticity. The most recent of these are "'Passionate Reciprocity': Love, Existentialism and Bodily Knowledge in *The French Lieutenant's Woman*" (*Partial Answers: Journal of Literary History of Ideas* 12.2), and "Goosebumps, Shivers, Visualization, and Embodied Resonance in the Reading Experience: *The God of Small Things*" (*Poetics Today* 38.2).

Edmund T. Rolls is at the Oxford Centre for Computational Neuroscience, Oxford, UK, and at the Department of Computer Science, University of Warwick, UK, where he is a Professor in Computational Neuroscience and is focusing on full-time research. He has served as Professor of Experimental Psychology at the University of Oxford and as Fellow and Tutor in Psychology at Corpus Christi College, Oxford. Rolls is a neuroscientist with research interests in computational neuroscience, including the operation of real neuronal networks in the brain involved in vision, memory, attention, and decision-making; functional neuroimaging of vision, taste, olfaction, feeding, the control of appetite, memory, and emotion; neurological disorders of emotion; psychiatric disorders including schizophrenia; and the brain processes underlying consciousness. These studies include investigations in patients and are performed with the aim of contributing to understanding the human brain in health and disease and of treating its disorders. He has published twelve books on neuroscience, and more than 560 full-length research papers on these topics.

Maureen Sie is Professor at the Institute of Philosophy, Leiden University, and Associate Professor of Meta-ethics and Moral Psychology at the Erasmus University, Rotterdam. From 2009 to 2014, she led a small research group enabled by a prestigious personal grant from the Dutch Organization of Scientific Research to explore the implications of findings in the behavioral, cognitive, and neuroscience fields to our concept of moral agency. Some publications include "Moral Hypocrisy and Moral Reasons," in *Ethical Theory and Moral Practice* (18(2), 2015); "Stereotypes and Prejudices, Whose Responsibility? Personal Responsibility vis-a-vis Implicit Bias" co-authored with Voorst Vader Bours in Brownstein's *Implicit Bias and Philosophy* (2016); "The Real Neuroscientific Challenge to Free Will," co-authored with Arno Wouters, in *Trends in Cognitive Science* (12(1), 2008); and "Self-Knowledge and the Minimal Conditions of Responsibility: A Traffic-Participation View on Human (Moral) Agency," in *Journal of Value Inquiry* (48(2), 2014).

Walter Sinnott-Armstrong is Chauncey Stillman Professor of Practical Ethics in the Department of Philosophy and the Kenan Institute for Ethics at Duke University. He holds secondary appointments in Duke's Law School, Center for Cognitive Neuroscience, and Center for Interdisciplinary Decision Sciences. He is a Partner Investigator at the Oxford Centre for Neuroethics and a Research Scientist with the Mind Research Network in New Mexico. He has published widely on ethics, empirical moral psychology and neuroscience, philosophy of psychiatry, philosophy of law, epistemology, philosophy of religion, and informal logic. His current work focuses on moral psychology and brain science, uses of neuroscience in legal systems, and moral artificial intelligence. He is writing books on scrupulosity, on freedom and moral responsibility, and on how arguments can overcome political polarization.

Peter Ulric Tse has been a Professor of Cognitive Neuroscience at Dartmouth College since 2001, returning to his alma mater, where he graduated in 1984 with a degree in math and physics. He has a PhD in Cognitive Psychology from Harvard University and was a postdoc at the Max Planck Institute for Biological Cybernetics. He wrote *The Neural Basis of Free Will: Criterial Causation* (2013) and is at work on a second book on free will titled *Imagining Brains: The Neural Sources of Human Freedom and Creativity*, from which his chapter here has been adapted.

Jennifer Cole Wright is Associate Professor of Psychology at the College of Charleston. Her area of research is moral development and moral psychology more generally. Specifically, she studies meta-ethical pluralism, the influence of individual and social "liberal vs. conservative" mindsets on moral judgments, and young children's early moral development. She has published papers on these and other topics in *Cognition, Mind & Language, Journal of British Developmental Psychology, Journal of Experimental Social Psychology, Oxford Studies in Experimental Philosophy, Journal of Moral Education, Philosophical Psychology, Journal of Cognition and Culture, Personality and Individual Differences, Social Development, Personality & Social Psychology Bulletin,* and *Merrill-Palmer Quarterly*. She co-edited, with Hagop Sarkissian, *Advances in Experimental Moral Psychology*, and she is currently co-authoring a book titled *Virtue Measurement: Theory and Application* with Nancy Snow. When she's not writing, she is usually busy warping young minds in the classroom, brainstorming experiments in her lab, or satisfying her lust for travel by backpacking across Europe or South East Asia—or sometimes just trekking (with the help of a fuel-efficient car) across the United States.

CHAPTER 1
Neuroexistentialism

Third-Wave Existentialism

OWEN FLANAGAN AND GREGG D. CARUSO

Jean Paul Sartre (1946/2007) was correct when he said existentialism is a humanism. Existentialisms are responses to recognizable diminishments in the self-image of persons caused by social or political rearrangements or ruptures, and they typically involve two steps: (a) admission of the anxiety and an analysis of its causes, and (b) some sort of attempt to regain a positive, less anguished, more hopeful image of persons. What we call *neuroexistentialism* is a recent expression of existential anxiety over the nature of persons. Unlike previous existentialisms, neuroexistentialism is not caused by a problem with ecclesiastical authority, as was the existentialism represented by Kierkegaard, Dostoevsky, and Nietzsche,[1] nor by the shock of coming face to face with the moral horror of nation state actors and their citizens, as in the mid-century existentialism of Sartre and Camus.[2] Rather, neuroexistentialism is caused by the rise of the scientific authority of the human sciences and a resultant clash between the scientific and the humanistic image of persons. Specifically, neuroexistentialism is twenty-first-century anxiety over the way contemporary neuroscience helps secure in a particularly vivid

---

This chapter includes some passages from Flanagan (2002, 2009) and Flanagan and Barack (2010).

1. See Kierkegaard (1843/1983, 1843/1992, 1844/2014, 1846/1971, 1849/1998), Dostoevsky (1866/2001, 1880/1976), and Nietzsche (1882/1974, 1883/1975, 1886/1989, 1887/1969).

2. See Sartre (1943/1992, 1946/2007), Camus (1942/1989, 1942/1991), de Beauvoir (1949/1989).

way the message of Darwin from 150 years ago, that humans are animals—not half animal, not some percentage animal, not just above the animals, but 100 percent animal, one kind of primate among the 200 or so species of primates. A person is one kind of fully material being living in a material world. Neuroexistentialism is what you get when *Geisteswissenschaften* reaches the stage where it finally and self-consciously exorcizes the *geist* and recommends that no one should take seriously the Cartesian myth of the ghost in the machine (Ryle 1949/2001).

In this introduction, we explain in Section 1 what neuroexistentialism is and how it is related to two earlier existentialisms. In Section 2, we explain how neuroexistentialism makes particularly vivid the clash between the humanistic and the scientific image of persons. In Section 3, we discuss the *hard problem* (Chalmers 1996) and the *really hard problem* (Flanagan 2007) and how they relate to neuroexistentialism. In Section 4, we inquire into the causes and conditions of flourishing for material beings living in a material world, whose self-understanding includes the idea that such a world is the only kind of world that there is and thus that the meaning and significance of their lives, if there is any, must be found in such a world. We conclude in Section 5 by providing a brief summary of the chapters to follow.

1. THIRD-WAVE EXISTENTIALISM

Neuroexistentialism is the third wave of existentialism, defined here as a zeitgeist that involves a central preoccupation with human purpose and meaning accompanied by the anxiety that there is none. Aristotle's biological teleology is all about purpose—humankind, like all kinds, has a proper function (e.g., reason and virtue), which can be seen, articulated, and secured. And when you achieve it or have it you are *eudaimon*, a person who flourishes. Existentialists in the West are all post-Aristotelians who respond to the idea that eudaimonia is not enough, there should be something more, something deeper and transcendental but who are honest about the difficulty of finding where or what this deeper, transcendental thing that would make sense of life and provide meaning is or even what it could possibly be.

Traditionally, religion—specifically monotheism in the West—played the role of supplying the something more, that which would make human life more significant than, say, Aristotle thought was significance enough. In some respects, now is a time when we are "Back to Aristotle," back to a time when secularists raise the question of what life means or could mean if there is nothing more than this world, this life. Is a picture of persons as gregarious, rational, embodied, social animals who seek to flourish enough to supply content and significance to what such flourishing could come to? Can the rational, embodied image of humans give us meaning?

1.1. The First Two Waves: Foundational Anxiety and Human Nature Angst

Several centuries after the Protestant Reformation began in 1517, after much blood was spilled for religious reasons, Europe entered a secular age. Charles Taylor (1989) characterizes what it means to live in a secular age in a useful way: it is to live in an age when atheism is a real and not simply a notional possibility—which it is even biblically, for example, in the *Psalms*, where we meet "the fool." The religious wars were all between true believers. Infidels, heretics, and atheists were just monikers applied to theists who held different—but often nearby—views of God and his nature. By the Enlightenment, there were not just some people who were atheists, but some of them were very smart, thoughtful, and morally decent. Hume, Voltaire, Diderot were such people.

Dostoevsky and Kierkegaard, both religious, and Nietzsche not, lived in this secular age, and each explored in his own gripping way the anxiety wrought by entertaining the possibility that there is no God who shores up and makes sense of the human predicament. Either God as traditionally conceived is insufficient to provide grounding for the human project or he is too far away for us to comprehend his being. Nietzsche's view is of the first sort, and, of course, he famously predicts that people are too milquetoast to accept this reality and to find meaning on their own, and so, as the message gets out, an age of nihilism will commence. Similarly, when Dostoevsky allows Ivan, one of the *Brothers Karamozov* to speak of the possibility of atheism, to speak out loud about his foundational doubts, this causes his brother Dmitri to express the horrifying thought that "if there is no God then everything is permitted." Meanwhile, Kierkegaard entertains the twin thoughts that the bureaucratic Church is corrupt and that, in any case, the divine is beyond human understanding and may, at its most compelling spiritual moments, as in God's demands on Abraham, ask for actions that are inexplicable in normal ethical terms and that even require the suspension of both reason and ethics. These twin assaults on religiosity, on the existence or intelligibility of the divine, together constitute the impetus behind the first wave of existentialism.

If first-wave existentialism can be characterized as the displacement of ecclesiastical authority and a consequent anxiety over how to justify moral and personal norms without theological foundations, second-wave existentialism was a response to an overly optimistic thought that emerged from the European enlightenment. The Enlightenment offered the idea that even if there is no God, we can count on human goodness and human rationality to make sense of meaning and morals. In fact, there was hope in the aftermath of various political revolutions in the eighteenth century that reason and goodness were already leading to good democratic and egalitarian polities, which can both ground and create the conditions for true fraternity, solidarity, and liberty. But this hope was dashed almost as soon as it was expressed by such horrifying realities as the scourge of colonialism, the fact that a Christian nation led by a democratically elected demagogue produced the Holocaust, and that the egalitarian projects of Stalin, Mao, and Pol Pot were as vicious and inhumane as the religious wars and Crusades. Second-wave existentialism culminated in the aftermath of the Second World War and expressed the genuine worry that humans might simply not be up to living morally or purposefully. Sartre, Camus, de Beauvoir, and Fanon maintain glimmers of hope in various liberatory projects at the same time as they worry that the quests for meaning, equality, gender justice, and racial justice may simply require ongoing revolutionary commitment. One cannot count on either God or human nature to secure these ends.

1.2. Third-Wave Existentialism

Both first- and second-wave existentialism continue to wash over modern consciousness, even as the precise nature and degree of skepticism over ecclesiastical and political authority fluctuates. The third wave, however, comes from a different source than the first two waves—it comes from science, rather than from questioning that undermines judgments about the honesty, goodness, and authority of religious and political leaders and institutions.

Conflicts between science and religion are familiar in the West—witness Galileo Galilei and Darwin, each undermining the authority of the Churches, but also even among nonbelievers by undermining a certain humanistic picture of persons. When one combines the neo-Darwinian picture of persons with advances in neuroscience, what one increasingly sees is the recognition in public consciousness that the mind is the brain and all mental processes just *are* (or are *realized in*) neural processes.[3] For certain intellectual elites, most philosophers, and many scientists, neo-Darwinism (including genetics, population genetics, etc.) combined with neuroscience (including cognitive and affective neuroscience, neurobiology, neurology, etc.) brings the needlepoint of detail to the picture of persons anticipated by and accepted in the physicalist or naturalist view of things—which, as such, has been avowed as the right metaphysical view ever since Darwin. But, for most ordinary folk and many members of the nonscientific academy, the idea that humans are animal

---

3. The claim that the "mind is the brain" should be understood in terms of what Eddy Nahmias calls *neuronaturalism* (see Chapter 14). As he describes it: "Neuronaturalism . . . is meant to be compatible with various forms of physicalism in philosophy of mind, including both non-reductive and reductive varieties (Stoljar 2009)." For instance, neuronaturalism does not commit one "to a reductionistic epistemological thesis that says the best explanations are always those offered by lower-level sciences (e.g., physics or neuroscience)" (Chapter 14, fn. 2).

and that the mind is the brain is destabilizing and disenchanting, quite possibly nauseating, a source of dread, fear, and trembling, sickness unto death even. Darwin's theory, on its own, has caused much dis-ease: witness the continuing debate in the United States about teaching Darwin's theory in schools without at least also teaching the allegedly equiplausible alternative(s), creationism or intelligent design. But neuroscience edges out the little space for the mind conceived as soul. And even if it does not turn out to be the case that the mind is, literally, the brain, plausible alternative views of the mind–brain relationship—such as "mind is a function of the brain" or "mind supervenes on the brain"—are no more likely to give comfort to those who wish to cling to a supernatural metaphysics. The official position of the Roman Catholic Church since the 1950s has been to accept Darwin with this caveat: When the speciation event(s) occurred that created *Homo sapiens*, then God, who had planned the whole thing, started inserting souls. This is considered a mature religious response to Darwin, but it is not. It is preposterous, and contemporary neuroscience shows why and how, every day in every way, as it removes all serious work that a soul might do—except, that is, the purported afterlife part. This scientific view results in the same feeling of drift and anchorless search for meaning that is a hallmark of all existentialisms and thereby constitutes the third wave of existentialism.

2. THE SCIENTIFIC AND MANIFEST IMAGES

Wilfrid Sellars famously wrote, "The aim of philosophy, abstractly formulated, is to understand how things in the broadest possible sense of the term hang together in the broadest possible sense of the term" (1963: 2). In this quote, we get the picture of the philosopher as a kind of synthesizer, or, if not that, one who keeps his eye on the whole so that the *Weltanschauung* of an age is not inconsistent, not fraught with incoherences. There is another image of the philosopher's vocation familiar from Socrates: the philosopher as gadfly. The two vocations can be linked up, especially since Plato's Socrates is all about the role of rational coherence and attention to destabilizing lacunae in the assumptions we make in living a good life overall.

Neuroexistentialism, like earlier existentialisms, is characterized by an anxiety arising from a clash between two or more sets of practices that contain internal to themselves certain commitments about the way things are, about metaphysics and ontology, and which are or at least seem inconsistent. The quickest way to understand the problem that is at the root of the cultural anxiety is to think once again about the conflict between the scientific image of persons and the humanistic image of persons.

The conflict between science and religion is well-known in the West. Galileo was imprisoned twice for his claim to have empirical evidence for

Copernicus's heliocentric theory and died under house arrest. Descartes suppressed *Le Monde*, his work on physics and astronomy, because of the treatment Galileo received. And Descartes's own work was put on the *Index* of the Roman Catholic Church thirteen years after his death, despite that fact that his Meditations contain two (still) famous proofs for the existence of God and three proofs for mind–body dualism, which he advertises as proofs for the immortality of the soul. The case of Darwin is the most familiar contemporary zone of this conflict, especially in America, where creationists and intelligent design advocates continue to argue about which theory is scientific and what should be funded by tax dollars and taught in schools. What the advocates of Darwin's theory of descent and modification by natural selection sometimes fail to see is that the opponents of the Darwinian view are right that there is a conflict between their antecedently held picture of persons and the one they ought epistemically to believe if Darwinians are right (i.e., if Darwin's theory is true). The stakes are extraordinarily high and pertain to how one understands oneself. The problem becomes understanding and facing directly the question of whether and how one is to find a conception of meaning and purpose for finite beings, literally animals, smart mammals, living in a material world.

Consider this list of commitments, which are typical of those who accept the humanistic picture of persons—which includes most of us. The *humanistic image* involves commitment to these beliefs:

- Free will
- Humans ≠ Animals
- Soul
- Afterlife
- Made in God's image
- Morality is transcendental
- Meaning is transcendental

The scientific image is a substantive one, not simply the negation of the humanistic image—one could read Darwin, Freud, contemporary naturalistic social science, philosophy, and neuroscience to get a feel for the positive picture—and as such it is an alternative to the humanistic image. But, for present contrastive purposes, it can be understood as denying the tenets that are constitutive of the humanistic image, and thus the *scientific image* asserts:

- No metaphysical free will
- Humans are completely animal
- No soul
- No afterlife
- Not made in God's image

- Morality is not transcendental
- Meaning is not transcendental

The scientific image is disenchanting and destabilizing for a number of familiar reasons. It denies that the mind is *res cogitans*, thinking stuff, and it denies that the mind conceived as brain could have any other fate than other smart mammals have: namely, death and decomposition.

It also rejects familiar conceptions of free will, such as the following one put forth by René Descartes in the seventeenth century:

> But the will is so free in its nature, that it can never be constrained.... And the whole action of the soul consists in this, that solely because it desires something, it causes a little gland to which it is closely united to move in a way requisite to produce the effect which relates to this desire. (Descartes 1649/1968)

And this conception held by Roderick Chisholm in the twentieth century:

> If we are responsible... then we have a prerogative which some would attribute only to God: each of us when we act, is a prime mover unmoved. In doing what we do, we cause certain things to happen, and nothing—or no one—causes us to cause those events to happen. (Chisholm 2002: 55–56)

Both of these quotes are expressing a libertarian conception of free will according to which we are capable of exercising *sui generis* kinds of agency and an unconditional ability to do otherwise. While such a conception of free will is often associated with dualistic and theistic thinking, second-wave existentialists like Sartre (no friend to theism) also embraced a libertarian conception of free will. In *Being and Nothingness* (1943/1992), Sartre rejects any and all forms of causal determinism—even the "psychological" determinism which finds the immediate causes of action and choice in the desires and beliefs of agents (see Morriston 1977). Sartre's existential freedom, or so-called radical freedom, maintains that *I* (as a responsible agent) am not simply another object in the world. As a human being, I am always open to (and engaged with) things in the world: that is what Sartre means by saying that I am a "being-for" itself (rather than a "being-in-self," which is when one allows oneself to be determined by facticity). According to Sartre, how I exist in the world is a function of my free decision to create meaning out of the facts with which I am confronted. Hence, for second-wave existentialists, the *existence of free will is disturbing* since I must take full responsibility for the meaning of the world in which I exist.

For third-wave existentialists, on the other hand, the reverse is the case: the possibility that we *lack libertarian free will* is what is disturbing and causes in us existential anxiety. As the brain sciences progress and we better understand the mechanisms that undergird human behavior, the more it becomes obvious that we lack what Tom Clark (2013) calls "soul control." There is no longer any reason to believe in a nonphysical self which controls action and is liberated from the deterministic laws of nature—a little *uncaused causer* capable of exercising counter-causal free will. While most naturalistically inclined philosophers, including most compatibilists, have long given up on the idea of soul control, eliminating such thinking from our folk psychological attitudes may not be so easy and may come at a cost for some. There is some evidence, for example, that we are "natural born" dualists (Bloom 2004) and that, at least in the United States, a majority of adults continue to believe in a nonphysical soul that governs behavior (Nadelhoffer 2014). To whatever extent, then, such dualistic thinking is present in our folk psychological and humanistic attitudes about free will and moral responsibility, it is likely to come under pressure and require some revision as the brain sciences advance and this information reaches the general public.[4]

The scientific image is also disturbing for other reasons. It maintains, for example, that the mind is the brain (see fn. 4), that humans are animals, that how things seem is not how they are, that introspection is a poor instrument for revealing how the mind works, that there is no ghost in the machine, no Cartesian theater where consciousness comes together, that our sense of self may in part be an illusion, and that the physical universe is the only universe that there is and it is causally closed. Many fear that if this is true, then it is the end of the world as we know it, or knew it under the humanistic regime or image. Neuroexistentialism is one way of expressing whatever anxiety comes from accepting the picture of myself as an animal (the Darwin part) and that my mind is my brain, my mental states are brain states (the neuro- part). Taken together, the message is that humans are 100 percent animal. One might think that that message was already available in Darwin. What does neuroscience add? It adds evidence, we might say, that Darwin's idea is true and that it is, as Daniel Dennett says, "a dangerous idea" (1995). Most people in the West still hold on to the idea that they have a nonphysical soul or mind. But as neuroscience advances, it becomes increasing clear that there is no place in the brain for *res cogitans* to be nor any work for it to do. The universe is causally closed, and the mind is the brain.

The next step, a consequence of the general undermining of the idea there is any nonphysical, nonnatural furniture in the universe, is the vertigo caused by the denial that morality, well-being, and life's meaning have anything outside

---

4. Predicting what revisions will be made is difficult. It is possible that relinquishing the humanistic idea of "soul control" and libertarian freedom will cause some to accept free will skepticism (see Pereboom and Caruso, Chapter 11). But it is also possible that some might adopt a *free-will-either-way* strategy causing them to accept compatibilism on pragmatic grounds, fearing the alternative.

the natural world to shore them up. Relinquishing the last reserve of an extra-bodily foundation for meaning and morality is the culmination of a process which started in the nineteenth century with the recognition of the inability of ecclesiastical authority to provide such a foundation and continued in the middle of the twentieth century with the rejection of the polity as such a source. If the soul does not exist, and it does not, then where do we derive our morals, our meaning, and our well-being? This problem is the "really hard problem," the special problem for those of us living in the age of brain science; of making sense of the nature, meaning, and purpose of our lives given that we are material beings living in a material world.

3. THE HARD PROBLEM AND THE REALLY HARD PROBLEM

The hard problem is ancient and turns on intuitions that, for centuries and across many different traditions, support dualism. Mind seems nonphysical, so it is. It is simply too hard to explain how agency, as it seems from the first-person perspective, could be analyzed as, or reduced to, physical processes. Here, the idea is that it is too hard to imagine how we could reduce mind to brain, so we can't. Thus we need metaphysical dualism.

In recent decades, as the physicalist view of the universe extends its reach to persons, and, despite dualist intuitions, mind-science advances under the guidance of the regulative idea that the mind is the brain, the intuition returns in two guises. First, there is the old intuition that mental events don't seem like brain events, followed by disbelief at the idea that some think they might be or in fact are brain events. So we are asked to wonder: How is consciousness possible in a material world? How could subjective experience arise/emerge from brain tissue? How could subjectivity arise from objective physical states of affairs? The questions are supposed to strike the audience as eternally bewildering and thus as questions that show that physicalism is not a view that we can really comprehend. Second, there is the intuition that, even if mental events are brain events, our concepts of the mental cannot be mapped onto or reduced to physical concepts, and this perhaps because mental concepts carry connotations of nonphysicality. Fair enough, but this conceptual problem is not a metaphysical problem. The morning star is the evening star, and it is not a star but, in fact, the planet Venus. All three concepts refer to the same heavenly body, but they mean different things. If my poem says that your eyes are like the morning star, I cannot replace those words with "evening star" and get the same meaning. So what? This explanatory or conceptual gap problem is commonplace when we are learning a new way of speaking. The various difficulties associated with treating the hard problem are to be expected when major conceptual change is called for, as it is by the scientific image of persons. From the perspective of the scientific image, the question of how subjectivity is realized in persons with brains is a problem for the human sciences, most especially neuroscience.

Assuming that the details of the answer to the question of how consciousness is realized is to be given, and is already being given, by neuroscience, a second problem remains—the really hard problem (see Flanagan 2007). It can be stated in these more or less equivalent forms: How—given that we are natural beings living in a material world and given that consciousness is a natural phenomenon—does human life mean anything? What significance, if any, does living our kind of conscious life have?

The really hard problem can be put more forcefully, in a way that enhances the already felt anxiety: Is there anything upbeat and truthful we can say in this post-Darwinian age about the meaning of life or about the meaning(s) of lives given that:

- We are short-lived animals.
- When we are gone, we are gone for good (i.e., forever).
- Even our species is likely to be short-lived, certainly not eternal.

One difference between the hard problem of consciousness and the really hard problem of meaning in a material world is that the first is a problem in science, whereas the second is a problem about how we humans can best understand our situation. Given that we are material beings living in a material world and given that we have every reason to believe that there is only this one life and then we are gone, gone for good, gone for all eternity, why and how does anything matter? This is a question that we are asked to answer with only the resources available, given a materialistic picture of things, but it is not itself a purely scientific question. It asks us what attitude, what philosophical attitude, we ought to adopt given what we think to be the true facts about our situation, our predicament.

4. THE NATURALISTS' RESPONSE TO THE NEUROEXISTENTIALIST PREDICAMENT

Historically, answers to questions of value and meaning were answered metaphysically and/or theologically. The humanistic image insists that humans are not animals, the mind is not the brain, and that meaning and morals need to be grounded—propped up—transcendentally. The scientific image says that humans are animals, the mind is the brain, and that there are no transcendental sources for meaning and morals. What there is, and all there is, is the natural world. Neuroexistentialism involves an acknowledgment of this conflict and a recognition of the anxiety it creates. It also involves an attempt to regain a positive, less anguished, more hopeful image of persons. While the contributors to this volume will likely disagree on the exact nature of that positive response, all share a fundamental commitment to naturalism and all hold that a proper response to our neuroexistentialist predicament should draw on insights from the behavioral, cognitive, and neurosciences.

During the Enlightenment, we saw the beginning of a movement toward naturalism, according to which morals and meaning are to be analyzed and understood psychologically—really in terms of history and the other human sciences more broadly, not metaphysically or theologically. Over the past few centuries, this movement has continued, and, most recently, we have seen the rise of moral psychology and other interdisciplinary attempts to understand moral development and human values, norms, judgments, and attitudes naturalistically. Contemporary moral psychology, for example, is methodologically pluralistic: it aims to answer philosophical questions about competing ethical perspectives, the structure of character, and/or the nature of moral reasoning, but in an empirically responsible way (see Doris and Stich 2006; Flanagan 1991, 2017). There is, in such an approach, a fundamental commitment to naturalism and the belief that moral philosophy should pay more attention to psychology and philosophy of mind (Flanagan 1991, 2017; Harman 2009).

If mind, morals, and the meaning of life are to be understood as problems inside the naturalistic view of things, not problems that require transcendental sources, then this three-part question arises: (1) How do we combine and harness the growing knowledge and insights of the human sciences with (2) the universal existential concern with meaning and flourishing in order to yield (3) a truthful, liberating, enlightened picture of our problems and our prospects as meaning-finders and meaning-makers. Understood this way, the central question becomes: Are there naturalistic resources that can quell the anxiety produced by the ascendancy of the scientific image generally and, specifically, the picture that comes from combining neo-Darwinism with neuroscience, which produces the new and nerve-racking anxiety associated with neuroexistentialism?

One promising approach is to pursue a kind of descriptive-normative inquiry into the causes and conditions of flourishing for material beings living in a material world whose self-understanding includes the idea that such a world is the only kind of world that there is and thus that the meaning and significance of their lives, if there is any, must be found in such a world. We can call such an inquiry *eudaimonics* (Flanagan 2007, 2009). Aristotle famously said that when he asked his fellow Greeks what they want (if anything) for its own sake, not for the sake of anything else, they all answered *eudaimonia*. Eudaimonia is best translated as *flourishing* or *fulfillment*, not as happiness. There are, of course, numerous ways one could go about developing a naturalistic eudaimonics, and this collection includes several different proposals on how we may be able to achieve eudaimonia and preserve meaning, morals, and purpose in a material world. Whether or not these proposals succeed, we leave it to the reader to decide. But we can say that neuroexistentialism, at least in its constructive stage, attempts to make use of the knowledge and insights of the behavioral, cognitive, and neurosciences to satisfy our existential concerns and achieve some level of flourishing and fulfillment.

In the following chapters, some of the world's leading philosophers, neuroscientists, cognitive scientists, and legal scholars tackle our neuroexistentialist predicament and explore what the mind sciences can tell us about morality, love, emotion, autonomy, consciousness, selfhood, free will, moral responsibility, law, the nature of criminal punishment, meaning in life, and purpose. The following section provides a brief summary of the chapters to come.

5. SUMMARY OF CHAPTERS

The book is divided into four main parts: Part I, Morality, Love, and Emotion; Part II, Autonomy, Consciousness, and the Self; Part III Free Will, Moral Responsibility, and Meaning in Life; and Part IV Neuroscience and the Law. While there is some overlap among the various sections—as would be expected in a collection like this—the four parts provide a rough and fairly accurate grouping of topics, one that identifies and highlights the key existential areas of concern.

Part I begins with Patricia Churchland exploring the impact of social neuroscience on moral philosophy. One tradition in moral philosophy depicts human moral behavior as unrelated to social behavior in nonhuman animals. Morality, on this view, emerges from a uniquely human capacity to reason. By contrast, recent developments in the neuroscience of social bonding suggest instead an approach to morality that meshes with ethology and evolutionary biology. According to Churchland, the basic platform for morality is attachment and bonding and the caring behavior motivated by such attachment. Churchland argues that oxytocin, a neurohormone, is at the hub of attachment behavior in social mammals and probably birds. Not acting alone, oxytocin works with other hormones, neurotransmitters, and circuitry adaptations. Among its many roles, oxytocin decreases the stress response, making possible the trusting and cooperative interactions typical of life in social mammals. Although all social animals learn local conventions, humans are particularly adept social learners and imitators. On Churchland's account, learning local social practices depends on the reward system because, in social animals, approval brings pleasure and disapproval brings pain. Subcortical structures, she argues, are the key to acquiring social values, and quite a lot is known about how the reward system works. Acquiring social skills also involves generalizing from samples so that learned exemplars can be applied to new circumstances. Problem-solving in the social domain gives rise to ecologically relevant practices for resolving conflicts and restricting within-group competition. Churchland argues that, contrary to the conventional wisdom that explicit rules are essential to moral behavior, norms are often implicit and picked up by imitation. This hypothesis connects to a different, but currently unfashionable tradition, beginning with Aristotle's ideas about social virtues and David Hume's eighteenth-century ideas concerning "the moral sentiment."

In Chapter 3, Maureen Sie builds on Churchland's account and argues that our nature as loving beings can explain our nature as moral beings. First, she points out that scientists have discovered the brain circuits and chemistry that are involved in not only regulating male and female sexuality and feelings of attachment but also in our sociability more broadly speaking, such as how we interact with strangers. Second, love and morality seem to be similar phenomena in many ways, and some of the properties that philosophers have traditionally struggled to understand in the case of morality seem much easier to explain when love is its source. She goes on to argue that if we can make sense of the claim that "love is the source of morality," then we would have a naturalized account of morality that leaves space for a variety of philosophical views. In an attempt to develop such an account, she distinguishes several kinds of loves and explains how they relate to different moral dimensions of our existence. She takes as her starting point C. S. Lewis's work on the subject. She elaborates on this framework in relation to the claim that love is the source of morality but completely abandons his Christian framework and renames his fourth kind of love "kindness." She argues that recent findings in affective neuroscience suggest that this fourth is a natural kind of love. She discusses the dynamics of Lewis's account, showing that each of the loves that he distinguishes requires the fourth love (kindness) to keep them from taking a nasty turn. She concludes by explaining why the fourth love that Lewis distinguishes actually fits the naturalist picture quite well if the recent finding that oxytocin is involved in our trusting interactions with strangers is correct.

In Chapter 4, Paul Henne and Walter Sinnot-Armstrong explore whether neuroscience undermines morality. Recent findings in neuroscience and psychology suggest that many kinds of moral judgments are deeply flawed—they are emotional, inconsistent, based on our distant evolutionary past, susceptible to racial and gender biases, and so on. Henne and Sinnot-Armstrong distinguish, analyze, and assess the main arguments for neuroscientific skepticism about morality and argue that neuroscience does not undermine *all* of our moral judgments. After quickly addressing several skeptical challenges, they focus the majority of their attention on one argument in particular—the idea that neuroscience and psychology might undermine moral knowledge by showing that our moral beliefs result from unreliable processes. They argue that the background arguments that are needed to bolster the main premise fail to support it in the way that is required for the argument to succeed. They conclude that the overall issue of neuroscience undermining morality is unsettled—we need more scientific research and philosophical reflection on this topic. Still, they contend, we can reach some tentative and qualified conclusions. First, neuroscience and psychology do not undermine all moral judgments as such, but they still might play an ancillary role in an argument that undermines some moral judgments. Second, they might lead us to think about moral judgments in new ways, such as by suggesting new divisions among moral judgments. Neuroscience is, then, "not a general underminer—but a trimmer and a categorizer." In these ways, "neuroscience can play a constructive role in moral theory, although not by itself. In order to make progress, neuroscience and normative moral theory must work together."

In Chapter 5, Edmund T. Rolls builds on evidence and theories he developed elsewhere about the neural base of emotions and explores what they can tell us about purpose, meaning, and morals. He begins by noting that one process to which "purpose" can refer is that genes are self-replicating. Another process to which "purpose" can apply, he contends, is that genes set some of the goals for actions. These goals are fundamental to understanding emotion. Another process to which "purpose" can apply is that syntactic multistep reason provides a route for goals to be set that are to the advantage of the individual, of the phenotype, and not of the genes. He proceeds to argue that meaning can be achieved by neural representations not only if these representations have mutual information with objects and events in the world, but also by virtue of the goals of the "selfish" genes and of the individual reasoner. This, he proposes, provides a means for even symbolic representations to be grounded in the world. He concludes by arguing that morals can be considered as principles that are underpinned by (the sometimes different) biological goals specified by the genes and by the reasoning (rational) system. Given that what is "natural" does not correspond to what is "right," he argues that these conflicts within and between individuals can be addressed by a social contract.

Jesse Prinz concludes Part I with his chapter on *moral sedimentation*. He begins by noting that existentialism is often regarded as a philosophy of radical freedom—that is, leading existentialists emphasized the human capacity for choice and self-creation. At the same time, there is a countercurrent in existentialist thought that calls freedom into question. This countercurrent draws attention to the ways in which behavior is determined by forces outside of our control. This is especially vivid in the moral domain. Prinz, for instances, explains that beginning with Nietzsche's claim that Christians are self-deceived and extending through feminist and decolonial perspectives within postwar existentialism, we find key authors pointing to ways in which deeply held values get shaped by social forces. Borrowing a term from phenomenology, Prinz calls this phenomenon "sedimentation." After tracing the idea of sedimentation and related concepts in existentialist thought, with special emphasis on the moral domain, Prinz argues that recent work in neuroscience, psychology, and other social sciences adds support to the thesis that we are vulnerable to sedimentation. He concludes by considering various tactics against sedimentation that have been proposed, arguing that some of the more prominent historical tactics are problematic while also pointing to some alternatives.

Part II begins with Neil Levy's chapter on "Choices Without Choosers: Toward a Neuropsychologically Plausible Existentialism." While existentialists are often accused of having painted a bleak picture of human existence, Levy contends that, in the light of contemporary cognitive science, there are grounds for thinking that the picture is not bleak enough. For second-wave existentialists, we live in a meaningless universe, condemned to be free to choose our own values, which have no justification beyond the fact that we have chosen them. But second-wave existentialists remained confident that there was someone, an agent, who could be the locus of the choice we each confront. Contemporary cognitive science shakes our faith even in the existence of this agent. Instead, it provides evidence that seems to indicate that there is no one to choose values; rather, each of us is a motley of different mechanisms and processes, each of which lack the intelligence to confront big existential questions and each pulling in a different direction. According to Levy, while there are grounds for thinking that the picture is in some ways bleaker than the existentialists suggested, he argues that it is not hopeless. The unified self that serves as the ultimate source of value in an otherwise meaningless universe may not exist, but we can each impose a degree of unity on ourselves. The existentialists were sociologically naïve in supposing a degree of distinction between agents and their cultural milieu that was never realistic. Agents are enculturated, and a realistic existentialist will recognize that. But they will also recognize that we are embodied and embedded agents: a biologically realistic picture will understand us as agents always already in process of unification but never achieving it, and always already in negotiation with values rather than choosing them. We are thrown beings: thrown into history, into culture, and into a biological and evolutionary history which we never fully understand and which we can do no more than inflect, all without foundations and lacking even the security of knowing the extent to which we choose or even what we choose. Existentialism must face up to an insecurity that is ontological and epistemological as much as it is axiological.

In Chapter 8, Shaun Gallagher, Ben Morgan, and Naomi Rokotnitz explore the notion of *relational authenticity*. They argue that to understand existential authenticity it will not do to return to the individuality celebrated by classical existentialism. Nor is it right to look for a reductionist explanation in terms of neuronal patterns or mental representations that would simply opt for a more severe methodological individualism and a conception of authenticity confined to proper brain processes. Rather, they propose, we should look for a fuller picture of authenticity in what they call the "4Es"—the embodied, embedded, enactive, and extended conception of mind. They argue that one requires the

4Es to maintain the 4Ms—mind, meaning, morals, and modality—in the face of reductionistic tendencies in neurophilosophy. The 4E approach, they contend, gives due consideration to the importance of the brain taken as part of the brain-body-environment system. It incorporates neuroscience in its explanations, but it also integrates important phenomenological-existentialist conceptions that emphasize embodiment (especially following the work of Merleau-Ponty) and the social environment. More specifically, they argue that phenomenological conceptions of intersubjectivity, or, in existentialist terms, being-with (*Mitsein*) and being-for-others, should play significant roles in our rethinking of authenticity.

In Chapter 9, Walter Glannon writes: "The existential angst of neuroscience is not the result of having to choose in the absence of religious or cultural models. Rather, the angst results from the idea that the subjectivity and conscious choice that presumably define us as persons can be completely explained—if not explained away—by neural and psychological factors to which we have no access." Neuroscience challenges our beliefs about agency and autonomy because it seems to imply that as conscious beings we have no control of our behavior. Most brain processes, for instance, are not transparent to us. We also have no direct access to the efferent system and only experience the sensorimotor consequences of our unconscious motor plans. Nevertheless, Glannon argues that the fact that unconscious processes drive many of our actions does not imply that conscious mental states have no causal role in our behavior and that we have no control over it. He argues that some degree of unconscious neural constraint on our conscious mental states is necessary to modulate thought and action and promote flexible behavior and adaptability to the demands of the environment. He maintains that a nonreductive materialist account of the mind–brain relation makes it plausible to claim that mental states can cause changes in physical states of the brain. He examines some psychiatric and neurological disorders and attempts to shows how the conscious mind can have a causal role in the etiology of these disorders as well as in therapies to control them and behavior more generally. He argues that lower level unconscious neural functions and higher level conscious mental functions complement each other in a constant process of bottom-up and top-down circular causal feedback that enables interaction between the organism and the external world. He concludes that the motivational states behind our actions and the meaning we attribute to them cannot be explained entirely by appeal to neural mechanisms. Although the brain generates and sustains our mental states, he argues that it does not determine them and leaves enough room for individuals to "will themselves to be" through their choices and actions.

In Chapter 10, Peter U. Tse describes various developments in neuroscience that reveal how volitional mental events can be causal within a physicalist paradigm and argues that two types of libertarian free will are realized in the human brain. He begins by attacking the logic of Jaegwon Kim's exclusion argument, which he specifies as maintaining that mental information cannot be causal and must be epiphenomenal because particle-level physical-on-physical causation is sufficient to account for apparent causation at all higher levels. Tse maintains that the exclusion argument falls apart if indeterminism is the case. He then proceeds to build an account of how mental events are causal in the brain. He takes as his foundation a new understanding of the neural code that emphasizes rapid synaptic resetting over the traditional emphasis on neural spiking. Such a neural code is an instance of "criterial causation," which requires modifying standard interventionist conceptions of causation. Tse argue that a synaptic reweighting neural code provides a physical mechanism that accomplishes downward information causation, a middle path between determinism and randomness, and a way for mind/brain events to turn out otherwise. This new view of the neural code, Tse argues, also provides a way out of self-causation arguments against the possibility of mental causation. Finally, Tse maintains that it is not enough to simply have "first-order free will." That is, only if present choices can ultimately lead to a chooser becoming a new kind of chooser—that is, only if there is a second-order free will or *meta-free will*—do brains have the capacity to both have chosen otherwise and to have meta-chosen otherwise. Tse concludes by discussing how the brain can choose to become a new kind of brain in the future, with new choices open to it than are open to it now.

Part III begins with Derk Pereboom and Gregg D. Caruso's chapter on *hard-incompatibilist existentialism*. In it, they explore the practical and existential implications of free will skepticism, focusing primarily on punishment, morality, and meaning in life. They begin by considering two different routes to free will skepticism. The first denies the causal efficacy of the types of willing required for free will and receives its contemporary impetus from pioneering work in neuroscience by Benjamin Libet, Daniel Wegner, and John-Dylan Haynes. The second, which is more common in the philosophical literature, does not deny the causal efficacy of the will but instead claims that, whether this causal efficacy is deterministic or indeterministic, it does not achieve the level of control to count as free will by the standards of the historical debate. They argue that while there are compelling objections to the first route, the second route to free will skepticism remains intact. They then go on to argue that free will skepticism allows for a workable morality and, rather than negatively impacting our personal relationships and meaning in life, may well improve our well-being and our relationships to others since it would tend to eradicate an often destructive form of moral anger. They conclude by arguing that free will skepticism allows for adequate ways of responding to criminal behavior—in particular, incapacitation, rehabilitation, and alternation of relevant social conditions—and that these methods are both morally justified and sufficient for good social policy. They present and defend their nonretributive alternative—the quarantine model, which is an incapacitation account built on the right to self-protection analogous to the justification for quarantine—and respond to recent objections to it by Michael Corrado, John Lemos, and Saul Smilansky.

In Chapter 12, Michael Gazzaniga tells us: "Let's face it. We are big animals with brains that carry out every single action automatically and outside our ability to describe how it works. We are a soup of dispositions controlled by genetic mechanisms, some weakly and some strongly expressed in each of us." Yet, he tells us there is some good news too: "We humans have something called the *interpreter*, located in our left brain, that weaves a story about why we feel and act the way we do. That becomes our narrative, and each story is unique and full of sparkle." He wonders, what's wrong with being that—just that? After all, being self-aware narrators is what brains do. Gazzaniga proceeds to explore the concepts of free will and moral responsibility in light of such facts, arguing that we all remain personally responsible for our actions because responsibility arises out of each person's interaction with the social layer we are embedded in. "Responsibility is not to be found in the brain," he concludes, rather it is "a needed consequence of more than one individual interacting with another."

In Chapter 13, Farah Focquaert, Andrea L. Glenn, and Adrian Raine return to the issue of free will skepticism and criminal behavior. They ask how we should, as a society, deal with criminal behavior in the current era of neuroexistentialism. They further ask if our belief in free will is essential to adequately addressing criminal behavior or if neurocriminology could offer a new way of addressing crime without the need to resort to backward-looking notions of *moral* responsibility and guilt. They begin by noting that the kind of free will that could justify retributive punishment based on a criminal's *moral* responsibility needs to be the "ultimate" kind—the kind which would allow an individual to behave differently given the exact same conditions. According to free will skepticism, however, we are not free in the sense that is required for *moral* responsibility (i.e., the basic desert sense), and we therefore lack the responsibility that is needed to justify any kind of punishment that draws on revenge or desert. They proceed to argue that what does remain is "moral answerability" and forward-looking claims of responsibility that focus on the moral betterment or moral enhancement of individuals who are prone to criminal behavior and on the realization of reparative measures toward victims. They go on to present a neurocriminology approach to criminal behavior and critically discuss the potential benefits and risks that may accompany such an approach. They argue that, whereas mass incarceration, severe sanctions, and stigmatization have resulted in more recidivism, adequate treatment programs that focus on increasing an individual's capacity to better control and change his future behavior have been linked to less recidivism. Such an approach can be placed within a broader public health perspective of human behavior and addresses both environmental and neurobiological risk factors of criminal behavior. Within this framework, neurocriminology approaches to criminal behavior may provide specific guidance within a broader moral enhancement framework. Hence, rather than undermining our current criminal justice practices, the free will skeptics' approach can draw on neurocriminological findings to reduce immoral behavior.

In Chapter 14, Eddy Nahmias defends a compatibilist account of free will and attempts to understand free will in the age of neuroscience. He begins by considering various reactions one could have to *neuronaturalism*—the thesis that, in imagining options, evaluating them, and making a decision, "each of those mental processes just *is* (or is *realized in*) a complex set of neural processes which causally interact in accord with the laws of nature." He diagnoses the different reactions one could have to this thesis and argues that the "natural reaction"—one that accepts neuronaturalism in stride and without any accompanying existential angst—is both common and correct. Focusing on free will, he offers reasons to think that a neuronaturalistic understanding of human nature does not take away the ground (or grounding) that supports most of our cherished beliefs about ourselves. While dualists and reductionists tend to think neuronaturalism conflicts with people's self-conception, Nahmias argues that most people are "theory-lite" and amenable to whatever metaphysics makes sense of what matters to them. He argues that even though we do not yet have a theory of how neural activity can explain our conscious experiences, such a theory will have to make sense of how those neural processes are crucial causes of our decisions about what to do. He concludes by suggesting that interventionist theories of causation offer the best way to see this.

In Chapter 15, Thomas Nadelhoffer and Jennifer Cole Wright investigate the relationship between free will beliefs (or the lack thereof) and existential anxiety. In an attempt to shed light on this relationship, they set out to test whether trait humility can serve as a "buffer" between the two—that is, are people who are high in dispositional humility less likely to experience existential anxiety in the face of skepticism about free will? Given the perspectival and attitudinal nature of humility, Nadelhoffer and Wright predict that humble people will be less anxious in the face of stories about the purported death of free will (or the reduction of the mind to the brain). In a series of four studies, they tested their hypothesis using various scales (e.g., the Free Will Inventory, the Humility Scale, the Existential Anxiety Questionnaire, the Existential Anxiety Scale, etc.) and primes designed to manipulate belief in free will. While they found some correlational support in Study 1 for their buffering hypothesis, their efforts were less successful than they had hoped since they were unable to push people's beliefs in free will sufficiently in Studies 2–4 to test the hypothesis further. This failure itself is instructional, however, since it tells us something important about the current use of primes in studies designed to manipulate people's belief in free will (usually to measure their pro- or antisocial effects). In this respect, they write, "our work should serve as a cautionary tale for philosophers, psychologists, and pundits who want to discuss the potential ramifications of the supposed death of free will. For while it is certainly possible for people to change their minds about free will, it is not clear that researchers have figured out effective, reliable, and stable methods for bringing these epistemic changes about (even temporarily)."

Physicist Sean M. Carroll closes out Part III with his chapter on purpose, freedom, and the laws of nature. He notes that the popular image of existentialism is associated with "philosophers sitting in cafes, smoking cigarettes and drinking apricot cocktails" and that this is at odds with the popular image of scientists decked out in lab coats. Despite these stereotypes, Carroll maintains that there is an undeniable connection between existentialism and science. This is perhaps easy to see with biology and neuroscience, but the connection goes beyond this. Carroll maintains that "An honest grappling with the questions of purpose and freedom in the universe must also involve ideas from physics and cosmology." He goes on to argue that if we want to create purpose and meaning at the scale of individual human lives, it behooves us to understand the nature of the larger universe of which we are a part. After discussing what modern physics can tell us about determinism, quantum mechanics, the arrow of time, and emergence, Carroll concludes by exploring the existential implications of these insights for freedom and meaning.

Part IV begins with Valerie Hardcastle's chapter on the neuroscience of criminality and our sense of justice. Taking the US courts as her stalking horse, Hardcastle analyzes appellate cases from the past five years in which a brain scan was cited as a consideration in the decision. After describing the methodology of her study, she presents the results of her analysis, focusing on how a defendant's race might be correlated with whether a defendant is able to get a brain scan, whether the scan is admitted into evidence, how the scan is used in the trial, and whether the scan changes the outcome of the hearing. Although she cautions against drawing any definitive conclusions until more studies are conducted, she identifies a trend indicating that brain scans of African-American defendants were less likely to be mitigating when used as evidence in court. She suggests one possible explanation for this that draws on Mark Alicke's culpable control model of blame (Alicke 2000, 2008) and recent work on implicit bias. She then provides a comparative analysis of the cases in which imaging data were successful in altering the sentence of defendants and those in which the data were unsuccessful. She concludes by pointing to larger trends in our criminal justice system indicative of more profound changes in how we as a society understand what counts as a just punishment.

The collection concludes in Chapter 18 with Stephen J. Morse arguing that neuroscience, for all its astonishing recent discoveries, raises no new challenges for the existence, source, and content of meaning, morals, and purpose in human life, nor for the robust conceptions of agency and autonomy that underpin law and responsibility. According to Morse, proponents of using the new neuroscience to revolutionize the law and legal system, especially criminal law, make two arguments. The first appeals to determinism and the specter of the person as simply a "victim of neuronal circumstances" (VNC) or "just a pack of neurons" (PON)—included here are those who argue that determinism and/or VNC/PON are inconsistent with responsibility. The second are those who defend "hard incompatibilism" (HI) (e.g., Pereboom and Caruso, in Chapter 11). Morse begins by reviewing the law's psychology, concept of personhood, and criteria for criminal responsibility. He then argues that neither determinism nor VNC/PON are new to neuroscience and that neither, at present, justifies revolutionary abandonment of moral and legal concepts and practices that have been evolving for centuries in both common law and civil law countries. He then turns to HI and argues that, although the metaphysical premises for responsibility or jettisoning it cannot be decisively resolved, the real issue should be the type of world we want to live in. He concludes by examining Pereboom and Caruso's quarantine proposal (Chapter 11) and argues that the hard incompatibilist vision is not normatively desirable, even it if is somehow achievable.

REFERENCES

Alicke, M. D. 2000. Culpable control and the psychology of blame. *Psychological Bulletin* 126: 556–574.
Alicke, M. D. 2008. Blaming badly. *Journal of Cognition and Culture* 8: 179–186.
Bloom, P. 2004. *Descartes' baby*. New York: Basic Books.
Camus, A. 1942/1989. *The stranger*. Tr. Matthew Ward. New York: Vintage.
Camus, A. 1942/1991. *The myth of Sisyphus*. Tr. Justin O'Brien. New York: Vintage.
Chalmers, D. 1996. *The conscious mind: In search of a fundamental theory*. New York: Oxford University Press.
Chisholm, R. 2002. Human freedom and the self. In R. Kane (Ed.), *Free will*, pp. 47–57. Malden, MA: Blackwell.
Clark, T. 2013. Experience and autonomy: Why consciousness does and doesn't matter. In G. D. Caruso (Ed.), *Exploring the illusion of free will and moral responsibility*, pp. 239–254. Lanham, MD: Lexington Books.
de Beauvoir, S. 1949/1989. *The second sex*. Tr. H. M. Parshley. New York: Vintage Books.
Dennett, D. C. 1995. *Darwin's dangerous idea: Evolution and meaning in life*. New York: Simon & Schuster.
Descartes, R. 1649/1968. *Passions of the soul*. In E. Haldane and G. Ross (Eds.), *The philosophical works of Descartes*. Vol. 1. Cambridge: Cambridge University Press.
Doris, J., and S. Stich. 2006. Moral psychology: Empirical approaches. *Stanford encyclopedia of philosophy*. Accessed online: https://plato.stanford.edu/entries/moral-psych-emp/
Dostoevsky, F. 1866/2001. *Crime and punishment*. Tr. Constance Garnett. New York: Dover.

Dostoevsky, F. 1880/1976. *The brothers Karamazov: The Constance Garnett translation revised by Ralph E. Matlaw*. New York: Norton.

Flanagan, O. 1991. *Varieties of moral personality: Ethics and psychological realism*. Cambridge, MA: Harvard University Press.

Flanagan, O. 2002. *The problem of the soul: Two visions of the mind and how to reconcile them*. New York: Basic Books.

Flanagan, O. 2007. *The really hard problem: Meaning in a material world*. Cambridge, MA: MIT Press.

Flanagan, O. 2009. One enchanted being: Neuro-existentialism and meaning. *Zygon: Journal of Science and Religion* 44(1): 41–49.

Flanagan, O. 2017. *The geography of morals: Varieties of moral possibility*. New York: Oxford University Press.

Flanagan, O., and D. Barack. 2010. Neuroexistentialism. *EurAmerica* 40(3): 573–590.

Harman, G. 2009. 1st Shearman lecture: Naturalism in moral philosophy. Delivered at Princeton University on May 19, 2009. Accessed online: https://www.princeton.edu/~harman/Papers/Naturalism.pdf

Kierkegaard, S. 1843/1983. *Fear and trembling*. Tr. Howard V. Hong and Edna H. Hong. Princeton: Princeton University Press.

Kierkegaard, S. 1843/1992. *Either/or*. Tr. Alastair Hannay. New York: Liveright.

Kierkegaard, S. 1844/2014. *The concept of anxiety: A simple psychologically oriented deliberation in view of the dogmatic problem of hereditary sin*. Tr. Alastair Hannay. New York: Liveright.

Kierkegaard, S. 1846/1971. *Concluding unscientific postscript*. Tr. David F. Swenson and Walter Lowrie. Princeton: Princeton University Press.

Kierkegaard, S. 1849/1998. *The sickness unto death*. Tr. Alastair Hannay. New York: Liveright.

Morriston, W. 1977. Freedom, determinism, and chance in the early philosophy of Sartre. *The Personalist* 58: 236–248.

Nadelhoffer, T. 2014. Dualism, libertarianism, and scientific skepticism about free will. In W. Sinnott-Armstrong (Ed.), *Moral psychology: Neuroscience, free will, and responsibility*, vol. 4, pp. 209–216. Cambridge, MA: MIT Press.

Nietzsche, F. 1882/1974. *The gay science*. Tr. Walter Kaufmann. New York: Vintage Books.

Nietzsche, F. 1883/1975. *Thus spoke Zarathustra*. In *The portable Nietzsche*. Tr. Walter Kaufmann. New York: Viking Press.

Nietzsche, F. 1886/1989. *Beyond good and evil*. Tr. Walter Kaufman. New York: Vintage Books.

Nietzsche, F. 1887/1969. *On the genealogy of morals*. Tr. Walter Kaufmann. New York: Vintage Books.

Ryle, G. 1949/2001. *The concept of mind (with introduction by Daniel C. Dennett)*. Chicago: University of Chicago Press.

Sartre, J.-P. 1943/1992. *Being and nothingness*. Tr. Hazel Barnes. New York: Washington Square Press.

Sartre, J.-P. 1946/2007. *Existentialism is a humanism*. Tr. Carol Macomber. New Haven: Yale University Press.

Sellars, W. 1963. *Science, perception, and reality*. London: Humanities Press.

Taylor, C. 1989. *Source of the self: The making of the modern identity*. Cambridge, MA: Harvard University Press.

PART I
Morality, Love, and Emotion

CHAPTER 2

The Impact of Social Neuroscience on Moral Philosophy

PATRICIA SMITH CHURCHLAND

An abiding puzzle is where moral motivation comes from.[1] Philosophers have long wanted to understand why we are not overwhelmingly and always completely selfish. You see me stumble on the staircase coming up from the subway. You, and strangers nearby, will probably be highly motivated to help. Why do you care about my plight? Why do we donate blood? It is neither fun nor lucrative. Why do we care about the fate of elephants?

Moral behavior comes in all sizes—from minor—for example, helping a short person put his luggage in an overhead bin—to momentous and far-reaching, such as traveling to Africa to treat patients infected with Ebola virus. This action is momentous because you, too, could catch the disease and die. Some doctors did, and some died.

In this context, by "moral motivation" I shall mean the willingness to incur a cost to oneself in order to benefit another. Imperfect though it is, that provisional characterization will suffice for the main points I wish to make here. Sometimes the costs incurred will be in terms of resources such as money or food, or time and energy, but they can also include opportunities foregone and reputational costs. Edward Snowden, who did what he did out of moral conviction, will likely never be allowed to return home and is despised by many Americans. The fact is, moral values and moral motivation loom large in human decision-making and behavior. The question is this: "Why?"

---

1. The central points in this chapter are drawn from my book, *Braintrust* (Churchland 2011). See also my paper, "The Neurobiological Platform for Moral Values" (2014).

1. VALUES IN THE NEURONAL CIRCUITRY

It is well understood that self-oriented values are built into the very structure of every nervous system. All animals, from worms to whales, are wired to be motivated to care for their own survival and well-being. If an animal were to lack self-regarding circuitry, it would not last long—it would become someone else's dinner. So, thanks to biological evolution, self-care is a basic value built right into the very structure of the brain. The prominent aspects of this circuitry in the brainstem of mammals are quite well understood, though many details are not yet nailed down. But caring for others? Sacrificing one's own interests for that of others? Where can that come from if biological evolution is driven by "me" and "my genes?"

Evolutionary biologists have tended to frame this question in terms of genes for altruism. The setting goes like this: if "altruism" genes happened to appear in some animals, they would not spread throughout the population. That is because the other selfish animals in the neighborhood would take advantage and out-compete the altruistic animals. They would not survive long enough to reproduce and spread their altruistic genes. One hypothesis, therefore, is that altruism probably is not a product of evolution; morality must be taught (Dawkins 1976).

Two major problems thwart this hypothesis. First, studies of preverbal infants as young as 3–6 months of age indicate that, by and large, the subjects are well able to distinguish between nice and nasty puppets and to recognize the difference between helpful puppets and obstructing puppets. They tend to prefer the helpful ones (Hamlin, Wynn, and Bloom 2007). There is no evidence that these infants have been grilled on the desirability of kindness over unkindness or, even if they had, that they would have understood. Much remains to be done to assess the significance of these studies, but, for present purposes, they do at the very least suggest that it might be useful to consider what infants early in their development have by way of moral sense.

The second problem concerns nonhuman mammals and birds. Some mammals, such as marmoset monkeys, are intensely social, the males taking a major role in the caring for the offspring and even alloparenting orphan infants. This has also been reported in five cases involving orphan chimpanzees adopted by males who are not their biological fathers.

Frans deWaal and colleagues have documented consolation behavior in chimpanzees, where one animal may console by touching and grooming another who has been defeated in a squabble, for example (deWaal 1989; DeWaal and van Roosmalen 1979). DeWaal has also documented prosocial choice in monkeys (deWaal, Leimgruber, and Greenberg 2008). In this experiment, one monkey is given two food items, and he can share one with a friend or keep them both. Prosocial choice is regularly, though not always, seen. These behaviors may not be as dramatic as joining Doctors Without Borders, but the important thing is that the monkey does show moral behavior suitable to a monkey—he does incur a cost to benefit his friend (deWaal 1996).

In her lab at the University of Chicago, Peggy Mason and colleagues showed altruism in rats that were cage mates (Ben-Ami Bartal, Decety, and Mason 2011). In this stunning experiment, one rat was placed in a tight, cylindrical plexiglas restrainer, with a door at one end that could be unlatched from the outside with ingenuity and effort. The trap and its rat were placed in the middle of an open space arena. A second rat was then placed in the arena. As Ben-Ami Bartal et al. explain, for the second rat to approach the trap, he has to overcome fear contagion since the trapped rat is making some squealing noises that broadcast his distress. Second, he has to overcome his fear of open spaces since rats, for safety, prefer to scurry along walls. The second rat approaches the trap, examines it thoroughly, and begins to dicker with the mechanism for opening the door. After much effort, the rat finally succeeds in unlatching the trap door, and his trapped friend escapes. Then there is some happy nuzzling. In a further test, Mason and colleagues put a cup of chocolate chips at some remove from the trapped rat. The friend enters the arena, recognizes that there is food, opens the trap, and then the two rats share the food.

Later, we shall want to consider this example of altruistic behavior of non-human animals in a wider framework of neurobiology and the evolution of nervous systems to determine whether, and how, it is relevant to addressing the origin of moral motivation.

One long-favored approach to the origin of morality sees religion as the watershed of moral values. Although popular, the approach faces a serious drawback. *Homo sapiens* have been on the planet for some 250,00 years, but organized religions, with the idea of divine law-givers and divine punishments are fairly recent—they appeared only after the advent of agriculture, about 10,000 years ago. Incidentally, reading and writing are even more recent inventions and seem to have been invented only about 5,000 years ago, implying that no leader of the Israelites could have been "given" the Ten Commandments much before Moses allegedly received them because God would have had to await the invention of Hebrew.

Hunter-gatherer groups, including those that still exist, such as the Inuit of the Arctic and the Piraha of the Amazon Basin, certainly have moral norms, but their "folk" religions do not share much with Christianity or Judaism. In particular, there is no place for God-the-Law-Giver-and-Punisher. Even so, the Inuit and the Piraha care, share, and defend each other. They have a social life. They have a range of within-tribe social norms to which individuals by and large adhere, such as norms against deceit and assault and norms regarding who decides where to hunt and what to do with poachers. Eastern religions such as Confucianism, though certainly very sophisticated morally on many topics, likewise do not espouse a divine Law-Giver. Confucius seems to have been a fairly wise man, but he made no claims about divinity. Going back much further—some two million years—it is reasonable to speculate that our ancestral hominins, such as *Homo erectus* and *Homo neanderthalis* more recently, had social arrangements that involved caring, sharing, and mutual defense.

The close links we now see between organized religion and morality that developed after the growth of stable farms and towns may be owed to the emergence of a priest class that adopted moral enforcement and moral education as its job and found a law-giving personal God to be helpful in these endeavors. A belief with these system features may have served to bind people together in a common moral framework, even when towns were large and not everyone knew everyone else (Purzycki et al. 2016). That a religion with these specific features is not essential to achieving such cohesion is suggestion by the great success of the Mongols under Ghengis Khan, the deity of whose religion was the wide open sky (Weatherford 2004). Nothing personal or law-giving in his conception of awesome power. The social institutions put in place by Ghengis Khan were remarkably sensible, including a kind of democracy and the realization that everyone needed to be able to read so that the laws were clearly understood throughout the community. Other successful cultural developments in China and India also suggest that people living in geographically wide areas can share a religion that involves no Law-Giving God. In any case, the main point is that organized religions seem not to be the fountainhead of moral motivation or basic moral norms themselves; rather, organized religions of diverse sorts seem to have built on what was already in place.

One popular tradition has it that pure reason is the source of moral decision-making as a person reflects on what it is to be genuinely human and to make genuinely moral decisions (Korsgaard 1996). In some versions, such as the Kantian one, emotions are best left out of moral decision-making. A common corollary of this view is that only humans have a faculty that we can recognize as being genuinely the faculty of "reason"; only humans reflect and have "self-government."

Notwithstanding the appeal of this neo-Kantian approach among philosophers, it is factually underpowered. Just as well-documented evidence of nonhuman altruism has provoked reconsideration of what social animals actually do without training, so well-documented examples of nonhuman animals solving difficult problems have also provoked reconsideration of the claim that only human can reason (Benson-Amram et al. 2016; Laurent and Balleine 2015; Smirnova et al. 2014; Rutz and St. Claire 2012). As we shall see in the next section, David Hume's deep insights concerning the role of moral emotions, or what he called "the moral sentiment," look plausible not only for the common observations about the emotional component of typical human moral choices, but they are also increasingly supported by what has been learned about sociality and the mammalian brain. So we return to our question: Where do moral values come from? Why are they often so powerful in generating our feelings and shaping our decisions?

2. LOOKING TO BIOLOGY

The human brain is a product of biological evolution. Every species is unique in some way relative to other species. To be sure, humans, too, are unique in some ways. In particular, we have more neurons than our closest relatives, chimpanzees. Nevertheless, we should avoid the temptation to suppose we are especially favored, or chosen, or "the best," or the "height of evolutionary progress," and so forth. When philosophers are taught that they can access a priori truth about the nature of the world, they should wonder how natural selection managed to make that possible for us. We are products of natural selection.

From everything we can tell now, the brains of *Homo sapiens* born now are much the same as the brain of *Homo sapiens* born 250,00 years ago. The genetic changes we know about concern noncognitive functions, such as changes in digestion and hair color, for example. So far as is known, there are no genetic changes that distinguish our brains from those of early *Homo sapiens*. Surprisingly perhaps, the cranial cavity in *Homo neanderthalis* is generally larger than in humans, implying that they typically had larger brains. Genetic analysis reveals that *Homo sapiens* and the Neanderthals did interbreed. Many humans of European descent, including myself, carry Neanderthal genes.

As we take pride in our rationality, one may find it sobering to note that, for about their first 100,000 years on the planet, *Homo sapiens* used only stone tools, without advancing even to fastening a sharp stone to a stick to make a spear or to make a bow to shoot an arrow. Body decorations and tool manufacture that is slightly more advanced than sharp stones took a long time to come—about 100,000 years. Because this contrasts so starkly with such feats as building a long-span bridge or doubling computational power every two years, we may tend to assume that human technological inclinations have always been robustly at the forefront. Probably not so. Such inclinations may be developed as a result of some successes that eventually gave rise to a culture of invention. The value of inventing technology could not have been just obvious.

3. FOOD AND CARING FOR OTHERS

The evolutionary changes that led to mammalian and bird styles of sociality were all about food—altruism, admirable though it may be, came along as a by-product. Here is the outline of the story. When warm-blooded animals first appeared among the dinosaurs, about 200 millions years ago, they enjoyed a masterful advantage over their cold-blooded competitors: they could forage at night when the warmth of the sun was absent, they could prosper in colder climates. A disadvantage, however, had to be overcome: gram for gram, the warm-blooded creature has to eat ten times as much as his cold-blooded cousins. The energy requirement is a massive evolutionary pressure on the evolution of homeotherms. How then to compete successfully? Being smart is one way to gain an advantage. Getting an advantage in cleverness happens mainly by ramping up the postnatal capacity to learn.

Two sets of altered genetic instructions enabled scaled-up learning: (1) make lots more neurons and thus make a bigger brain, and (2) ensure that the neurons learn at the right time—namely, after the offspring are born. The first led to the emergence of the neocortex, a structure overlying the older motivational and homeostatic circuitry and that is unique to mammals and birds. Exactly how the neocortex evolved from earlier brains is not well understood. The second was achieved by scheduling infants to be born with highly immature brains. Thus, a baby turtle hops right out of its broken shell and marches off to the sea, after which it learns only a modest amount. By contrast, a newborn rat pup is completely dependent on its mother. Luckily, it has at least the capacity to reflexively struggle toward any warmish thing and suck on it. Once latched on, it learns voraciously from its environment.

For mammals and birds, immaturity was a boon since their learning-ready brains could tune themselves up to whatever causal circumstance they happened to be born into. Once mature, they were much smarter than their cold-blooded competition. As an evolutionary strategy, Big Learning was a game-changer. In brief, mammals and birds ended up with the neocortex, a kind of soft-tissue computer that connects with the ancient structures embodying motivation, drives, and emotions. Mammals and birds are very smart—and flexibly smart—relative to reptiles and fish. In some mammals, such as primates, and some birds, such as corvids, the neocortex expanded to a remarkable degree.

A major problem had to be solved concurrently, however, because the downside of this strategy for expanding cleverness is that infant mammals are pitifully dependent and easy prey. The solution to enabling their survival? Rig the brain wiring of a mature smart animal so it cares for the infants until they can fend for themselves. Modifying parental brains to be caring brains was also a game changer.

Making parents *parental* was achieved by changing the existing self-survival mechanisms so that care of *me* extended to *me-and-mine* (Rilling and Young 2014). Just as the mature rat is wired to care for her own food and safety and warmth, so she is wired to care for the food and safety and warmth of her pups. Both mother and babies feel pain when separated and pleasure when they are together. They are bonded, and the bonding is embodied in neural circuitry. Is the love we feel *real*? Yes indeed. It is as real as anything the brain does, such as remembering where home is, seeing the moon, or deciding to hide rather than run. Notice, too, that social instincts, built on this edifice, will spread throughout the population. That is because parental animals who neglect and abandon their offspring will not pass on their genes.

Oxytocin is an ancient body-and-brain molecule found in reptiles as well as birds and mammals. It has a sibling, vasopressin, and the two peptides probably had a common origin many hundreds of millions of years ago. These two peptides, oxytocin and vasopressin, are at the crux of the intricate neural adaptations sustaining mammalian and bird sociality. In reptiles, oxytocin facilitates egg ejection, and, in mammals, oxytocin is important in sperm ejection and in ovulation. It is also essential in milk ejection. Evolution found it convenient to extend the role of oxytocin and vasopressin to adjust circuitry in the brain to facilitate parental behavior.

In mammals, oxytocin is released in the brains during mother–infant cuddling, bonding them strongly to each other. The brain-made opioids and cannabinoids also play a crucial role so that mother and baby feel pleasure when they are together, pain when separated. The pathways for physical pain and social pain appear to be very closely interconnected, suggesting that social pain is an evolutionary extension of the physical pain system (Eisenberger 2012, 2015). This bonding pattern, regulated by oxytocin and a palette of other neurochemical and neurohormones working in their proprietary circuitry, is the basic platform for morality—for caring for others. These changes were probably achieved via rather modest genetic changes affecting self-care circuitry. It is not known whether mammals and birds had a common ancestor whose brain was organized for parental care or whether parental care circuitry in mammals and birds is a case of convergent evolution.

What does seem evident is that the cluster of characteristics—homeothermy, Big Learning, and parental care—were linked in the evolution of the mammalian and avian brains and that the result was highly successful. Mammals and birds thrived, and many species emerged. At this stage of our knowledge, it appears probable that circuitry supporting this cluster of behaviors is the neural platform for morality. Outstanding empirical questions certainly remain, such as how oxytocin and the endogenous opioids and endocannabinoids interact and how environmental stressors or learning can alter their function.

How does this platform scale to get caring beyond that of mothers and babies, caring that we more readily recognize as typical morality? The outlines of the answer took shape with a discovery about the social behavior of voles and the neural elements that enable this behavior (Insel and Hulihan 1995).

Voles are rodents, and there are many species of voles, including prairie voles that live on the open grasslands and montane voles that inhabit rocky mountainous areas. Montane voles behave in a way that we tend to suppose is typical: the males and female meet, they mate, and they go their separate ways. Prairie voles, however, are remarkably different. They meet, they mate, and they are bonded for life (Williams et al. 1994). Biology being what it is, there is variability, and not every prairie vole shows this behavior. Nevertheless, long-term pair-bonding is common among prairie voles.

Prairie voles like to be together, they are distressed when separated, and the male guards the nest and helps rear the pups. Most, though not all, of their reproductive action is with their mate. They live in communities of kin and friends, and siblings also tend the younger pups (Ahern, Hammock, and Young 2011). Remarkably perhaps, they also show consolation behavior to a stressed friend (Burkett et al. 2016). So, here we have monogamous voles and promiscuous voles: What is the difference in the brain to explain the difference in social behavior?

The surprising answer depends on oxytocin and its receptors—the sites on nerve cells where a specific neurotransmitter can fit precisely into a dock and modify the responses of the neurons. Prairie voles tend to have a high density of oxytocin receptors in a very specific region of the reward system, the nucleus accumbens, and they have a high density of receptors for vasopressin in another, highly specific region of the reward system, the ventral pallidum (Williams et al. 1994 Young 1999). Montane voles have a mere smattering of oxytocin and vasopressin receptors in those respective regions. If you experimentally block the oxytocin receptors in the prairie voles, they behave like montane voles. A similar neural organization is also seen in other pair-bonding animals such as marmosets, titi monkeys, and deer mice, though there are also significant anatomical differences (Beery, Lacey, and Francis 2008; Freeman et al. 2014).

What about humans? Roughly speaking, in our behavior, we are more like prairie voles and marmoset monkeys than like solitary species such as montane voles and wolverines. We humans are intensely social. We like to be together, we find social inclusion pleasurable and social exclusion painful. And what of our brains? For technical reasons, much regarding the human oxytocin story is not yet satisfactorily nailed down (Leng and Ludwig 2016). Still, it is known from postmortem studies that humans have many receptors for oxytocin in the reward system, just like prairie voles, and we seem to have oxytocin receptors in many areas of the cortex as well (Boccia et al. 2013). It is noteworthy that a particular region of the cortex—the anterior insula—is highly responsive to many aspects of the state of the body, such as injury or trauma, but it is also highly responsive to social exclusion. This can be seen in brain imaging studies (Craig 2015). One intriguing finding is that the anterior insula, important for assessing the physiological state of the body, contains a high density of one class of opioid receptors (μ receptors), suggesting also a link to pleasurable aspects of sociality (Baumgartner et al. 2006). It is possible that left and right anterior insula may have different degrees of sensitivity to, inclusion and exclusion, respectively, but on this topic, too, more questions stretch ahead of us.

Stress hormones such as corticotrophin-releasing factor also play an important role in social behavior, as they do in addiction (Lim, Nair, and Young 2005). When oxytocin levels rise, stress hormone levels decrease and vice versa (Panksepp 2003). When animals are in high alert against danger, when they are preparing to fight or flee, stress hormones are high and oxytocin levels are low. When the threat has passed and the animal is among friends, hugging and chatting, stress hormones back off and oxytocin levels surge. Among its many roles, oxytocin decreases the stress response, making possible the friendly, trusting interactions typical of life in social mammals. I can let my guard down when I know I am among trusted family and friends. It has also been suggested that oxytocin plays a role in allowing animals in a group to tolerate each other. For a casual observation, consider that when you are stressed, other people's habits and "ways" can rather suddenly become extremely annoying. Normally, you might easily overlook or just not care about someone's habitual punning or constant sniffling, but, under stress, these fidgets tend to rub you raw. In general, animals under stress are more easily irritated by intrusive others, and an important role of oxytocin seems to be to dampen some of this irritation response from neurons, thus permitting the social instead of the solitary life.

With small genetic changes, friends, and sometimes strangers, came to be embraced in the sphere of *me-ness*; we nurture them, fight off threats to them, keep them safe. My brain knows these others are not *me*, but if I am attached to them, their plight fires up caring circuitry, motivating me to incur a cost to benefit the other. Different social species have somewhat different social arrangements, likely reflecting the ecology that shaped their evolution. For example, in baboons, the young bachelors leave the troop and search for a new home troop; but in chimpanzees, it is the adolescent females that leave the natal troop. In baboons, the matrilines of mothers, daughters, granddaughters, and so forth are important for strength of attachment, and the matrilines within a troop are recognized to rank from highest to lowest. Chimpanzees do not have matrilines. Crows live together in flocks, but bald eagles tend to live with just the bonded pair and their young until independence. The variations abound and are assumed to reflect the ecological conditions in which the animal makes its living. Given the basic platform for caring, small genetic changes in circuitry can facilitate whether attachment and caring extend to mates or kin or friends and so forth.

A by-product of basic sociality anchored by parental and kin attachment is the emergence of group life, which is highly advantageous. Cooperation in hunting and defense gives benefits galore. These advantages may have also favored genes for extending attachment to friends. Cooperation feels good, especially if it brings success. This, too, helps us extend caring to friends.

When cooperative behavior yields rewards, cooperation becomes the favored practice for solving problems, whether they concern battling off intruders or building a bridge across a river. The cooperative behavior seen between ravens and wolves, for example, seems to have been learned at some point and seems to be picked up by each generation. For very social animals such as wolves and bonobos, the bonds to others in the group are very strong, and something akin to human grieving can be seen in the wolf pack when a mother wolf is killed or when a chimpanzee mother loses her baby (https://www.youtube.com/watch?v=jzrige2nqqw).

What of norms and rules, which are endemic to human morality? They are important because self-caring values did not disappear with the emergence of the platform for mammalian willingness to incur a cost to benefit another. Self-caring must coexist with other-caring values. We must compete as well as cooperate with our siblings and friends. And sometimes the selfish urges do get the upper hand. Group living often involves norms that keep selfish behavior in check. Social motivation for caring connects to learning norms via the reward system. Because mammals care intensely about approval and disapproval, very young brains are highly disposed to internalize the social practices and norms of the family and, eventually, the wider group.

The workhorse here is the mammalian reward system, a system evolution updated from the ancient structures of reptiles and that were linked with the new cortical mantle in mammals and birds—the old basal ganglia with the new frontal cortex. In this part of the story, dopamine is the critical neurochemical, but, as before, the endogenous opioids and cannabinoids are extremely important as well, along with serotonin and likely many other neurochemicals as well. Pleasure and pain are central to motivation and to learning. As with evolutionarily older animals, the basal ganglia allow mammals to develop habits and skills that enhance their ability to compete. In mammals, however, some of these habits and skills structure *social interactions*, with the upshot that certain socially significant plans are inhibited—such as stealing or deceiving or killing within the group (Bhanji and Delgado 2014; Graybiel 2008; Xiang, Lohrenz, and Montague 2013). On the other hand, the individual may be powerfully motivated to overcome a fear of injury to rescue a brother or a friend. This is seen not only in humans, but also in rodents, birds, and monkeys.

Generally, approval for an action is rewarding and feels good, whereas disapproval feels bad. The emotions are profoundly engaged. We pick up appropriate social behavior by imitating, sometimes quite unconsciously, those around us, thereby facilitating social harmony. We also learn by trial and error, and a little disapproval goes a long way, especially in the young. As conditions change, solutions to social dilemmas may also change. The need for new solutions ignites new problem-solving, necessity being the mother of invention.

Powerful dispositions—intuitions—concerning what is right and what is wrong emerge in the developing human as its brain solves social problems and internalizes social norms. Roughly, a complex neural organization takes shape, consisting of a blend of emotions and cognition, against a background of a powerful desire to belong to the group. This organization is recruited in evaluating a social situation and sketches relevant predictions about what will or could happen; it assesses what others are probably thinking and feeling and decides what should be done. Many of the neurobiological details concerning exactly how this works are still unclear, but they will become clearer, at least concerning the general features of the mechanism, if not in every mechanistic detail.

4. CONCLUSION

As a science, neurobiology can help us understand why humans typically have a moral conscience, but neuroscience per se does not adjudicate specific norms or laws that make up the social superstructure on the neurobiological platform. Nevertheless, as Flanagan et al. (2014) point out, data from neuroscience and psychology can be part of the conversation concerning the ends of a good life and what in the conditions of life are conducive to well-being. In such conversations, it will always be relevant to know such things as the toll on the brain of early abandonment and neglect or of systematic social exclusion (Beyer, Münte, and Krämer 2013; van Harmelen et al. 2014). On the other hand, it is quite obvious that neuroscience cannot tell us whether it is morally appropriate to use drones to kill terrorists, or whether a flat tax is better than a graduated income tax, or whether CRISPR-cas9 should be used to alter the human genome in sperms and eggs. For those decisions, we, as a diverse group of humans with shared interests, still need negotiation, compromise, good sense, and good will. Neuroscientists, as concerned citizens, should of course be part of that conversation, along with farmers, welders, and baseball players. For these kinds of questions, we do know at least this much: there is no algorithm for generating answers; there is no metaphysical entity to appeal to and no magical faculty of pure reason. As an anonymous wag once put it, there is no JUSTICE; there is just us.

REFERENCES

Ahern T. H., E. A. Hammock, and L. J. Young. 2011. Parental division of labor, coordination, and the effects of family structure on parenting in monogamous prairie voles (Microtus ochrogaster). *Developmental Psychobiology* 53(2): 118–131.

Bhanji J. P., and M. R. Delgado. 2014. The social brain and reward: Social information processing in the human striatum. *Wiley Interdisciplinary Reviews: Cognitive Science* 5(1): 61–73.

Baumgartner, U., H.G. Buchholz, A. Bellosevich, W. Magerl, T. Siessmeier, R. Rolke, et al. 2006. High opiate receptor binding potential in the human lateral pain system. *Neuroimage* 30: 692–699.

Beery, A. K., E. A. Lacey, and D. D. Francis. 2008. Oxytocin and vasopressin receptor distributions in a solitary and a social species of tuco-tuco (Ctenomys haigi and Ctenomy). *Journal of Comparative Neurology* 507: 1847–1859.

Ben-Ami Bartal, I., J. Decety, and P. Mason. 2011. Empathy and pro-social behavior in rats. *Science* 334(6061):1427–1430. doi: 10.1126/science.1210789.

Benson-Amram, S., B. Dantzer, G. Stricker, E.M. Swanson, and K.E. Holekamp. 2016. Brain size predicts problem-solving ability in mammalian carnivores. *Proceedings of the National Academy of Science, USA* 113: 2532–2537.

Beyer, F., T. F. Münte, and U. M. Krämer. 2013. Increased neural reactivity to socio-emotional stimuli links social exclusion and aggression. *Biological Psychology* 96: 102–110.

Boccia, M. L., P. Petrusz, K. Suzuki, L. Marson, and C. A. Pedersen. 2013. Immunohistochemical localization of oxytocin receptors in human brain. *Neuroscience* 253: 155–164.

Burkett, J. P., E. Andari, Z. V. Johnson, D. C. Curry, F. B. de Waal, and L. J. Young. 2016. Oxytocin-dependent consolation behavior in rodents. *Science* 351(6271): 375–378.

Churchland, P. S. 2011. *Braintrust*. Princeton, NJ: Princeton University Press.

Churchland, P. S. 2014. The neurobiological platform for moral values. *Behaviour* 151: 283–296.

Craig, A. D. 2015. *How do you feel? An interoceptive moment with your neurobiological self*. Princeton, NJ: Princeton University Press.

Dawkins, R. 1976. *The selfish gene*. Oxford: Oxford University Press.

de Waal, F. B. 1989. *Peacemaking among primates*. Cambridge, MA: Harvard University Press.

de Waal, F. B. 1996. *Good natured: The origins of right and wrong in humans and other animals*. Cambridge, MA: Harvard University Press.

de Waal, F. B., K. Leimgruber, and A. R. Greenberg. 2008. Giving is self-rewarding for monkeys. *Proceedings of the National Academy of Science USA* 105(36): 13685–13689.

de Waal, F. B., and van Roosmalen. 1979. Reconciliation and consolation among chimpanzees. *Behavioral Ecology and Sociobiology* 5: 55–66.

Eisenberger, N. L. 2012. The pain of social disconnection: Examining the shared neural underpinnings of physical and social pain. *Nature Reviews Neuroscience* 13: 421–434.

Eisenberger, N. I. 2015. Social pain and the brain: Controversies, questions, and where to go from here. *Annual Review of Psychology* 66: 601–629.

Flanagan, O., A. Ancell, S. Martin, and G. Steenbergen. 2014. Empiricism and normative ethics: What do the biology and psychology of morality have to do with ethics? *Behaviour* 151: 209–228.

Freeman, S. M., H. Walum, K. Inoue, A. L. Smith, M. M. Goodman, K. L. Bales, and L. J. Young. 2014. Neuroanatomical distributions of pxytocin and vasopressin 1a receptors in the socially monogamous coppery titi monkey (Callicebus cupreus). *Neuroscience* 273: 12–23.

Graybiel, A. 2008. Habits, rituals, and the evaluative brain. *Annual Review of Neuroscience* 31: 359–387.

Hamlin, J. K., K. Wynn, and P. Bloom. 2007. Social evaluation by preverbal infants. *Nature* 450: 557–560.

Insel, T. R., and T. J. Hulihan. 1995. A gender-specific mechanism for pair bonding: oxytocin and partner preference formation in monogamous voles. *Behavioral Neuroscience* 109: 782–789.

Korsgaard, C. M. 1996. *The sources of normativity*. Cambridge, MA: Cambridge University Press.

Laurent, V., and B. W. Balleine. 2015. Factual and counterfactual action-outcome mappings control choice between goal-directed actions in rats. *Current Biology* 25(8): 1074–1079.

Leng, G., and M. Ludwig. 2016. Intranasal oxytocin: Myths and delusions. *Biological Psychiatry* 79(3): 243–250.

Lim, M. M., H. P. Nair, and L. J. Young. 2005. Species and sex differences in brain distribution of corticotropin-releasing factor receptor subtypes 1 and 2 in monogamous and promiscuous vole species. *Journal of Comparative Neurology* 487: 75–92.

Panksepp, J. 2003. Feeling the pain of social loss. *Science* 302(5643): 237–239.

Purzycki, B. G., C. Apicella, Q. D. Atkinson, E. Cohen, R. A. McNamara, A. K. Willard, D. Xygalatas, A. Norenzayan, and J. Henrich. 2016. Moralistic gods, supernatural punishment and the expansion of human sociality. *Nature* 530: 327–330.

Rilling, J. K., and L. J. Young. 2014. The biology of mammalian parenting and its effect on offspring social development. *Science* 345(6198): 771–776.

Rutz, C., and J. J. St. Claire. 2012. The evolutionary origins and ecological context of tool use in New Caledonian crows. *Behavioral Processes* 89(2): 153–165.

Smirnova, A., Z. Zorina, T. Obozova, and E. Wasserman. 2014. Crows spontaneously exhibit analogical reasoning. *Current Biology* 25 (2): 256–260.

St Clair, J. J., and C. Rutz. 2013. New caledonian crows attend to multiple functional properties of complex tools. *Philosophical Transactions of Royal Society London, B Biological Sciences* 368(1630): 20120415.

van Harmelen, A. L., K. Hauber, B. Gunther Moor, P. Spinhoven, A. E. Boon, E. A. Crone, and B. M. Elzinga. 2014. Childhood emotional maltreatment severity is associated with dorsal medial prefrontal cortex responsivity to social exclusion in young adults. *PLoS One* 8/9(1): e85107. doi: 10.1371/journal.pone.0085107.

Weatherford, J. 2004. *Genghis Khan and the making of the modern world*. New York: Broadway Books.

Williams, J. R., T. R. Insel, C. R. Harbaugh, and C. S. Carter. 1994. Oxytocin administered centrally facilitates formation of a partner preference in prairie voles (*Microtus ochrogaster*). *Journal of Neuroendocrinology* 6: 247–250.

Xiang, T., T. Lohrenz, and P. R. Montague. 2013. Computational substrates of norms and their violations during social exchange. *The Journal of Neuroscience* 33(3): 1099–1108.

Young, L. J. 1999. Frank A. Beach Award. Oxytocin and vasopressin receptors and species-typical social behaviors. *Hormones and Behavior* 36 (3): 212–221.

CHAPTER 3

All You Need Is Love(s)

Exploring the Biological Platform of Morality

MAUREEN SIE

> Does it look like a pair of pyjamas,
> Or the ham in a temperance hotel?
> Does its odour remind one of llamas,
> Or has it a comforting smell?
> Is it prickly to touch as a hedge is,
> Or soft as eiderdown fluff?
> Is it sharp or quite smooth at the edges?
> O tell me the truth about love.
> —W. H. Auden, *Oh Tell Me the Truth About Love*

This chapter explores the claim that our nature as loving beings can explain our nature as moral beings. Are we able to make sense of such a claim and, if so, how? This chapter aims to offer a possible philosophical answer by distinguishing among several kinds of loves and relating them to what I call different moral dimensions of our existence. The moral dimensions that I distinguish correspond to dimensions that philosophers have traditionally applied to explain our nature as moral beings. If my answer is appealing, then we have a naturalized account of morality that leaves space for a variety of philosophical views.

The claim that our nature as loving beings can explain our nature as moral beings is interesting for two reasons (hereafter, I shorten this claim to "love is the source of morality"). First, scientists have discovered the brain circuits and chemistry that are involved in not only regulating male and female sexuality and feelings of attachment (mothers' urge to care for their offspring) but also in our sociability more broadly speaking, such as how we interact with strangers. This might suggest that we have discovered "the biological sources of friendship and love," as expressed by Jaak Panksepp (1998: 246), the founder of affective neuroscience, and the biological platform of morality, as argued by Patricia Churchland (2011). If such is indeed the case, future "love" research might offer new insights into morality, building on what we know from the animal research about these brain circuits of friendship and love and from more recent work on behavioral economics and psychology (Churchland 2011, 2013; Chapter 2, this volume).

Second, love and morality seem to be similar phenomena. This is of interest because some of the properties that philosophers have traditionally struggled to understand in the case of morality seem much easier to explain when love is its source. Both love and morality entail caring for and attention to others. When we are in a loving relationship with someone or something, this person or object limits our self-centered strivings, behaviors, and actions. Often, morality seems concerned with limiting such self-centered strivings as well. Moreover, both phenomena possess a natural motivational authority in our lives. Suppose that you suddenly have to cancel an appointment with someone because your boyfriend needs you. Although you have to cancel "last minute," why you do so is clear and will probably not be held against you. Loving someone implies or consists of an authoritative motivational state to do certain things for that person. Not only do you tend to desire to take care of those you love (motivational state), you also take that desire as a directive (having authority); it justifies your action.

Morality has a similar character. Under normal conditions, you "promised something" tends to be enough to explain your attempt to keep the promise, all other things being equal. Although you might desire to do something else, your promise exerts authority over your behavior; people tend to be motivated to keep their promises. This is also the case for other moral commitments and considerations. If you judge it to be morally wrong to steal from, lie to, or harm another person, you will be inclined to refrain from such acts.

This is not to claim that nothing could undermine the motivational authority of love or morality since, evidently, we occasionally transgress moral expectations or those of our loved ones. We are motivationally complex beings, and other considerations might take the upper hand or conditions might be such that the force of what we (normally) care about leaves us unmoved. However, we would be surprised if someone did not treat moral considerations or their loving relationships with at least some default authority.

The advantage of explaining morality as springing from love is that to explain our nature as moral beings, we do not need a metaphysically extravagant story that conflicts with our scientific view of the world, such as the existence of an independent moral reality with queer moral properties.[1] They are

---

1. I use "queer" instead of "peculiar" because of Mackie's famous argument from queerness.

"queer" because the perception of those properties somehow has an intrinsic connection to motivation or the existence of an independent moral authority.[2] Love seems a natural enough phenomenon; it is easy to perceive how it can play a role in our reproductive practices and survival, or does it not? By now, psychologists have widely published works on love after a decline in interest in the phenomenon during the behaviorist period in psychology. Romantic love,[3] specifically, its phase that we call "falling in love," has been established as a universal phenomenon that can be detected in our brains and bodies with hormones correlating to its different constitutive ingredients (Fisher 2014: 80). Romantic love, whose first phase is called "limerence,"[4] tends to bind people together, at least for the amount of time required to provide care for their offspring until the latter are able to survive on their own (Fisher 1994).[5] However, it is exactly this phase of romantic love that is also often "gloriously amoral and a-rational."[6] It notoriously makes people act immorally, steal, lie, kill, and even lead whole armies astray.[7] Hence, the question is whether the love that makes evolutionary sense is the kind that can plausibly be the source of morality.

Therefore, perhaps we need to focus on a different kind of love when we are interested in it as a source of morality. In her book entitled *Braintrust*, Churchland (2011) defends the thesis that love is the source of morality. What makes our human society a moral one is our ability to trust one another and to extend this trust to people to whom we are not related and even to strangers. As Churchland (2011) argues, it is exciting that recent findings show trust as the basis for our human sociality and, similar to romantic love, that trust is also connected to the hormones oxytocin and vasopressin.[8]

---

2. Of course, there are other philosophical views that are defended as compatible with our scientific view of the world, such as sentimentalist and neo-Kantian rationalist views.

3. By "romantic love," I refer to the love between adult human beings that starts with an episode of falling in love.

4. See de Sousa (2015: 3), who takes over the term from psychologist Dorothy Tennov and argues that this phase makes it appropriate to view love as a syndrome.

5. The story is a bit more complicated than this. According to Fisher, romantic love might be capricious by nature's design, leading to what this author calls "two complementary reproductive strategies in tandem," taking care of socially sanctioned marriage and reproduction, on one hand, while reproducing with clandestine lovers, on the other hand (2004: 151).

6. Nick Zangwill (2013), among others, has recently argued this in a paper of that title.

7. This is portrayed in Ondaatjes's *The English Patient*, to the dismay of some philosophers (see Jollimore 2011: 148).

8. However, Churchland (2013) warns us that the popular idea that oxytocin is a "love" or "cuddle" hormone is misperceived. As she wittily writes, playing off a famous Mae West remark, in the case of oxytocin, "too much of a good thing is not wonderful, it can be catastrophic" (Churchland 2013: 78). So far, not only are the experimental results obtained from oxytocin weak, but they also show that administration of the hormone can have a result opposite to a loving/cuddling response (Bartz et al. 2011; Churchland 2013: 102).

When we speak of love in the sense of a trusting and caring relationship with other human beings, what we have in mind is far removed from the romantic love discussed earlier. The trust and care that are extended even to strangers seem more like a mix of what philosophers have traditionally referred to as *agape* and the Greek *storge*, now often referred to as "affection." *Agape* is a kind of universal love that we feel for other human beings regardless of who they are; affection refers to a special feeling of attachment to our children and parents, for example. The hypothesis that morality springs from love seems to make more sense if we have a kind of caring and trusting attitude in mind. It seems hard to imagine someone who is very friendly, attentive, and loving in this sense yet completely immoral. Perhaps this was what made Tony Soprano, the character played by James Gandolfini in the hit series *The Sopranos*, such an interesting figure. It is very likely that this kind of love also makes evolutionary sense. After all, as mammalian species, we are born immature and helpless and only survive when others take care of us, a fact that Panksepp (1998) and Churchland (2011) emphasize. It is comprehensible for us to be the kind of beings that can become attached to others in a way that facilitates parental care.

Nevertheless, if the kind of love that explains morality is different from the romantic one, then the claim that love enables us to understand our nature as moral beings is ambiguous. Alternatively, is it perhaps the case that the discovery of oxytocin and vasopressin shows us that romantic love and the caring and attentive kind of love actually share a common nature? Are we perhaps mistaken in arguing that they are phenomenologically distinct? People do terrible deeds to protect their children as well. Moreover, as many parents will remember, our emotions toward our newborn feel a lot like falling in love. Perhaps these kinds of loves are alike after all; only our interpretation, the stories we tell about them, and our theories differ. It seems that, to make headway on the interesting claim that love is the source of morality, we should deal with "this thing called love" with some clarity.

For me, such clarity is derived from reading a short book called *The Four Loves* by C. S. Lewis, the Christian apologist, famous writer, and scholar of English literature and Medieval and Renaissance English. He distinguishes among four kinds of loves, three of which are natural in his view, with the fourth one being Christian love, called "charity." I elaborate on this framework in relation to the claim that love is the source of morality, although I completely abandon his Christian framework and rename the fourth kind of love "kindness." I also argue that recent findings in affective neuroscience suggest that this fourth kind is a natural kind of love as well. In Section 1, I elaborate on the basics of Lewis's account of the natural loves. In Section 2, I explain how I understand the nature of his enterprise and in what sense love can be understood as the source of morality. In Section 3, I discuss the dynamics of Lewis's account, showing that each of the loves that he distinguishes requires the fourth love (kindness) to keep them from taking a nasty turn. I conclude by explaining why the fourth love that Lewis distinguishes actually fits the naturalist picture quite well if the recent finding that oxytocin is involved in our trusting interactions with strangers is correct.

1. AFFECTION, FRIENDSHIP, AND *EROS*

What makes Lewis's account appealing for our purposes does not involve the kinds of loves that he distinguishes. Each of these has been discussed by many other scholars, several of whom have already been mentioned in the introduction. He distinguishes among (1) affection, connected to the Greek word *storge*; (2) *philia* or friendship; (3) *eros*, referred to as romantic love in the introduction (hereafter, I use these labels interchangeably); and (4) Christian charity, connected to the Greek agape: "a sort of indiscriminate, universalized, and sexless storge," as de Sousa (2015: 2) describes it. It is the way in which Lewis distinguishes among these kinds of loves that is special, allowing us to relate them to the moral dimensions of our existence, on one hand, and to understand love as a natural phenomenon, on the other hand. Moreover, the distinction enables us to comprehend love as a dynamic, messy, and imperfect phenomenon. In my view, this is just as it should be. We start life by being completely dependent on our caregivers; hence, from an early age onward, we need to anticipate the behaviors of others, be sensitive to their expectations of us, and make them take care of us (cf. Churchland 2011: 61; cf. Herman, forthcoming). These early relationships and everything that can go wrong in them are bound to determine our adult relationships, an important insight from attachment theory.[9] Furthermore, as we grow up, our relationships with our parents, peers, friends, and lovers change, as do our insights on how such relationships should be or on what we need in them. An adequate philosophical account of love should accommodate such messiness and imperfection, as well as its dynamics. Before I elaborate on how Lewis's account does so exactly, let me make an observation on how to understand it against the background of contemporary discussions on love as the source of morality.

As far as I can see, Lewis's arguments differ from contemporary discussions on love as the source of morality by not trying to capture the "essence of love" as either evaluational or not. The controversy in contemporary philosophical discussions revolves around the question of whether we have reasons to love (evaluational) or whether love rather "happens to us" and can even be

---

9. In turn, these adult relationships comprise a prominent feature for explaining immoral (criminal) behavior and impaired moral development (Van Ijzerdoorn 1997).

compatible with a negative evaluation of the beloved (Zangwill 2013: 307).[10] If love itself is not evaluational (or rational), we might think that it would be difficult to argue that it is related to morality or could even be its source.[11] As I will show, the kind of argument that Lewis enables us to raise is of an altogether different nature and does not require us to claim that love is either evaluational and easy to square with morality or not. The same is true with another theme that seems to be at the core of many philosophical arguments about love as the source of morality. This is the idea that, in the case of love, we are concerned with particular persons, singled out for their individual characteristics, and their specific relation to ourselves, whereas, in the case of morality, we are required to ignore individual differences and particular relationships. Jollimore (2011: 147) summarizes this position, "Love . . . is precisely the sort of emotion that morality is intended to govern." Certainly, other philosophers argue against this view, denying that morality requires us to ignore individual differences or the idea that love requires preferential treatment for those with whom we have special relationships. As I will show, Lewis does not defend the position that love is either immoral/amoral or the opposite. Love can go wrong (or right) in many ways. Nevertheless, it makes sense to argue that it is the source of morality as well. Now, let us turn to the account itself to explain this.

Lewis starts his discussion of love by defining affection as the most instinctive love and the one without which we cannot live. Although for Lewis this is a general observation about the strength of our desire to be loved in an affectionate way, he actually locates the origin of affection in parental love and filial love (1960/2012: 39). Hence, his observation on the importance of affection for our psychological survival is quite in line with the fact that, as mammalian species, we only survive when taken care of in the first years of our existence. Lewis argues that affection is distinct from friendship and *eros* for its ability to make appreciations possible, "which, but for it, might never have existed" (1960/2012: 44). The reason for this is that he considers affection to be indiscriminate at its core.

Affection knows no boundaries; it does not know barriers of age, sex, class, and education, not even of species. Lewis points out that affection can exist between a nurse and a student, a dog and its master, and a dog and a cat. In contrast to the case of *eros* or friendship, Lewis observes that we do not always

---

10. Harry Frankfurt and Bernard Williams have famously argued against such evaluative accounts of love. See Schaubroeck (2011) for an insightful discussion of the relation between morality and love.

11. This is especially true when our view on morality relates it to rationality, to what is reasonable, and to observing rules that hold for everyone and can be grasped by everyone. Probably for this reason, someone such as Churchland goes to the trouble of arguing against rule-based accounts of morality and understands herself as arguing against Kantian accounts of morality (2011: 173–175).

know when we love someone affectionately; neither do we necessarily take pride in it. This is why Lewis calls affection a modest and secretive kind of love, even "shame faced" (1960/2012: 42). It happens to us, seeps through our lives, often without us noticing its exact beginning, and it often even "needs absence or bereavement to set us praising to whom only affection binds us" (Lewis 1960/2012: 42). Due to its sneaky nature, this kind of love can unite those who share little in common, who find themselves together accidentally, living in the same neighborhood and sharing a community (Lewis 1960/2012: 45). Lewis argues that affection can open and broaden our minds, not to what is already similar to or cared for by us, but to the unfamiliar and unknown.

Hence, in Lewis's description, affection possibly expands our horizons and confronts us with the reality and possible value of worlds and perspectives that are unknown to us. In other words, affectionate love binds us to people who are not our caregivers, siblings, or like-minded peers.[12] The moral dimension of affectionate love is that it puts us in immediate relationships with other people and things. It attaches us to them, making us sensitive to their responses to us, their well-being, desires, and projects.

Now, let us turn to friendship.

According to Lewis, friendship needs commonalities, a common focus on the world, a focus or interest that, prior to the friendship, felt as something quite unique to each individual involved in the friendship. In contrast to affection, Lewis considers friendship the least natural of the four loves, something that we do not need to survive. Nonetheless, perhaps we do need the "structure of friendship" or companionship (Lewis 1960/2012: 78) for survival—that is, people with whom we spend time with and share a common focus, a hobby, or an interest. For Lewis, though, friendship is more than companionship; it is a "meeting of minds," where each individual can be himself or herself and feels reassured of his or her view on things by discovering that the other (others) shares (share) it. Friends come to know one another and one another's minds by sharing experiences and discovering their common outlook and way of responding to the world.

In Lewis's description, friendship is an energizing and empowering relationship; it connects like-minded individuals and combines their strengths to pursue a common goal and effect change. It is also the relationship that makes people feel good about themselves as "rational" individuals, with all kinds of ideas about the world that do not necessarily align exactly with those of others (their caregivers, educators, and those current in their culture). In Lewis's

---

12. If we bear in mind the insights from developmental psychology and affective neuroscience, it might well be that our first affectionate relationships (those with our caregivers) also determine or shape our horizons and value of the world, as Peter Hobson argues (2002).

understanding, friends affirm one another's perspectives on things and ways of relating to the world, freely and spontaneously, without in any way being required to do so. Hence, friendship can be said to open up the moral dimension of our existence, often referred to as "autonomy" or "authenticity"—the fact that each of us has a special way of being in and responding to the world that distinguishes us from others. We are the kind of beings that do not only care about others but also tend to care about the kind of people that we are.

Now, let us turn to romantic love, the kind that has a negative reputation when its relation with morality is concerned. On the other hand, some scholars have also argued that romantic love truly opens us up to the reality of the other person and makes us rise above ourselves. For example, Jollimore argues that it is this aspect that makes love deeply moral (2011: 149–151). This is not exactly how Lewis understands it, although he agrees with Jollimore and many other scholars that what is special about romantic love is our preoccupation with a particular person. Lewis observes that although *eros* for the evolutionist is something that grows out of our sexual nature, this is not what most often happens in the individual's consciousness (1960/2012: 113). Often, what comes first is a preoccupation with another person "in the totality" of this person (1960/2012: 113). When you are in love, you do not want just a woman or a man but a particular woman or man. Neither do you simply want the pleasure that this person can give you, but, somehow, you want the other person as such (Lewis 1960/2012: 115).

Ironically, this longing is inextricably linked with a physical appetite despite its transcendent appearance (Lewis 1960/2012: 123). Scientific findings suggest that falling in love makes our dopamine levels rise, which can stimulate the release of testosterone, the hormone of sexual desire (Fisher 2014: 83). In the act of love, Lewis claims that "we are not merely ourselves. We are also representatives. It is here no impoverishment but an enrichment to be aware that forces older and less personal than we work through us" (Lewis 1960/2012: 125). Nonetheless, it is not lust that is most characteristic of *eros*, as Lewis understands it. Such distinction between love and lust is also corroborated by scientific findings on the hormones involved, as well as the pattern of brain activity involved (Fisher 2014: 80).

On the phenomenological level, Lewis observes that without eros, sexual desire is "a fact about us" (1960/2012: 116). It is us who "lust after someone," as they say, but when we are in love, our sexual desires are about the one we love. We are in love with someone. In this sense, our love confronts us with a reality outside ourselves, something objective, moreover, something that merits our love and longing. In other words, we are directed at our beloved; our relation to him or her is not about satisfying our longing or the pleasure that fulfilling it would give us (Lewis 1960/2012: 116). Moreover, eros does not aim at happiness; it will not disappear or dissuade people from pursuing it even when, clearly, no good or happiness will come of it. "For it is the very mark of Eros that when he is in us we had rather share unhappiness with the Beloved than be happy on any other terms" (Lewis 1960/2012: 130).

Certainly, Lewis's point is not that romantic love might require sacrifices, as Jollimore (2011: 150) understands him. Rather, the moral dimension that romantic love opens up is the existence of something that has value regardless of our needs, even regardless of our happiness or that of our beloved.

According to Lewis, what characterizes romantic love, which brings us to its moral dimension, is that it "wonderfully transforms what is *par excellence* a need-pleasure into the most *appreciative* of pleasures" (1960/2012: 115–116).[13] *Eros*—though its origin is our sexual desire, undoubtedly connected to our urge to reproduce—for human beings, opens up or connects us to a world of values, something that in and of itself is of value regardless of the pleasure, satisfaction, and even the happiness that it brings us or our beloved. In Lewis's account, the moral nature of romantic love has nothing to do with enabling us to bridge a gap between us and other human beings[14] but with enabling us to care for things unrelated to our needs and happiness or to those of people whom we love. In the next section, I elaborate on what exactly Lewis's point is in my view and take stock of how he enables us to understand love as the source of morality.

2. LOVE AND MORALITY

Lewis is aware that his view on friendship is quite out of vogue. His chapter on friendship is mainly devoted to arguing against all kinds of misperceptions about this kind of relationship (1960/2012: 69–109). Hence, he does not mean his description to capture what we call friendship in this day and age (Lewis 1960/2012: 70). More generally, he clarifies that most of our relationships will be a mixture of the kinds of loves that he distinguishes.[15] In a real-life friendship, affection and romantic love often occur together or develop

---

13. Lewis distinguishes among three modes in which we love—"need love" (as most children love their parents), "gift love" (as most parents love their children), and "appreciative love" (as many of us love certain things or persons for their own sake, regardless of their relation to us as individuals or of our needs). Certainly, these are not exclusive modes and might even never appear distinct from one another in our love relationships, and Lewis does not mean to claim that they are. "In actual life [they] mix and succeed one another, moment by moment" (Lewis 1960/2012: 21). The only one that might "ever exist alone, in "chemical" purity . . . is need love," which might be because "[n]othing about us except our neediness is . . . permanent."

14. In her forthcoming work, *Love and Morality* (or *Love's Complexities*), Barbara Herman criticizes philosophical accounts of love for wrongfully ignoring the obvious fact that we start life not as individual atoms that later in life occasionally bridge a gap between us and others in our special love relations, as friends or lovers.

15. This is similar to the modes in which we love (see note 13).

from or into each other (Lewis 1960/2012: 43). Hence, although it is easy to misinterpret him, he does not offer ideals of existing relationships. In my understanding, the point he makes is that our relationships with others can differ qualitatively and still be instances of and experienced as loving relationships. We can value others' company, attention, and opinions of us without necessarily longing to be with them sexually or feeling the need to take care or be taken care of by them (friendship). We can want and long to be with people sexually or otherwise, without sharing their interests and views on the world (*eros*). We can care a lot about others, want them to take care of us, and desire to take care of them without sharing their interests or longing to be with them sexually (affection). All these cases are experiences of loving others, and all bring authoritative motivational states, as mentioned in the introduction. In an analogy with these kinds of loves, perhaps originating from the same neurochemical processes, we can also love things, places, our hometown, a character in a television series, a particular seaside in Wales, our country, nature, or a poem about the red color of the Jewish bride's dress in Rembrandt's painting. Moreover, in an analogy with the kinds of loves that Lewis distinguishes, we can love these things in different ways, that is, in an (mixture of) affectionate, friendship-like, or romantic way.

With these qualitative distinctions in place, Lewis's taxonomy of loves enables us to understand the idea that our nature as loving beings is the source of morality, not by taking sides in the philosophical discussion on the "a-rational" or amoral nature of love. Such a discussion might very well have no answer if we are pluralists with respect to love—that is, if we believe with Lewis that different kinds of loves exist. Perhaps love is sometimes evaluative and rational (in the case of friendship), and, at other times, it is not (in the case of affection and *eros*); maybe some aspects of love are evaluative, while others are not. Sometimes, we fall out of love with someone when she or he has become a different person. Apparently, the friendship in that relationship was a more important determining feature than affection. Many parents love their children regardless of the kinds of persons they become; in such relationships, affection might be the most important characteristic. Some romantic relationships develop into strong affective ones, strong friendships, or a combination of the two. The relationships that evolve into strong affective ones are probably less evaluational, if at all, more "a-rational," if at all, than strong friendships. However, as Lewis observes, most relationships are mixtures of several kinds of loves; as I explain in the next section, their relation to morality is multifaceted as well. Hence, what do I mean by stating that our taxonomy can make sense of the idea that love is the source of morality?

Understood along Lewis's line of argument, our nature as loving beings opens up dimensions of our existence—to which philosophers have taken recourse—to explain our nature as moral beings, that is, our nature as caring and/or valuing beings, able to take the perspective of others and see the value of autonomy. Our affective responses to one another make us connect to other living beings and other things regardless of our individual rational nature (our specific views, evaluations, and experiences of things). On the other hand, friendship strengthens our relation with the world as individual rational beings with a particular perspective on it and a specific way of responding to it. Finally, romantic love makes us rise above ourselves as beings primarily motivated by the avoidance of pain and suffering and by the pursuit of happiness or the satisfaction of desires (ours and/or of those close to us).

As far as I can see, this account does not enable or require us to choose among particular ethical or meta-ethical views. Someone might still argue that only one of these dimensions is prototypical for morality and accordingly defend a Kantian, rationalist, or sentimentalist account of morality. Alternatively, someone might argue that an Aristotelian or another pluralist view best captures morality as we understand it. It is beyond the scope of this chapter to elaborate on this point. However, we can further discuss how each of these kinds of loves brings its own dangers, how Lewis introduces the fourth kind of love, and how this love relates to recent insights from affective neuroscience.

3. "TOO MUCH OF A GOOD THING" AND KINDNESS

Each of the natural loves that Lewis distinguishes has features and a certain dynamic that can do harm as well. This is how he arrives at a point where he proposes a fourth love that is needed to keep the other loves from becoming what he calls "demons." Christians refer to this love as "charity," which sounds to my non-Christian ears as a loving attitude of "care and attention" without it being directed (solely) at other persons because of their particular characteristics or relation to us. This kind of indiscriminate care and attention is not based on individual characteristics of the giver or the receiver; that is, it can be addressed to strangers. As mentioned, I ignore Lewis's Christian perspective on the fourth love and rename it "kindness."

According to Lewis, kindness starts with "'decency and common sense,' later being revealed as goodness" (1960/2012: 141), which is required to keep the other three loves "healthy" or prevent them from doing harm. Our relationship with others tells us when to reject or withdraw and make no concessions to our romantic or affective feelings, our friendship, or the person with whom we have this loving relationship (Lewis 1960/2012: 149). This kind of love also counterbalances the particularity of the other kinds although not by pitting universal rules and reason against the pull of our loving inclinations.[16]

---

16. Although rules and reason might tell us what kindness comprises, in certain situations, it is what tells us *how* to be kind.

Lewis observes that friendship—remember that his understanding of this relationship does not fit our modern understanding—has always been regarded with distrust by figures of authority (teachers and politicians), among others. The reason for this is the same as what makes friendship valuable from the perspective of the individuals involved. At the core of friendship is a withdrawal from the larger community, an affirmation of each individual involved in the friendship. This can create strength and enable a group to start something new and to change something for the better, but it can just as well be put to use for something wrongful. The withdrawal from the larger community can also originate from a shared and harmful purpose. Both Lewis (1960/2012: 95) and Churchland (2013: 125–126) remind us about the pleasure of strong social bonds brought about by shared hatred and grievance, which many of us had known as kids. Furthermore, Lewis points out that friendship in itself is a selective, undemocratic, and arrogant relationship by nature. Due to mutual admiration, friendship strengthens the individuals involved, but the downside is that this makes them inclined to seek each other's opinions (Lewis 1960/2012: 96), which might easily lead to general deafness and arrogance with respect to the opinions of the broader world. Kindness, an attentive and open attitude to others who do not belong to our in-group and to other things that do not belong to our field of interest, is needed to prevent friendship from taking this negative turn.

The danger with *eros* and with romantic lovers is not so much that they will withdraw from the community (turning their union into a mutual admiration society, as in the case of friendship)—that is, not as long as *eros* characterizes the relationship. According to Lewis, the danger is not to idealize the beloved or the relationship with her or him; rather, it is to idolize *eros* itself—to treat it as an authority without reservation. When people say they do things "out of love" in a romantic setting, Lewis observes (as many scholars have done before him) that they do not tend to mean this as an excuse; "they say the word love, not so much pleading 'extenuating circumstances' as appealing to an authority" (1960/2012: 136; cf. Jollimore 2011: 147–148; Zangwill 2013). This feature of love makes so many philosophers doubt that it can be a source of morality. Although *eros* has an unconditional nature, Lewis points out that its nature is fleeting and temporary. The first phase of *eros*, falling in love (as already noted), has a certain time span—luckily enough, for it is an exhausting state. As Lewis argues, the relationship that starts with romantic love needs the other loves to make true its promises of eternity. Each of us has qualities and characteristics that are not lovable; hence, we all need to show and receive kindness. The same is true for affectionate relationships and for similar reasons.

Lewis observes that since the mark of affection is that it is spontaneous and requires nothing, all of us expect to be loved by and to love those close to us (typically our family). Since not everything about us is lovable all the time, a battlefield of disappointment is just waiting to happen. Although this is not how Lewis states it, when we express it this way, it is reminiscent of the insight of attachment theory that many things can go wrong in these first years of our development in which we completely depend on our caregivers.[17] Moreover, the difficulty of affective relationships—the discussion now is about close relationships somewhat later in life—is that their proximity and ease require delicacy and subtlety to know what liberties we can and cannot take with one another (Lewis 1960/2012: 53–55). The rituals that we perform in public life regarding what we can or cannot say and do are often absent in the proximity of family life. In this respect also, the affectionate relationships that we have known from our early lives can easily go wrong. Last, Lewis observes a paradox in the affectionate love of caregivers. Humans not only need the care of others; they also need to care (Lewis 1960/2012: 40). Although the proper aim of caring for offspring is to put them in a state in which they no longer need it, this is somehow contrary to its very instinct (Lewis 1960/2012: 62). By this, he means that our caring impulse itself is directed at the good of the other but in relation to the care we provide. Hence, there is the danger of keeping the objects of this love needy or inventing imaginary needs; which, I guess, is easily recognized by many adult children.[18]

Many of the examples that Lewis provides are from family life, and, as noted before, he is quite aware that this kind of love derives from our mammalian nature (1960/2012: 49). Lewis believes, I think rightfully so, that our solid and long-lasting happiness is closely intertwined with the presence of affectionate love in our lives (1960/2012: 65–66). And he is very sensitive, again I think rightfully so, to the many ways in which affectionate relationships can go wrong. Although he does not discuss it in so many words, within the framework he offers, it is easy to see how the loving relationships with those who (are supposed to) take care of us in the first years of our lives are bound to turn up in our adult relationships.

More generally, Lewis believes that nearly all of affection's characteristics are ambivalent (Lewis 1960/2012: 48). Let me start with two of affection's more mundane characteristics. First, due to its immediacy, it can easily trump more important considerations. This of course can make actions done with affection harmful for others or even morally wrong. Second, affection is conservative by nature. Since it relies on or often comprises what is "old" and "familiar," whenever there is strong affection between people, jealousy is close

---

17. One of the important insights that Herman attempts to put central stage in our discussion of the relation between love and morality; also, she emphasizes that these first years might very well be crucial to the development of the capacities required for moral agency (Herman FC). See also note 9.

18. According to Lewis, other care relations that invite the same dynamic are that of master and protégé and the "pampering" relation that some people have with their pets.

by as well. In a sense, we don't want familiar faces and places to change—not even for the better (Lewis 1960/2012: 56). In both cases, we need common sense, decency, give and take, patience, self-denial, humility, and, above all, kindness to prevent affection, just like the other kinds of loves, from taking a bad turn (Lewis 1960/2012: 68).

At the psychological level, Lewis mentions two other dangers that characterize affection. He observes that it is easy to experience affection as the least selfish of our loves, especially when it makes us take care of others. This makes it easy to "use" affectionate relationships to silence doubts about our own importance and the quality of our feelings. We all know extreme cases of people sacrificing their lives in the service of others (children, parents, and relatives), but, Lewis warns us, this should not mislead us to the everyday nature of this temptation for each and everyone of us (1960/2012: 66–67). The same is true for the—closely related—second danger of affectionate love: the self-satisfying grievance and resentment at being "taken for granted," of not being appreciated enough. Here, too, we all know examples of people who cannot be loved enough, and here, too, this should not blind us to the fact that each and every one of us knows the "pleasure of resentment" (1960/2012: 68).

Hence, although "[a]ffection at its best wishes never to wound, nor to humiliate nor to domineer" (Lewis 1960/2012: 54), the dangers of doing so are very much part of the dynamic of our affective relationships. As with the other kinds of love, affection also needs kindness to be kept in check; that is, an attentive and caring attitude, even at moments or with respect to features of our loved ones we do not particularly care for. Likewise, we need to realize and accept that such kindness is shown to us as well, as part of our affectionate relationships. That might sound much easier than it actually is because it requires us to allow that even that most "immediate" and, in many senses, basic ingredient of our loving relationships to others is deeply ambivalent.

This concludes my discussion of Lewis's account of love. The moral dimensions opened up by the kinds of loves that he distinguishes neatly line up with the capacities put on central stage by recent scientific interest in love as the source of morality or, more precisely, in oxytocin and vasopressin. As noted in the introduction, according to Churchland, these hormones play a role in reproduction, attachment, and, more broadly, our sociability. This sociability includes our capacity to anticipate (evaluate and interpret) the behaviors and feelings of others and a "neural reward-and-punishment system linked to internalizing social practices and applying them suitably" (Churchland 2011: 61; see also Chapter 2 this volume). This capacity might express itself as what I have called "kindness." Thus far, I have discussed kindness in relation to the other loves. Attending to, considering, and caring for and about others, regardless of their individual characteristics or relation to ourselves, also characterize our friendly, trusting, and cooperative interactions with strangers.

As noted in the introduction, over the past decades, scientists have identified the brain mechanisms and hormones that seem to be involved in the care of dependent offspring and in sexual reproduction as well as in our social interactions with strangers. This has even led to the discussion about oxytocin as the "love" or "cuddle" molecule. It is exciting that oxytocin can be administered to human beings via nasal spray. Hence, it can and is used in numerous experiments to influence cooperative processes; for example, in behavioral economics and social psychology. In developmental psychiatry as well, scientists have investigated how oxytocin affects individuals with social impairments. For these reasons, it has been argued that our growing knowledge about the brain mechanisms involving these hormones is exciting and concerns nothing less than the biological sources of friendship, love (Panksepp 1998: 246), and morality (Churchland 2011, Chapter 2 this volume). The fact that oxytocin and vasopressin have effects on people who lack the natural capacity to anticipate (evaluate and interpret) the behaviors and feelings of others, as well as influence how we interact with strangers, suggests that kindness is also a natural enough kind of love. As mentioned, Churchland and other scholars warn us that, so far, the experimental results obtained with oxytocin are weak and difficult to interpret. This, of course, might suggest that scientists are on the wrong track in believing that they have identified the biological platform of morality. On the other hand, it might also imply that, similar to the other kinds of loves, kindness is dynamic and complex as well.[19]

4. CONCLUSION

In the picture proposed by Lewis, love is a dynamic phenomenon. It can easily accommodate the fact that our everyday love relationships are messy, subject to change, can take a bad turn, and might do harm and make us behave in ways that we disapprove of or are even immoral. Nonetheless, it can also accommodate the idea that our nature as loving beings explains our nature as moral beings, be it in a broad and general manner; that is, without deciding on a specific view on morality. The moral dimensions of our existence that are opened up by the kinds of loves that Lewis distinguishes are exactly those that philosophers have traditionally adopted to explain morality. These include our natural inclination to care about and for others, the ability to take the perspective of others, our nature as rational beings with specific individual perspectives and ways of responding to the world, and our nature as valuing beings. Many opinions can and have been expressed about these explanations.

---

19. Lewis has worthwhile things to say about the darker sides of his fourth kind of love—charity—as well, and many of those things are of interest to a discussion of kindness. This will have to wait for another opportunity.

However, the idea that our nature as loving beings explains our moral nature is appealing, regardless of the specific moral view that someone favors. At least, I have attempted to make this perspective plausible, with the help of Lewis.

ACKNOWLEDGMENTS

Thanks to Gregg D. Caruso, Arno Wouters, and Nicole van Voorst Vader Bours for comments on earlier drafts of this chapter.

REFERENCES

Bartz, J. A., J. Zaki, N. Bolger, and K. N. Ochsner. 2011. Social effects of oxytocin in humans: Context and person matter. *Trends in Cognitive Science* 5(7): 301–309.

Churchland, P. 2011. *Braintrust*. Princeton, NJ: Princeton University Press.

Churchland, P. 2013. *Touching a nerve. Our brains, our selves*. New York: Norton.

De Sousa, R. 2015. *Love: A very short introduction*. New York: Oxford University Press.

Fischer, H. E. 1994. *Anatomy of love: A natural history of mating, marriage, and why we stray*. London: Ballantine Books.

Fisher, H. E. 2014. The tyranny of love: Love addition—an anthropologist's view. In L.C. Feder and K. Rosenberg, eds. *Behavioral addictions: Criteria, evidence and treatment*. Elsevier Press.

Herman, B. Forthcoming. Love and morality (or Love's complexities). Paper delivered at Utrecht University, September 18, 2012.

Hobson, P. 2002. *The cradle of thought: Exploring the origin of thinking*. Basingstoke and Oxford: Pan Books.

Jollimore, T. A. 2011. *Love's vision*. Princeton, NJ: Princeton University Press.

Lewis, C. S. 1960/2012. *The four loves*. London: Collins.

Panksepp, J. 1998. *Affective Neuroscience: The foundations of human and animal emotions*. New York: Oxford University Press.

Schaubroeck, K. 2011. Moeten we altijd doen wat goed is? Harry Frankfurt en de foundering van de moraal [Should we always be good? Harry Frankfurt and the foundation of morality]. In K. Schaubroeck and T. Nys (Eds.), *Vrijheid, noodzaak en liefde. Een kritische inleiding tot de filosofie van harry frankfurt* [*Freedom, necessity and love: A critical introduction to Harry Frankfurt's philosophy*], pp. 163–182. Kapellen/Kampen: Pelckmans/Klement.

Van IJzendoorn, M. H. 1997. Attachment, emergent morality, and aggression: Toward a developmental socioemotional model of antisocial behaviour. *International Journal of Behavioral Development* 21(4): 703–728.

Zangwill, N. 2013. Love: Gloriusly amoral and arational. *Philosophical Explorations* 16(3): 298–314.

CHAPTER 4

Does Neuroscience Undermine Morality?

PAUL HENNE AND WALTER SINNOTT-ARMSTRONG

Imagine that you have an opportunity to kill your neighbor and steal her money. You hate your neighbor, and you need money. There is almost no chance that you will be caught, and you won't suffer any bad consequences, so it is in your interest to kill her and steal her money. Still, most people think that it would be morally wrong for you to kill and steal from her. This common moral judgment stops you from killing and stealing. Here, your moral judgment matters.

The same point applies in other cases. Suppose your neighbor is drowning in her pool so that you do not have to kill her—only let her die. If you dislike her, you might let her drown—if you did not think that doing so is immoral. And suppose you don't have to kill her or let her die; you can steal from her without getting caught, and she has so much money that she won't even miss what you steal. Then you would be more likely to steal—if you did not think stealing is morally wrong. Similarly for lying, cheating, breaking promises, and so on, not only to our neighbors but also to our spouses, bosses, friends, and family. Our moral judgments often prevent us from misbehaving when tempted.

The problem is that the seemingly stable underpinning of those ordinary moral judgments that keep us in line seem to conflict with recent findings in neuroscience and psychology. Many studies suggest that many kinds of moral judgments are deeply flawed—they're emotional, inconsistent, based on a distant evolutionary past, susceptible to racial and gender biases, and so on. If this research shows that moral judgments are unreliable, then it makes them difficult to believe. And if we do not believe them, then we might be more likely to misbehave. That's the problem.

But neuroscience does not undermine all of our moral judgments. Hence, we do not have to eschew or retract all of them. And we can trust many of our ordinary moral judgments—such as that it is wrong to kill your neighbor for her money. Here, we will show why.

1. WHICH CHALLENGE?

Our opponents have claimed that neuroscience and psychology undermine morality in a number of distinct ways. Here are a few:

(A) If neuroscience undermines free will and moral responsibility, then it indirectly undermines all moral judgments that require such freedom or responsibility. Specifically, suppose that acts are morally wrong only when agents who do them intentionally deserve blame and punishment, but nobody deserves blame or punishment unless they have free will. Then any science that shows that agents have no free will indirectly show that nothing agents do is morally wrong. In that case, it is morally permissible for you to kill and to steal from your neighbor when you hate her and need the money. So, why not do it? (cf. Pereboom 2001; Pereboom and Caruso, Chapter 11 this volume)

(B) Even if neuroscience does not undermine moral wrongness judgments, some people still might believe that it does. This false belief might then lead people to commit certain immoral actions because they think that they will not be responsible (cf. Baumeister, Masicampo, and DeWall 2009; Vohs and Schooler 2008). Here, neuroscience undermines moral behavior, even if not moral beliefs.

(C) Some have proposed that all (or almost all) moral reasoning is actually post hoc rationalization (Haidt 2001). On this account, discoveries in psychology and neuroscience show us that people's moral beliefs and judgments are not based on reasoning. Rather, they rely on intuition and emotional processing to produce moral judgments. These conclusions would seem to undermine morality on any view that sees morality as essentially rational.

(D) Any theory of moral judgments as a whole seems to assume that morality is a unified field. In the physical sciences, there cannot be an adequate theory that applies to all and only metals unless metals are a unified class. Similarly, if moral judgments are not a unified class—because nothing really distinguishes moral judgments from religious, conventional, legal, and aesthetic judgments—then there cannot be any adequate theory of morality. One (but not the only) way for moral judgments to be unified would be for a distinctive neural system to produce them. Thus, if neuroscience shows that moral judgments are not unified as such, it might undermine all theories of moral judgment as such (see Sinnott-Armstrong and Wheatley 2014).

(E) Neuroscience might also show that humans don't detect real moral properties in the world but instead project their sentiments, attitudes, and beliefs onto the world. Some people seem more beautiful than others, but that appearance is only in the eye of the beholder (see Serling 1960). If neuroscience explains moral judgments in the same way as we explain judgments of beauty, without assuming that anything really is morally right or wrong, then—given Occam's razor—neuroscience might undermine any realist moral metaphysics (see Greene 2003).

(F) Even if there are moral properties independent of us, there is a question whether or not people can know them. We might think that people do not attribute knowledge to beliefs when they result from unreliable processes. If my color vision, for example, is unreliable in dim light at long distances, and I see your car only in dim light at a long distance, then, even if your car seems red to me, and even if it is red, I do not know that it is red. Similarly, neuroscience and psychology might undermine moral knowledge if they showed that our moral beliefs result from unreliable processes. Moreover, they might suggest that moral judgments are unreliable if they showed that moral judgments are affected by factors that everyone takes to be morally irrelevant (like race). In this way, neuroscience could undermine moral epistemology (see Sinnott-Armstrong 2006, ch. 7).

While A–E are worthy of discussion and will be addressed in other work, we will make only a very few all-too-quick responses to them here—simply to show that they are answerable:

(A*) Some defenders of morality try to meet challenge (A) by proposing compatibilist theories of free will and moral responsibility along with theories of mental causation (see Sinnott-Armstrong 2012). Additionally, some free will skeptics argue that rejecting moral responsibility does not eschew morality generally. For example, Pereboom and Caruso argue in Chapter 11 of this volume that "free will skepticism can accommodate judgments of moral goodness and badness, which are arguably sufficient for moral practice" (see also Caruso 2012; Pereboom 2001, 2014).

(B*) Denials of free will might lead to some minor immoral acts in some circumstances without making people kill and steal from their neighbors and families (see Evans 2013). Many also doubt the findings of Baumeister and Vohs (see Caruso 2017: 12–13).

(C*) A number of psychologists and philosophers have suggested that moral reasoning is not always post hoc and that this accounts for many moral judgments (e.g., Paxton and Greene 2010).

(D*) We might be able to develop theories about subsets of moral judgments or about all normative judgments (including nonmoral normative judgments) even if we cannot defend any theory of morality per se (see Sinnott-Armstrong and Wheatley 2014).

(E*) Even if our moral judgments are associated with or engendered by emotions, that does not show that these judgments merely project those emotions onto a morally neutral world. Moreover, whether we can explain moral judgments or other phenomena without assuming that any moral judgments are true depends on what makes moral judgments true; so, moral truth might be needed for explanations according to some theories of moral judgment (e.g., Copp 2008).

Of course, each of these challenges deserves much more attention than we can give here.

In this short chapter, we will focus on Challenge (F). Given our reading of (F), this basic argument that neuroscience undermines morality runs like this:

(1) Discoveries in neuroscience and psychology show that moral judgments result from processes that are unreliable.
(2) If a judgment results from an unreliable process (that the believer knows or should know is unreliable), then the agent doesn't know that the judgment is correct.
(3) Therefore, nobody (who knows or should know about the discoveries in (1))[1] knows whether any moral judgment is correct.

This argument is valid, but we doubt that it is sound. Specifically, we will show that background arguments that are supposed to bolster Premise (1) fail to support that crucial premise in the way that is required for the argument to succeed.[2]

---

1. The parenthesis in Conclusion (3) is needed if the parenthesis is included in Premise (2), and that parenthesis in Premise (2) is needed in order to make Premise (2) plausible. We will not discuss this complication because we want to focus on Premise (1).

2. There also might be reason to doubt Premise 2, such as in Turri (2016). That discussion is beyond the scope of this chapter.

2. THE DISUNITY OF MORALITY

Before we examine the support for the premise, it is unclear whether Premise (1) is about *all* moral judgments or only *some* moral judgments. If it is about only some moral judgments, then Conclusion (3) about "any" moral judgment does not follow. So, Premise (1) needs to be about all moral judgments:

(1*) Discoveries in neuroscience and psychology show that *all* moral judgments result from processes that are unreliable.

But then the problem is that neuroscientists and psychologists have tested only a small sample of the vast array of moral judgment types. As a result, "discoveries in neuroscience and psychology" do not come close to undermining all moral judgments, so Premise (1) is false.

Opponents might reply that researchers have tested a representative sample of moral judgment types, so they can generalize to all moral judgments. Such generalizations, however, fail when there are relevant differences within the class of judgments, and that is exactly what we find among moral judgments. For example, Greene et al. (2001) found that personal moral dilemmas produced increased activity in areas associated with social and emotional processing: medial frontal gyrus, posterior cingulate gyrus, and bilateral superior temporal sulcus. Impersonal and nonmoral dilemmas produced increased activity in areas associated with working memory: dorsolateral prefrontal and parietal areas. Notably, they also found comparatively little difference between the impersonal-moral and nonmoral conditions, suggesting that impersonal moral judgment has less in common with personal moral judgment than with certain kinds of nonmoral practical judgment.

Parkinson et al. (2011) found further evidence suggesting that different kinds of moral judgments result from distinct neural processes. In their study, they looked at brain activation during moral judgments across three types of moral transgressions: physical harm, dishonesty, and sexual disgust. They found that different areas of the brain were activated for the different types of moral transgression. For physical harm transgressions, areas associated with action understanding activated. For dishonesty, areas associated with representing other people's beliefs were activated. And for sexual disgust, areas associated with affective processing were activated. In fact, the only brain area active across different moral transgressions was the dorsomedial prefrontal cortex, which is activated for all social conflicts, whether moral or not.

These studies show that research on a subset of moral judgments cannot by itself be enough to support a universal conclusion about all moral judgments. The differences among moral judgments preclude such hasty generalization.

Another reply is that experiments showing unreliability in a subset of moral judgments can still shift the burden of proof onto those who claim that other moral judgments are reliable. Consider again a person who has found that he is often mistaken about cars' colors in dim light at a distance but now claims to be sure about the colors of Ford cars in dim light from a distance. Even if his vision has been tested only on cars that are not Fords, these findings extend to other cars unless he gives some reason why his vision is better for Fords than for other cars. Of course, he still might be a reliable detector of colors in bright light at short distances, but that needs to be shown in light of his deficits in so many situations. Analogously, even if our moral judgments have been tested only in certain contexts, these findings might support the claim that other moral judgments are also unreliable unless there is some reason to think that these other moral judgments are more reliable than those tested.

All this shows is that people who believe that the processes behind some moral judgments are reliable need some reason to believe that those particular processes or moral judgments are reliable. None of what we have claimed so far implies that they cannot carry that burden of proof. They still might be able to use coherence with other moral and nonmoral judgments and emotions to corroborate certain moral judgments and show that they depend on reliable processes (see Sinnott-Armstrong 2008, ch. 10). We often have special reasons to believe in reliability. For instance, some moral judgments seem especially obvious, are widely accepted, or persist after careful reflection and discussion. Our moral beliefs might be justified in such cases, even if they are unjustified in other circumstances. If so, then neuroscience cannot undermine all moral judgments. Consequently, we must limit our first premise to:

(1**) Discoveries in neuroscience and psychology show that *some* moral judgments result from processes that are unreliable.

Of course, the conclusion (3) will need to be limited in a parallel fashion in order to keep the argument valid. Thus, (3) becomes:

(3**) Therefore, nobody (who knows or should know about the discoveries in (1**)) knows whether the judgments in (1**) are correct.

3. CONSERVING AND REVISING

According to this revised argument, neuroscience still might undermine some large and significant classes of moral judgments. Then the question is how to distinguish the moral judgments that are undermined from the ones that are not.

Many philosophers seem to believe that this distinction cannot be drawn without substantive assumptions about which moral judgments are true. Without knowing what is true, we can't determine which processes reliably lead to truth, so we can't determine reliability if we can't know what is true about controversial moral issues—or so they say.

Another option is to appeal to higher order principles that are neutral on substantive moral issues but still distinguish reliable from unreliable judgments. Consider a normative principle about genuine ordering effects (Horne and Livengood 2015). Suppose subjects are given two pieces of evidence: E_1 and E_2. Half of the subjects are given the evidence in order E_1 then E_2, while the other half is given the same two bits of evidence in the reverse order. Then subjects are asked to give judgments of whether a certain act is morally wrong. If the judgments of subjects in the two groups differ, there is a genuine order effect. Such order effects reveal unreliability because the truth about whether or not an act is wrong cannot change just because of the order of the evidence (unless your view allows for such inconsistency). Different people might encounter evidence in different orders, and one person might encounter it in different orders on different days, but our moral judgments, one might think, need to be consistent in order to be reliable. Suppose that Framish judges a lie to be immoral when told that the agent lied to his friend and then told that it occurred during a business deal. But then suppose that Framish reaches the opposite judgment when told that an agent engaged in a business deal and then was told that the agent lied to his friend (cf. Haidt and Baron 1996). Framish's judgment cannot be correct on both occasions, so Framish is correct at most half of the time. The processes that lead to such conflicting judgments are, to this extent, unreliable.

Such cases lead to a normative principle like this:

(OEUR)[3] If anyone's moral judgments exhibit genuine order effects, then these moral judgments are unreliable.

This principle applies when moral judgments are subject to genuine order effects, so we can know that the processes that lead to moral judgments are unreliable at least to the extent that they are subject to such order effects (Sinnott-Armstrong 2008: 2011).[4] Thus, this principle would support the first

---

3. *OEUR*, Order Effects Undermine Reliability.
4. Horne and Livengood (2015) claim that the order effects discussed in Sinnott-Armstrong (2008) involve updating effects instead of genuine order effects. We disagree, but we will not reply to them here because our points here work even on their restricted understanding of order effects.

premise (1**) of the main argument and provide a way to eschew such moral judgments as unreliable. That would undermine some moral judgments.

Of course, this kind of argument will not work when order effects do not occur, and similar arguments will not work if we do not find true inconsistencies. They do not occur, for instance, in obvious cases like killing and stealing from your neighbor. Even in areas where they do occur, only a subset of people are subject to order effects (Demaree-Cotton 2014). And even those who are subject to order effects might be able to correct their moral judgments in light of reflection and reasoning. Thus, we will be left with a lot of moral judgments that cannot be undermined in this way—at least directly. Nonetheless, many other moral judgments still might be unreliable for reasons other than order effects.

Another source of unreliability in moral judgment, according to many commentators, is emotion. Their claim is not that all emotions always produce unreliability. Instead, they claim (or need to claim) only that certain emotions make moral judgments less reliable in certain circumstances. Consider disgust. Although controversy remains, some experiments suggest that many moral judgments (such as judgments that homosexuality is immoral) are based on disgust. Some philosophers argue that disgust is an unreliable basis for moral judgment (Kelly and Morar 2014; Nussbaum 2004). However, others reply that disgust is reliable at least in some areas (Kass 1997; Plakias 2013).

In order to assess this dispute, we can contrast disgust in response to features of an act from disgust triggered by the environment in which the moral judgment was made. Consider a study where participants were hypnotized to feel disgust whenever they read a particular word (Wheatley and Haidt 2005). When the participants were asked to rate acts in a vignette that included the stimulus word, they judged more harshly. In a few cases, they even judged an act to be immoral that nobody who was not hypnotized saw as immoral. Everyone agrees that whether the person who made the judgment was hypnotized cannot make any difference to whether another person's act is immoral, so the study shows that disgust leads to unreliability here (but see Huebner, Dwyer, and Hauser 2009; May 2014). To show unreliability in this kind of case, we don't need to appeal to any substantive moral judgments. The conclusion follows from a widely held view that moral wrongness of an act cannot depend on feelings in the person who is judging the act. We do not dispute such easy cases of unreliability.

In contrast, there are other cases where a person makes a moral judgment on the basis of feeling disgust at some feature of the act being judged, such as when someone judges that incest or political corruption is immoral because it is disgusting. When disgust results from facts of the case, it is more difficult to determine whether moral judgments based on such disgust are reliable. How can we tell whether this kind of disgust is or is not a reliable source? And how much and what kind of unreliability is enough to undermine moral judgments in the sense of showing that we should not trust them?

One's answer to these questions depends one's higher order inclinations. It is useful to distinguish being *conservative* or *revisionist* about moral judgments (Knobe and Doris 2010; Kumar and Campbell 2012; Vargas 2004). Conservatives think that ordinary moral judgments are for the most part right. Hence, we should accept these first-order moral judgments and develop our moral theories to accord with them. Revisionists, on the other hand, think that moral judgments often deviate from what theories give us reason to believe is correct, and then moral judgments should be revised to bring them in line with theory. Of course, many people waver or are intermediate between these extremes.

Consider a case where experiments suggest that moral judgments depend on disgust. Some revisionists will be more inclined to distrust or even reject those moral judgments because disgust often leads us astray. In contrast, some conservatives will be more inclined to trust those moral judgments despite knowing that disgust sometimes leads us astray because disgust also sometimes prevents us from making mistakes (see Strohminger 2014: 480). How one reacts to the finding that the moral judgment depends on disgust thus depends on one's higher order inclinations or second-order beliefs. Some people prioritize the avoidance of false positives (thinking an act is wrong when it is not), whereas others prioritize the avoidance of false negatives (thinking an act is not wrong when it is). Given these second-order inclinations, the former will be revisionist about pretheoretical judgments that an act is morally wrong—favoring the moral theory—and the latter will be conservative about these judgments—maintaining them even when they conflict with moral theory.

The same problem arises with regard to empathy. Sometimes empathy leads to altruistic actions that we see as morally good (Batson 2011). However, sometimes empathy leads us to break moral rules or favor well-off neighbors over needy strangers (Bloom 2014; Prinz 2011). It is unclear how we should respond if we find that a certain class of moral judgments depends on empathy. Higher order beliefs about the priority of avoiding false positives or false negatives will lead some to be revisionists and others to be conservatives.

Similarly, sometimes fear is a reliable indicator of danger, perhaps when that fear is based on experience (Allman and Woodward 2008). For instance, an abused wife might reliably be afraid of her husband. But fear can also mislead us. Just as dogs can be frightened by balloons, so we might be frightened of scientific progress far beyond any real danger. Fear can lead to questionable moral judgments (as in the case of cognitive enhancers) (Buchanan 2009).

Any single emotion—disgust, empathy, fear, and so on—can be unreliable in some cases but not others. Such emotions react to various features of the case, so our beliefs about which features are relevant will affect our claims about what is reliable or unreliable. If we discover that some feature we take to be morally irrelevant (such as race) influences a process involved in producing moral judgments, then we will have grounds for taking that process to be unreliable. At a higher level, our beliefs about whether it is more important to avoid false positives or false negatives will also affect whether a certain degree or kind of unreliability is too much and whether a certain moral judgment is too unreliable to be trusted. Thus, how we react to the scientific news about the processes that produce our moral judgments will depend on our background assumptions and our proclivities toward revisionism or conservatism.

What does this show about the main argument? Recall the first premise:

(1**) Discoveries in neuroscience and psychology show that *some* moral judgments result from processes that are unreliable.

We have argued that which moral judgments are the *some* moral judgments in (1**) is unclear. Even if we can eliminate some judgments (like the easy cases discussed earlier) as unreliable, we cannot determine which other classes of judgments satisfy this premise without appealing to substantive moral principles. Which moral principles one is willing to assume depends on one's beliefs about which facts are morally relevant as well as one's higher order inclinations toward revisionism or toward conservatism.

A similar point applies to the second premise:

(2) If a judgment results from an unreliable process (that the believer knows or should know is unreliable), then the agent doesn't know that the judgment is correct.

Here again second-order beliefs matter. Whether one accepts or rejects this premise depends on how much unreliability is shown, but it also depends on one's higher order inclinations. Revisionists will deny moral knowledge when experiments reveal significant unreliability, but conservatives will continue to claim moral knowledge despite acknowledging the same degree of unreliability. For example, Demaree-Cotton (2014) concludes, "We . . . have some reason to suspect that . . . modest framing effects . . . , which tended to lead to approximately 80 percent reliability, are particularly powerful, and that they do not support the claim that moral intuitions in general are unreliable." Although 80 percent seems enough to satisfy Demaree-Cotton, 80 percent will not satisfy other philosophers with more revisionist inclinations.

The literature includes moderates (Kumar and Campbell 2012) as well as extreme conservatives (Doris, Knobe, and Woolfolk 2007) and extreme revisionists (Vargas 2004). The problem is that we see no adequate way to adjudicate among these positions. When a moral judgment seems correct, but we know that it comes from a source that is somewhat unreliable in somewhat similar circumstances, we have few resources to help us decide whether or not to trust it. This decision cannot be made easier with reflection and reasoning about the judgments. As we have mentioned, higher order inclinations do a lot of work here; so, reflection and reasoning will lead different people to different conclusions. The issue is largely a matter of how we prioritize kinds of errors, which will vary with context and practical concerns. Our second-order beliefs about what is and is not morally important will lead us to revise or to conserve our judgments. These beliefs and inclinations will differ drastically among thinkers.

This particular dispute between conservatism and revisionism needs to be resolved in order to assess Premises (1**) and (2). Our inability to resolve the dispute thus renders the central argument at least questionable. We cannot, then, determine which moral judgments are reliable enough. Hence, we see no adequate reason to claim or deny that neuroscience undermines morality in the relevant sense.

4. WHAT IS LEFT FOR NEUROSCIENCE TO DO?

We will close by identifying a different role that neuroscience can play in moral theory: Neuroscience can affect how we subdivide morality and moral judgments.

On the one hand, neuroscience can show us that subsets of moral judgments that seem very different actually resemble each other in unexpected ways. For example, Hsu et al. (2008) show that activity in parts of anterior insula predicts aversion to inequity. These same areas are related to disgust at bodily fluids, incest, and homosexuality in conservatives. Of course, much more work, including replication and extension, is needed on this topic. Nonetheless, this experiment suggests how neuroscience can lead us to think in new ways about moral judgments. These new ways of subdividing might even affect which moral judgments we take to be based on reliable processes. If the same processes lead to judgments about inequity that lead to moral judgments about homosexuality, then we cannot hold that one of these kinds of moral judgment is based on a reliable process but the other is not (as many political liberals and conservatives claim)—although truth is a different story.

Conversely, neuroscience can show us that a natural grouping of moral judgments is not really as unified as it seems. Some neuroscience suggests that moral judgments about harm depend on a variety of processes. Greene et al. (2001) distinguish personal and impersonal dilemmas. Heekeren et al. (2005) show that bodily harm is processed differently than other kinds of harm, and Clifford et al. (2015) suggest that physical and emotional harm as well as physical harm to persons versus animals are processed differently. If these results stand up, they might show that philosophers and psychologists could make more progress and achieve more precision if they stopped trying to construct moral theories about harm in general and instead focused on one kind of harm at a time.

5. CONCLUSION

The overall issue of neuroscience undermining morality is unsettled. We need more scientific research and philosophical reflection on this topic. Still, we can reach some tentative and qualified conclusions. Neuroscience and psychology do not undermine all moral judgments as such, but they still might play an ancillary role in an argument that undermines some moral judgments. They also might lead us to think about moral judgments in new ways, such as by suggesting new divisions among moral judgments (see, e.g., Chituc et al. 2016). Neuroscience is, then, not a general underminer, but a trimmer and a categorizer. In these ways, neuroscience can play a constructive role in moral theory, although not by itself. In order to make progress, neuroscience and normative moral theory must work together.

ACKNOWLEDGMENTS

We thank MAD Lab—and Aaron Ancell and Michael Campbell, in particular—as well as Gregg Caruso for their comments on this draft.

REFERENCES

Allman, J., and J. Woodward. 2008. What are moral intuitions and why should we care about them? A neurobiological perspective. *Philosophical Issues* 18: 164–185.
Batson, C. D. 2011. *Altruism in humans*. New York: Oxford University Press.
Baumeister, R. F., E. J. Masicampo, and C. N. DeWall. 2009. Prosocial benefits of feeling free: Disbelief in free will increases aggression and reduces helpfulness. *Personality and Social Psychology Bulletin* 35(2): 260–268.
Bloom, P. 2014. Against empathy. *Boston Review*. Available online: http://bostonreview.net/forum/paul-bloom-against-empathy
Buchanan, A. 2009. Moral status and human enhancement. *Philosophy & Public Affairs* 37(4): 346–381.
Caruso, G. 2012. *Free will and consciousness: A determinist account of the illusion of free will*. Lanham, MD: Lexington Books.
Caruso, G. 2017. Free will skepticism and its implications: An argument for optimism. In E. Shaw and D. Pereboom (Eds.), *Free will skepticism in law and society*. Cambridge University Press.
Chituc, V., P. Henne, W. Sinnott-Armstrong, and F. De Brigard. 2016. Blame, not ability, impacts moral "ought" judgments for impossible actions: Toward an empirical refutation of "ought" implies "can." *Cognition* 150: 20–25.

Clifford, S., V. Iyengar, R. Cabeza, and W. Sinnott-Armstrong. 2015. Moral foundations vignettes: A standardized stimulus database of scenarios based on moral foundations theory. *Behavior Research Methods* 47(4): 1178–1198.

Copp, D. 2008. Darwinian skepticism about moral realism. *Philosophical Issues* 18(1): 186–206.

Demaree-Cotton, J. 2014. Do framing effects make moral intuitions unreliable? *Philosophical Psychology*: 1–22.

Doris, J. M., J. Knobe, and R. L. Woolfolk. 2007. Variantism about responsibility. *Philosophical Perspectives* 21(1): 183–214.

Evans, J. 2013. The moral psychology of determinism. *Philosophical Psychology* 26(5): 639–661.

Greene, J. 2003. From neural 'is' to moral 'ought': What are the moral implications of neuroscientific moral psychology? *Neuroscience* 4: 847–852.

Greene, J. D., R. B. Sommerville, L. E. Nystrom, J. M. Darley, and J. D. Cohen. 2001. An fMRI investigation of emotional engagement in moral judgment. *Science* 293(5537): 2105–2108.

Haidt, J. 2001. The emotional dog and its rational tail: A social intuitionist approach to moral judgment. *Psychological Review* 108: 814–834.

Haidt, J., and J. Baron. 1996. Social roles and the moral judgement of acts and omissions. *European Journal of Social Psychology* 26(2): 201–218.

Heekeren, H. R., I. Wartenburger, H. Schmidt, K. Prehn, H. P. Schwintowski, and A. Villringer. 2005. Influence of bodily harm on neural correlates of semantic and moral decision-making. *Neuroimage* 24(3): 887–897.

Horne, Z., and J. Livengood. 2015. Ordering effects, updating effects, and the specter of global skepticism. *Synthese*: 1–30.

Hsu, M., C. Anen, and S.R. Quartz. 2008. The right and the good: Distributive justice and neural encoding of equity and efficiency. *Science* 320: 1092–1095.

Huebner, B., S. Dwyer, and M. Hauser. 2009. The role of emotion in moral psychology. *Trends in Cognitive Sciences* 13(1): 1–6.

Kass, L. R. 1997. The wisdom of repugnance: why we should ban the cloning of humans. *New Republic* 216(22): 17–26.

Kelly, D., and N. Morar. 2014. Against the yuck factor: On the ideal role of disgust in society. *Utilitas* 26(02): 153–177.

Knobe, J., and J. Doris. 2010. Strawsonian variations: Folk morality and the search for a unified theory. *The handbook of moral psychology*, pp. 321–354. New York: Oxford University Press.

Kumar, V., and R. Campbell. 2012. On the normative significance of experimental moral psychology. *Philosophical Psychology* 25(3): 311–330.

May, J. 2014. Does disgust influence moral judgment? *Australasian Journal of Philosophy* 92(1): 125–141.

Nussbaum, M. 2004. *Hiding from humanity: Disgust, shame and the law*. Princeton, NJ: Princeton University Press.

Parkinson, C., W. Sinnott-Armstrong, P. E. Koralus, A. Mendelovici, V. McGeer, and T. Wheatley. 2011. Is morality unified? Evidence that distinct neural systems underlie moral judgments of harm, dishonesty, and disgust. *Journal of Cognitive Neuroscience* 23(10): 3162–3180.

Paxton, J., and J. Greene. 2010. Moral reasoning: Hints and allegations. *Topics in Cognitive Science* 2: 511–527.

Pereboom, D. 2001. *Living without free will*. New York: Cambridge University Press.

Pereboom, D. 2014. *Free will, agency, and meaning in life*. Oxford: Oxford University Press.

Plakias, A. 2013. The good and the gross. *Ethical Theory and Moral Practice* 16(2): 261–278.

Prinz, J. 2011. Against empathy. *The Southern Journal of Philosophy* 49(1): 214–233.

Sinnott-Armstrong, W. 2006. *Moral skepticisms*. New York: Oxford University Press.

Sinnott-Armstrong, W. 2008. Framing moral intuitions. In W. Sinnott-Armstrong (Ed.), *Moral psychology* (vol. 2): *The cognitive science of morality*, pp. 47–76. Cambridge, MA: MIT Press.

Sinnott-Armstrong, W. 2012. Free contrastivism. In M. Blaauw (Ed.), *Contrastivism in philosophy* pp. 134–153. Routledge.

Sinnott-Armstrong, W., and T. Wheatley. 2014. Are moral judgments unified? *Philosophical Psychology* 27(4): 451–474.

Serling, R. (Writer), and D. Heyes (Director). 1960. The eye of the beholder. In R. Serling (Executive Producer), The Twilight Zone. CBS.

Strohminger, N. 2014. Disgust talked about. *Philosophy Compass* 9: 478–493.

Turri, J. 2016 A new paradigm for epistemology from reliabilism to abilism. *Ergo* 3 (8): https://quod.lib.umich.edu/e/ergo/12405314.0003.008/--new-paradigm-for-epistemology-from-reliabilism-to-abilism?rgn=main;view=fulltext

Vargas, M. 2004. Responsibility and the aims of theory: Strawson and revisionism. *Pacific Philosophical Quarterly* 85(2): 218–241.

Vohs, K. D., and J. W. Schooler. 2008. The value of believing in free will encouraging a belief in determinism increases cheating. *Psychological Science* 19(1): 49–54.

Wheatley, T., and J. Haidt. 2005. Hypnotic disgust makes moral judgments more severe. *Psychological Science* 16(10): 780–784.

CHAPTER 5

The Neuroscience of Purpose, Meaning, and Morals

EDMUND T. ROLLS

In this chapter, I build on detailed evidence and theories about the neural bases of emotion (Rolls 2013, 2014a, 2014b, 2016a) and their implications (Rolls 2012b) that are described elsewhere and further develop the ideas that arise about purpose, meaning, and ethics.

I emphasize that a great deal of evidence is available in the sources cited and that evidence provides the foundation for the ideas considered further here.

1. THE NEUROSCIENCE OF PURPOSE

There are a number of ways in which the notion of "purpose" can be approached in neuroscience.

1.1. Gene Replication and Purpose

One biological sense of purpose is that life is kept going by the self-replicating mechanisms of reproduction. The reproduction can be asexual, with evolution driven mainly by gene mutation, or sexual, which has the added advantage that different genes can be brought together in new combinations, which facilitates local hill-climbing in the high dimensional space that genetics can search (Rolls 2014a, 2016a). Some events, of course, have to facilitate the start of the whole process, but once self-replicating genes have become possible, the whole process of evolution by the mechanisms of variation and Darwinian natural selection provides a basis for understanding the design of organisms. This is one sense in which the notion of "purpose" can be considered in neurobiology.

1.2. Seeking Gene-Identified Goals, Emotion, and Purpose

1.2.1. Emotions as States Elicited by Instrumental Reinforcers (Rewards and Punishers)

Emotions can usefully be defined (operationally) as states elicited by rewards and punishers that have particular functions (Rolls 1999, 2005a, 2014a). The functions are defined herein and include working to obtain or avoid the rewards and punishers. A reward is anything for which an animal (which includes humans) will work. A punisher is anything that an animal will escape from or avoid. An example of an emotion might thus be the happiness produced by being given a particular reward, such as a pleasant touch, praise, or winning a large sum of money. Another example of an emotion might be fear produced by the sound of a rapidly approaching bus or the sight of an angry expression on someone's face. We will work to avoid such stimuli, which are punishing. Another example would be frustration, anger, or sadness produced by the omission of an expected reward or the termination of a reward, such as the death of a loved one. Another example would be relief produced by the omission or termination of a punishing stimulus, such as the removal of a painful stimulus or sailing out of danger. These examples indicate how emotions can be produced by the delivery, omission, or termination of rewarding or punishing stimuli and go some way to indicate how different emotions could be produced and classified in terms of the rewards and punishers received, omitted, or terminated. A diagram summarizing some of the emotions associated with the delivery of a reward or punisher or a stimulus associated with them, or with the omission of a reward or punisher, is shown in Figure 5.1.

I consider elsewhere a slightly more formal definition than rewards or punishers, in which the concept of reinforcers is introduced, and it is shown that emotions can be usefully seen as states produced by instrumental reinforcing stimuli (Rolls 2005a). Instrumental reinforcers are stimuli that, if their occurrence, termination, or omission is made contingent upon the making of a response, alter the probability of the future emission of that response. Some stimuli are unlearned reinforcers (e.g., the taste of food if the animal is hungry, or pain), whereas others may become reinforcing by associative learning, because of their association with such primary reinforcers, thereby becoming "secondary reinforcers."

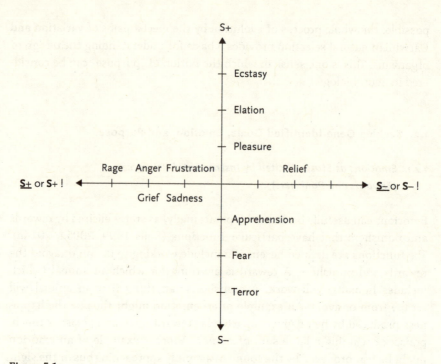

Figure 5.1:
Some of the emotions associated with different reinforcement contingencies are indicated. Intensity increases away from the center of the diagram, on a continuous scale. The classification scheme created by the different reinforcement contingencies consists of (1) the presentation of a positive reinforcer (S+), (2) the presentation of a negative reinforcer (S−), (3) the omission of a positive reinforcer (S+) or the termination of a positive reinforcer (S+!), and (4) the omission of a negative reinforcer (S−) or the termination of a negative reinforcer (S−!). It should be understood that each different reinforcer will produce different emotional states: this diagram just summarizes the types of emotion that may be elicited by different contingencies, but the actual emotions will be different for each reinforce (Rolls 2013, 2014a).

This foundation has been developed (Rolls 2005a, 2014a) to show how a very wide range of emotions can be accounted for as a result of the operation of a number of factors, including the following:

1. The *reinforcement contingency* (e.g., whether reward or punishment is given, or withheld) (see Figure 5.1).
2. The *intensity* of the reinforcer (see Figure 5.1).
3. Any environmental stimulus might have a *number of different reinforcement associations*. (For example, a stimulus might be associated both with the presentation of a reward and of a punisher, allowing states such as conflict and guilt to arise.)
4. Emotions elicited by stimuli associated with *different primary reinforcers* will be different.
5. Emotions elicited by *different secondary reinforcing stimuli* will be different from each other (even if the primary reinforcer is similar).

6. The emotion elicited can depend on whether an *active or passive behavioral response* is possible. (For example, if an active behavioral response can occur to the omission of a positive reinforcer, then anger might be produced, but if only passive behavior is possible, then sadness, depression, or grief might occur.)

By combining these six factors, it is possible to account for a very wide range of emotions (Rolls 2005a, 2014a). It is also worth noting that emotions can be produced just as much by the recall of reinforcing events as by external reinforcing stimuli; that cognitive processing (whether conscious or not) is important in many emotions because very complex cognitive processing may be required to determine whether or not environmental events are reinforcing. Indeed, emotions normally consist of cognitive processing which analyzes the stimulus, then determines its reinforcing valence, and then produces an elicited mood change if the valence is positive or negative. I note that a mood or affective state may occur in the absence of an external stimulus, as in some types of depression, but that normally the mood or affective state is produced by an external stimulus, with the whole process of stimulus representation, evaluation in terms of reward or punishment, and the resulting mood or affect being referred to as emotion (Rolls 2014a).

1.2.2. Gene-Defined Goals Provide a Purpose for Action

Given that emotions can be considered as states elicited by goals for action, we may ask what the evolutionary adaptive value is of emotions. It turns out that the design of brains to seek rewards and avoid punishers is highly adaptive and provides another neurobiological approach to purpose in that it is adaptive for animals to be designed by evolution to seek goals, as described next.

The most important function of emotion is that it is related to seeking goals and obtaining or not-obtaining the goals, as follows, and as described more fully elsewhere (Rolls 2012b, 2013, 2014a). This is the first function of emotion.

1.2.3. Emotions and Goals that Provide a Purpose for Action

Emotional (and motivational) states allow a simple interface between sensory inputs and action systems, which allow for flexibility of behavioral responses to reinforcing stimuli. The essence of this idea is that goals for behavior are specified by reward and punishment evaluation. When an environmental stimulus has been decoded as a primary reward or punishment, or (after previous stimulus-reinforcer association learning) as a secondary rewarding or punishing stimulus, then it becomes a goal for action. The human can then perform any action to obtain the reward or to avoid the punisher. (Instrumental learning typically allows any action to be learned, though some actions may be more easily learned than others; Lieberman 2000; Pearce 2008.) Thus there is flexibility of action, and this is in contrast with stimulus-response, or habit, learning in which a particular response to a particular stimulus is learned. The emotional route to action is flexible not only because any action can be performed to obtain the reward or avoid the punishment, but also because the human can learn in as little as one trial that a reward or punishment is associated with a particular stimulus, in what is termed "stimulus-reinforcer association learning."

To summarize and formalize, two processes are involved in emotional behavior. The first is stimulus-reinforcer association learning: emotional states are produced as a result (Rolls 2014a). This process is implemented in structures such as the orbitofrontal cortex and amygdala (Figure 5.2) (Grabenhorst and Rolls 2011; Rolls 2014a; Rolls and Grabenhorst 2008).

The second is instrumental learning of an action made to approach and obtain the reward or to avoid or escape from the punisher. This is action-outcome learning and involves brain regions such as the cingulate cortex

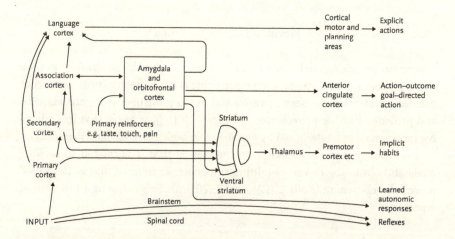

Figure 5.2:
Multiple routes to the initiation of actions and other behavioral responses in response to rewarding and punishing stimuli (Rolls 2016a). The inputs from different sensory systems to brain structures, such as the orbitofrontal cortex and amygdala, allow these brain structures to evaluate the reward- or punishment-related value of incoming stimuli or of remembered stimuli. One type of route to behavior may be implicit and includes the anterior cingulate cortex for action-outcome, goal-related, learning, and the striatum and rest of the basal ganglia for stimulus–response habits. Another type of route is via the language systems of the brain, which allow explicit (verbalizable) decisions involving multistep syntactic planning to be implemented. Outputs for autonomic responses can also be produced using outputs from the orbitofrontal cortex and anterior cingulate cortex (some of which are routed via the anterior insular cortex) and amygdala.

when the actions are being guided by the goals and the striatum and rest of the basal ganglia when the behavior becomes automatic and habit-based (i.e., uses stimulus-response connections; see Figure 5.2) (Rolls 2005a, 2009; Rolls 2014a; Rushworth et al. 2011). Emotion is an integral part of this, for it is the state elicited in the first stage by stimuli which are decoded as rewards or punishers, and this state has the property that it is motivating. The motivation is to obtain the reward or avoid the punisher (the goals for the action), and animals must be built to obtain certain rewards and avoid certain punishers. Indeed, primary or unlearned rewards and punishers are specified by genes which effectively specify the goals for action.

This is the solution that natural selection has found for how genes can influence behavior to promote their fitness (as measured by reproductive success) and for how the brain could interface sensory systems to action systems, and it is an important part of Rolls's theory of emotion (2005a, 2014a).

The implication is that operation by animals (including humans) using reward and punishment systems tuned to dimensions of the environment that increase fitness provides a mode of operation that can work in organisms that evolve by natural selection. It is clearly a natural outcome of Darwinian evolution to operate using reward and punishment systems tuned to fitness-related dimensions of the environment if arbitrary actions are to be made by the animals, rather than just preprogrammed movements such as tropisms, taxes, reflexes, and fixed action patterns. This view of brain design in terms of reward and punishment systems built by genes that gain their adaptive value by being tuned to a goal for action offers, I believe, a deep insight into how natural selection has shaped many brain systems and is a fascinating outcome of Darwinian thought (Rolls 2005a, 2011a, 2014a).

The point being made then is that another sense in which behavior can be described as purposive is that when genes specify rewards and punishers, they provide the goals for action, that is, the purposes for actions. We are built to be goal-seeking machines in the interests of our selfish genes, which find the specification of rewards and punishers an efficient way to guide our "purposive" behavior for the genes' reproductive success.

Selecting between available rewards with their associated costs and avoiding punishers with their associated costs is a process that can take place both implicitly (unconsciously) and explicitly using a language system to enable long-term plans to be made. These many different brain systems, some involving implicit evaluation of rewards and others explicit, verbal, conscious, evaluation of rewards and planned long-term goals, must all enter into the selector of behavior (see Figure 5.2).

Other functions of emotion are described elsewhere (Rolls 2012b, 2013, 2014a).

1.3. Goal-Seeking, Reasoning, the Individual, and Purpose

I have put forward a position that, in addition to the gene-based goal system for emotion described earlier, there is a separate rational (i.e., reasoning) system that can plan ahead and work for what are sometimes different, long-term goals (Rolls 1997a, 2003, 2004, 2005a, 2005b, 2007a, 2007b, 2008b, 2011b, 2012b, 2013, 2014a). This type of processing involves multistep trains of thought, as might be required to formulate a plan with many steps. Each step has its own symbols (e.g., a word to represent a person), and so syntactic linking (binding) is needed between the symbols within each step and some syntactic (relational) links must be made between symbols in different steps. I have argued that when we correct such multistep plans or trains of thought, we need to think about these first-order thoughts, and the system that does this is thus a higher order thought system (in that it is thinking about first-order thoughts).

There is a fundamentally important distinction here: working for a gene-specified reward, as in many emotions, is performed for the interests of the "selfish" genes. Working for rationally planned rewards may be performed in the interest of the particular individual (e.g., the person, the phenotype), and not in the interests of the genotype (Rolls 2011b).

It is suggested that this arbitrary symbol manipulation using important aspects of language processing and used for planning but not in initiating all types of behavior is close to what consciousness is about. In particular, consciousness may *be* the state which arises in a system that can think about (or reflect on) its own (or other people's) thoughts; that is, in a system capable of second-order or higher order thoughts (Carruthers 1996; Dennett 1991; Gennaro 2004; Rolls 1995, 1997a, 1997b, 1999, 2004, 2005a, 2007a, 2014a; Rosenthal 1986, 1990, 1993, 2004, 2005).

It is of great interest to comment on how the evolution of a system for flexible planning might affect emotions. Consider grief, which may occur when a reward is terminated and no immediate action is possible (see Rolls 1990, 2005a). It may be adaptive by leading to a cessation of the formerly rewarded behavior and thus facilitating the possible identification of other positive reinforcers in the environment. In humans, grief may be particularly and especially potent because it becomes represented in a system which can plan ahead and understand the enduring implications of the loss (Rolls 2016c).

The question then arises of how decisions are made in animals such as humans that have both the implicit, direct reward-based and the explicit, rational, planning systems (see Figure 5.2) (Rolls 2008a). One particular situation in which the first, implicit system may be especially important is when rapid reactions to stimuli with reward or punishment value must be made, for then structures such as the orbitofrontal cortex may be especially important (Rolls 2005a, 2014a). Another is when there may be too many factors to be taken into account easily by the explicit, rational, planning system, when the implicit system may be used to guide action. In contrast, when the implicit system continually makes errors, it would then be beneficial for the organism to switch from automatic habit, or from action-outcome goal-directed behavior, to the explicit conscious control system which can evaluate with its long-term planning algorithms what action should be performed next. Indeed, it would be adaptive for the explicit system to regularly be assessing performance by the more automatic system and to switch itself in to control behavior quite frequently because, otherwise, the adaptive value of having the explicit system would be less than optimal.

It may be expected that there is often a conflict between these systems in that the first, implicit system is able to guide behavior particularly to obtain the greatest immediate reinforcement, whereas the explicit system can potentially enable immediate rewards to be deferred and longer term, multistep plans to be formed that may be in the interests of the individual, not the genes. For example, an individual might decide not to have children but instead to devote him- or herself to being a creative individual or to enjoying opera, or the like. This type of conflict will occur in animals with a syntactic planning ability; that is, in humans and any other animals that have the ability to process a series of "if . . . then" stages of planning. This is a property of the human language system, and the extent to which it is a property of nonhuman primates is not yet fully clear. In any case, such conflict may be an important aspect of the operation of at least the human mind because it is so essential for humans to correctly decide, at every moment, whether to invest in a relationship or a group that may offer long-term benefits or whether to directly pursue immediate benefits (Rolls 2005a, 2008a, 2011b).

The point being made here is that another sense in which behavior can be described as purposive is that actions in humans (and perhaps in related animals) may be performed where the goal is the interest of the individual, the phenotype, using a reasoning system that can calculate using multistep reasoning the advantages and costs to the individual of performing an action to obtain a goal. A goal for the individual might be to live a long, intellectually productive, and healthy life, and such a person might decide, for example, to forgo a gene-identified goal such as the delicious taste of sweet and the flavor (mouth-feel) of fat (such as in an ice cream) in order to remain healthy. The point has been developed elsewhere that the implicit and explicit systems that define goals for action, with their somewhat different interests, are likely to remain across a population more or less in balance because a strong gene-defined emotional system will facilitate reproduction, whereas a rational system may not do this but may provide rewards for the individual.

It should be noted that the evolution of the rational system may have occurred because it conferred significant advantages in terms of reproductive success but that, nevertheless, once such a system evolved, the properties of this system enabled it to compute goals that might be more in the interests of the individual than of the genes.

2. THE NEUROSCIENCE OF MEANING

2.1. Content and Meaning in Representations: How Are Representations Grounded in the World?

One sense of "meaning" is "purpose," and this was considered in Section 1. Another sense of meaning is how it is that representations in the brain have "meaning" and content. I now describe what I understand by representations being grounded in the world, which addresses how representations, and even the symbols used in language, have meaning. These concepts were developed as part of the much larger issue of the nature and functions of consciousness (Rolls 2012b, 2014a, 2016a).

It is possible to analyze how the firing of populations of neurons encodes information about stimuli in the world (Rolls 2016a; Rolls and Treves 2011). For example, from the firing rates of small numbers of neurons in the primate inferior temporal visual cortex, it is possible to know which of twenty faces has been shown to the monkey (Rolls, Treves, and Tovee 1997; Rolls and Treves 2011). Similarly, a population of neurons in the anterior part of the macaque temporal lobe visual cortex has been discovered that has a view-invariant representation of objects (Booth and Rolls 1998; Rolls 2012a). From the firing of a small ensemble of neurons in the olfactory part of the orbitofrontal cortex, it is possible to know which of eight odors was presented (Rolls, Critchley, and Treves 1996; Rolls et al. 2010). From the firing of small ensembles of neurons in the hippocampus, it is possible to know where in allocentric space a monkey is looking (Rolls et al. 1998). In each of these cases, the number of stimuli that is encoded increases exponentially with the number of neurons in the ensemble, so this is a very powerful representation (Aggelopoulos, Franco, and Rolls 2005; Franco et al. 2004; Rolls 2016a; Rolls et al. 2004; Rolls et al. 1997; Rolls and Treves 2011).

What is being measured in each example is the mutual information between the firing of an ensemble of neurons and which stimuli are present in the world. In this sense, one can read off the code that is being used at the end of each of these sensory systems (Rolls 2016a; Rolls and Treves 2011).

However, what sense does the representation make to the animal? What does the firing of each ensemble of neurons "mean"? What is the content of the representation? In the visual system, for example, it is suggested that the representation is built by a series of appropriately connected competitive networks operating with a modified Hebb-learning rule (Rolls 2012a, 2016a). Now, competitive networks categorize their inputs without the use of a teacher (Rolls 2016a, 2016b). So which particular neurons fire as a result of the self-organization to represent a particular object or stimulus is arbitrary. What meaning, therefore, does the particular ensemble that fires to an object have? How is the representation grounded in the real world? The fact that there is mutual information between the firing of the ensemble of cells in the brain and a stimulus or event in the world (Rolls 2016a; Rolls and Treves 2011) partly, but does not fully, answers this question.

One answer to this question is that there may be meaning in the case of objects and faces that it *is* an object or face and not just a particular view. This is the case in that the representation may be activated by any view of the object or face. This is a step suggested to be made possible by a short-term memory in the learning rule that enables different views of objects to be associated together (Rolls 2012a, 2016a). But it still does not provide the representation with any meaning in terms of the real world. What actions might one make, or what emotions might one feel, if that arbitrary set of temporal cortex visual cells was activated?

This leads to one of the answers I propose. I suggest that one type of meaning of representations in the brain is provided by their reward (or punishment) value: activation of these representations is the goal for actions. In the case of primary reinforcers such as the taste of food or pain, the activation of these representations would have meaning in the sense that the animal would work to obtain the activation of the taste of food neurons when hungry and to escape from stimuli that cause the neurons representing pain to be activated. Evolution has built the brain so that genes specify these primary reinforcing stimuli and so that their representations in the brain should be the targets for actions (Rolls 2014a, 2016a). In the case of other ensembles of neurons in, for example, the visual cortex that respond to objects with the color and shape of a banana—and which "represent" the sight of a banana in that their activation is always and uniquely produced by the sight of a banana—such representations come to have meaning only by association with a primary reinforcer, involving the process of stimulus-reinforcer association learning.

The second sense in which a representation may be said to have meaning is by virtue of sensory-motor correspondences in the world. For example, the touch of a solid object such as a table might become associated with evidence from the motor system that attempts to walk through the table resulting in cessation of movement. The representation of the table in the inferior temporal visual cortex might have "meaning" only in the sense that there is mutual information between the representation and the sight of the table until the table is seen just before and while it is touched, when sensory-sensory association between inputs from different sensory modalities will be set up that will enable the visual representation to become associated with its correspondences in the touch and movement worlds. In this second sense, meaning will be conferred on the visual sensory representation because of its associations in the sensory-motor world. Related views have been developed by the philosopher Ruth Millikan (1984). Thus, it is suggested that there are two ways by which sensory representations can be said to be grounded (i.e., to have meaning) in the real world.

It is suggested that the symbols used in language become grounded in the real world by the same two processes, as follows (Rolls 2016a). In the first, a symbol such as the word "banana" has meaning because it is associated with primary reinforcers such as the flavor of the banana and with secondary reinforcers such as the sight of the banana. These reinforcers have "meaning" to the animal in that evolution has built animals as machines designed to do everything that they can to obtain these reinforcers, so that they can eventually reproduce successfully and pass their genes on to the next generation. The fact that some stimuli are reinforcers but may not be adaptive as goals for action is no objection. Genes are limited in number and cannot allow for every eventuality, such as the availability to humans of [non-nutritive] saccharin as a sweetener. The genes can just build reinforcement systems the activation of which is generally likely to increase the fitness of the genes specifying the reinforcer [or may have increased their fitness in the recent past].) In this sense, obtaining reinforcers may have life-threatening "meaning" for animals, though of course the use of the word "meaning" here does not imply any subjective state, just that the animal is built as a survival-for-reproduction machine. This is a novel, Darwinian approach to the issue of symbol grounding.

In the second process, the word "table" may have meaning because it is associated with sensory stimuli produced by tables, such as their touch, shape, and appearance, as well as with other functional properties, such as, for example, being load-bearing and obstructing movement if they are in the way (Rolls 2016a).

These points are relevant for my *higher order syntactic thought* (HOST) theory of consciousness, by addressing the sense in which the thoughts can be grounded in the world. The HOST theory holds that the thoughts "mean" something to the individual in the sense that they may be about the survival of the individual (the phenotype) in the world, which the rational, thought system aims to maximize (Rolls 2012b, 2016a).

3. A NEUROSCIENCE APPROACH TO MORALS

3.1. Biological Underpinnings

In this section, I consider the neurobiological underpinnings of ethics, with a much fuller account provided by Rolls (2012b).

Rolls (2012b, 2014a) has argued that much of the foundation of our emotional behavior arises from specification by genes of primary reinforcers that provide goals for our actions. We have emotional reactions in certain circumstances, such as when we see that we are about to suffer pain, when we fall in love, or if someone does not return a favor in a reciprocal interaction. What is the relation between our emotions and what we think is right; that is, our ethical principles? If we think something is right, such as returning something that has been on loan, is this a fundamental and absolute ethical principle or might it have arisen from deep-seated biologically based systems shaped to be adaptive by natural selection operating in evolution to select genes that tend to promote the survival of those genes?

Many principles that we regard as ethical principles *might* arise in this way. For example, guilt might arise when there is a conflict between an available reward and a rule or law of society. Jealousy is an emotion that might be aroused in a male if the faithfulness of his partner seems to be threatened by her liaison with another male. In this case, the reinforcement contingency that is operating is produced by a punisher, and it may be that males are specified genetically to find this punishing because it indicates a potential threat to their paternity and parental investment. Similarly, a female may become jealous if her partner has a liaison with another female because the resources available to the "wife" useful to bring up her children are threatened. Again, the punisher here may be gene-specified. Such emotional responses might influence what we build into some of the ethical principles that surround marriage and partnerships for raising children.

Many other similar examples can be surmised from the area of evolutionary psychology (Buss 2015; Ridley 1993, 1996). For example, there may be a set of reinforcers that are genetically specified to help promote social cooperation and even reciprocal altruism and that might thus influence what we regard as ethical, or at least what we are willing to accept as ethical principles. Such genes might specify that emotion should be elicited, and behavioral changes should occur, if a cooperating partner defects or "cheats" (Cosmides and Tooby 1999). Moreover, the genes may build brains with genetically specified rules that are useful heuristics for social cooperation, such as acting with a strategy of "generous tit-for tat," which can be more adaptive than strict "tit-for-tat" in that being generous occasionally is a good strategy to help promote further cooperation that has failed when both partners defect in a strict "tit-for-tat" scenario (Ridley 1996). Genes that specify good heuristics to promote social cooperation may thus underlie such complex emotional states as feeling forgiving. There are many other examples, including kin altruism, which has a genetic basis.

The situation is clarified by the ideas I have advanced about a rational syntactically based reasoning system and how this interacts with an evolutionarily older emotional system with gene-specified rewards. The rational system enables us, for example, to defer immediate gene-specified rewards and make longer term plans for actions that in the long term may have more useful outcomes. This rational system enables us to make reasoned choices and to reason about what is right. Indeed, it is because of the linguistic system that the naturalistic fallacy becomes an issue. In particular, we should not believe that what is right is what is natural (*the naturalistic fallacy*), because we have a rational system that can go beyond simpler gene-specified rewards and punishers that may influence our actions through brain systems that operate at least partly implicitly (i.e., unconsciously).

The suggestion I make is that, in all these cases and in many others, there are biological underpinnings that determine what we find rewarding or punishing, designed into genes by evolution to lead to appropriate behavior that helps to increase the fitness of the genes. When these implicit systems for rewards and punishers start to be expressed explicitly (in language) in humans, the explicit rules, rights, and laws that are formalized are those that set out in language what the biological underpinnings "want" to occur.

Clearly, in formulating the explicit rights and laws, some compromise is necessary in order to keep the society stable. When the rights and laws are formulated in small societies, it is likely that individuals in that society will have many of the same genes, and rules such as "help your neighbor" (but "make war with 'foreigners'") will probably be to the advantage of one's genes. However, when the society increases in size beyond a small village (in the order of 1,000 individuals), then the explicitly formalized rules, rights, and laws may no longer produce behavior that turns out to be to the advantage of an individual's genes. In addition, it may no longer be possible to keep track of individuals in order to maintain the stability of "tit-for-tat" cooperative social strategies (Dunbar 1996; Ridley 1996). In such cases, other factors doubtless come into play to additionally influence what groups hold to be right. For example, a group of subjects in a society might demand the "right" to free speech because it is to their economic advantage.

Thus, overall, it is suggested that many aspects of what a society holds as right and moral, and of what becomes enshrined in explicit "rights" and laws, are related to biological underpinnings, which have usually evolved because of the advantage to the individual's genes, but that, as societies develop, other factors also start to influence what is believed to be "right" by groups of individuals, related to socioeconomic factors. In both cases, the laws and rules of the society develop so that these "rights" are protected but often involve compromise in such a way that a large proportion of the society will agree to, or can be made subject to, what is held as right.

Society may set down certain propositions of what is "right." One reason for this is that it may be too difficult on every occasion, and for everyone, to work out explicitly what all the payoffs are for each rule of conduct. A second reason is that what is promulgated as "right" could actually be to someone else's advantage, and it would not be wise to expose this fully. One way to convince members of society not to do what is apparently in their immediate interest is to promise a reward later. Such deferred rewards are often offered by religions (Rolls 2012b). The ability to work for a deferred reward using a one-off plan in this way becomes possible, it is suggested, with the evolution of the explicit, propositional, system.

3.2. Ethical Principles Arising from Advantages for the Phenotype Versus for the Genotype

Many of the examples described earlier of the biological underpinnings to ethical behavior—what we have come to regard as right and just—reflect advantages to the genotype. These underpinnings related to advantages to the genotype are present in many nonhuman animals, though developed to greater or lesser extents. Examples include kin altruism (common in nonhuman animals), reciprocal altruism including tit-for-tat exchanges and forgiveness, and stakeholder altruism (Rolls 2012b).

However, I wish to introduce here a new concept related to the underlying biology, that some aspects of what we regard as ethical and right are related to advantage to the phenotype. Phenotypic advantage may not necessarily be to the advantage of the genotype, and indeed we can contrast the "selfish phene" with the "selfish gene" (Rolls 2011b, 2012b). The concept of phenotypic advantage in relation to ethics is that, by reasoning, the multiple step syntactic thought brain system may lead people to agree to rules of society that are to the advantage of their bodies, even when there may be no genetic advantage (Rolls 2012b). Consider "Thou shalt not kill" and "Thou shalt not steal." Both rules could be to the advantage of the individual person who may wish to stay alive and not be robbed of all the things that she or he enjoys, even though that may not necessarily be to the advantage of genes. Indeed, in the animal kingdom, conflict and killings even within species that are driven by genotypic advantage are common, as in intrasexual competition as part of sexual selection (Rolls 2012b) and a lion killing the cubs of his new lioness by a previous father. In this context, the rules "Thou shalt not kill" and "Thou shalt not steal" may not be to the genetic advantage of an individual, but may well be to the advantage of the individual person who may wish to stay alive to enjoy life and property, even beyond the age of reproduction.

Helping and care for the elderly is another example where individuals may contract into a society where help and care for the elderly is valued and is even a "right" because it may be to their own advantage later on in life, and they are willing to pay some cost (e.g., not stealing from others) to be part of the society in which they can enjoy "rights" such as not having their possessions stolen. This social contract, agreeing to the rules of a society, has similarities with a contract with an insurance company, in which there is a cost, but also a potential benefit to an individual, and where that advantage may not necessarily be genetic and may never have been selected for genetically.

The concept here is as follows. At least humans (and possibly some other animals) have evolved to have a brain that can reason, and that has many advantages in enabling them to pursue long-term goals rather than immediate gene-defined goals. However, once one has a reasoning system, it can reason that some of the things that one has been built to enjoy (perhaps for genetic reasons originally, such as wealth, power, good food) might be enjoyed even when there is no genetic advantage, for example, by prolonging life into old age, which becomes a "right" where it is possible.

This results in ethical values and "rights" which have their (biological) basis in the good of the phenotype and not necessarily the good of the genotype (though they are not mutually exclusive and may often work together). This phenotypically underpinned system of rights that becomes enshrined in a social contract is somewhat like the "right" to a pension and to medical care when elderly (and beyond the age at which genetic potential is likely to be enhanced by longer life), which allows the individual person or phenotype to benefit by having obeyed the rules of the society earlier by contributing to its insurance provisions. Such a social contract has similarities with a contract with an insurance company: it costs a bit, but may protect your body in the long term. A pension is a bit like such a social contract. One agrees rationally to a cost, as it may benefit you (and not necessarily your genes) in the long term.

3.3. The Social Contract

The view that one is led to is that some of our moral beliefs may be explicit, verbal formulations of what may reflect factors built genetically by kin selection into behavior; namely, a tendency to favor kin because they are likely to share some of an individual's genes. In a small society, this explicit formulation may be "appropriate" (from the point of view of the genes) in that many members of that society will be related to that individual. When the society becomes larger, the relatedness may decrease, yet the explicit formulation of the rules or laws of society may not change. In such a situation, it is presumably appropriate for society to make it clear to its members that its rules for what is acceptable and "right" behavior are set in place so that individuals can live in safety and with some expectation of help from society in general.

It is argued that the second biological underpinning of ethics is the evolution in (at least) humans of a reasoning system, which leads humans to values based on phenotypic advantage, which may not always correspond to genetic advantage.

The operation of this reasoning system may encourage acceptance of a social contract, in part because it may be judged to be useful for the individual's kin and genetic fitness and also for the individual's phenotype, which may, for example, accept a cost of not harming others and benefitting from that in order not to be harmed.

Indeed, the view to which this approach based on neuroscience and the evolution of the brain leads is that there may be no absolute rights or god-given rights (Rolls 2012b) but that, instead, there may be rules and laws of a society, and indeed perhaps "universally" across societies, that may be agreed by a social contract, and these rules and laws may imply "rights." Justice may in this approach be used to refer to the implementation of these laws and rules (Hobbes 1651; Locke 1689; Rawls 1971). For example, Thomas Hobbes, beginning from a mechanistic understanding of human beings and the passions, postulates what life would be like without government, a condition which he calls the state of nature. In that state, each person would have a right, or license, to everything in the world. This, Hobbes argues, would lead to a "war of all against all" (*bellum omnium contra omnes*), and thus lives that are "solitary, poor, nasty, brutish, and short." This is close to my argument about what would represent the interests of the genotype. In this context, Hobbes then argues that, to escape this state of war, men in the state of nature accede to a social contract and establish a civil society, a commonwealth. John Locke (1689) continued the development of Hobbes's argument, though more from a starting point that humans are rational, reasoning people.

Other factors that can influence what is held to be right might reflect socioeconomic advantage to groups or alliances of individuals. It would be then in a sense up to individuals to decide whether they wished to accept the rules, with the costs and benefits provided by the rules of that society in a form of Social Contract. Individuals who did not agree to the social contract might wish to transfer to another society at a different place on the continuum of costs and potential benefits to the individuals or to influence the laws and policies of their own society, but while acting within its laws. Individuals who attempt to cheat the system or break the laws that operate within a society would be expected to pay a cost in terms of punishment meted out by the society in accordance with its rules and laws, what it considers to be "right" and "just," with the underpinnings in genotypic and phenotypic advantage described here and in more detail by Rolls (2012b).

4. CONCLUSION

One process to which "purpose" can refer is that genes are self-replicating. Another process to which "purpose" can apply is that genes set some of the goals for actions. These goals are fundamental to understanding emotion.

Another process to which "purpose" can apply is that syntactic multistep reason provides a route for goals to be set that are to the advantage of the individual, the phenotype, and not of the genes.

Meaning can be achieved by neural representations not only if these representations have mutual information with objects and events in the world, but also by virtue of the goals just described of the "selfish" genes and of the individual reasoner. This, it is suggested, provides a means for even symbolic representations to be grounded in the world.

Morals can be considered as principles that are underpinned by (the sometimes different) biological goals specified by the genes and by the reasoning (rational) systems. Given that what is "natural" does not correspond to what is "right," it is suggested that these conflicts within and between individuals can be addressed by a social contract.

REFERENCES

Aggelopoulos, N. C., L. Franco, and E. T. Rolls. 2005. Object perception in natural scenes: Encoding by inferior temporal cortex simultaneously recorded neurons. *Journal of Neurophysiology* 93: 1342–1357.
Booth, M. C. A., and E. T. Rolls. 1998. View-invariant representations of familiar objects by neurons in the inferior temporal visual cortex. *Cerebral Cortex* 8: 510–523.
Buss, D. M. 2015. *Evolutionary psychology: The new science of the mind, 5th edition.* New York: Pearson.
Carruthers, P. 1996. *Language, thought and consciousness.* Cambridge: Cambridge University Press.
Cosmides, I., and J. Tooby. 1999. Evolutionary psychology. In R. Wilson and F. Keil (Eds.), *MIT encyclopedia of the cognitive sciences*, pp. 295–298. Cambridge, MA: MIT Press.
Dennett, D. C. 1991. *Consciousness explained.* London: Penguin.
Dunbar, R. 1996. *Grooming, gossiping, and the evolution of language.* London: Faber and Faber.
Franco, L., E. T. Rolls, N. C. Aggelopoulos, and A. Treves. 2004. The use of decoding to analyze the contribution of the information of the correlations between the firing of simultaneously recorded neurons. *Experimental Brain Research* 155: 370–384.
Gennaro, R. J. 2004. *Higher order theories of consciousness.* Amsterdam: John enhamins.
Grabenhorst, F., and E. T. Rolls. 2011. Value, pleasure, and choice in the ventral prefrontal cortex. *Trends in Cognitive Science* 15: 56–67.
Hobbes, T. 1651. *Leviathan, or the matter, forme, and power of a commonwealth, ecclesiasticall and civil.*
Lieberman, D. A. 2000. *Learning.* Belmont, CA: Wadsworth.
Locke, J. 1689. *The two treatises of civil government.*
Millikan, R. G. 1984. *Language, thought, and other biological categories: New foundation for realism.* Cambridge, MA: MIT Press.
Pearce, J. M. 2008. *Animal learning and cognition, 3rd edition.* Hove, UK: Psychology Press.
Rawls, J. 1971. *A theory of justice.* Oxford: Oxford University Press.

Ridley, M. 1993. *The red queen: Sex and the evolution of human nature*. London: Penguin.
Ridley, M. 1996. *The origins of virtue*. London: Viking.
Rolls, E. T. 1990. A theory of emotion, and its application to understanding the neural basis of emotion. *Cognition and Emotion* 4: 161–190.
Rolls, E. T. 1995. A theory of emotion and consciousness, and its application to understanding the neural basis of emotion. In M. S. Gazzaniga (Eds.), *The cognitive neurosciences*, pp. 1091–1106. Cambridge, MA: MIT Press.
Rolls, E. T. 1997a. Consciousness in neural networks? *Neural Networks* 10: 1227–1240.
Rolls, E. T. 1997b. Brain mechanisms of vision, memory, and consciousness. In M. Ito, Y. Miyashita, and E. T. Rolls (Eds.), *Cognition, computation, and consciousness*, pp. 81–120. Oxford: Oxford University Press.
Rolls, E. T. 1999. *The brain and emotion*. Oxford: Oxford University Press.
Rolls, E. T. 2003. Consciousness absent and present: A neurophysiological exploration. *Progress in Brain Research* 144: 95–106.
Rolls, E. T. 2004. A higher order syntatic thought (HOST) theory of consciousness. In R. J. Gennaro (Ed.), *Higher-order theories of consciousness: An anthology*, pp. 137–172. Amsterdam: John Benjamins.
Rolls, E. T. 2005a. *Emotion explained*. Oxford: Oxford University Press.
Rolls, E. T. 2005b. Consciousness absent or present: A neurophysiological exploration of masking. In H. Ogmen and B. G. Breitmeyer (Eds.), *The first half second: The microgenesis and temprial dynamics of unconscious and conscious visual processes*, pp. 89–108. Cambridge, MA: MIT Press.
Rolls, E. T. 2007a. The affective neuroscience of consciousness: Higher-order linguistic thoughts, dual routes to emotion and action, and consciousness. In P. Zelazo, M. Moscovitch, and E. Thompson (Eds.), *Cambridge handbook of consciousness*, pp. 831–859. Cambridge: Cambridge University Press.
Rolls, E. T. 2007b. A computational neuroscience approach to consciousness. *Neural Networks* 20: 962–982.
Rolls, E. T. 2008a. *Memory, attention, and decision-making: A unifying computational neuroscience approach*. Oxford: Oxford University Press.
Rolls, E. T. 2008b. Emotion, higher order syntatic thoughts, and consciousness. In L. Weiskrantz and M. K. Davies (Eds.), *Frontiers of consciousness*, pp. 131–167. Oxford: Oxford University Press.
Rolls, E. T. 2009. The anterior and midcingulate cortices and reward. In B. A. Vogt (Ed.), *Cingulate neurobiology and disease*, pp. 191–206. Oxford: Oxford University Press.
Rolls, E. T. 2011a. A neurobiological basis for affective feeling and aesthetics. In E. Schellekens and P. Goldie (Eds.), *The aesthetic mind: Philosophy and psychology*, pp. 116–165. Oxford: Oxford University Press.
Rolls, E. T. 2011b. Consciousness, decision-making, and neural computation. In V. Cutsuridis, A. Hussain, and J. G. Taylor (Eds.), *Perception action cycle: Models, algorithms and systems*, pp. 287–333. Berlin: Springer.
Rolls, E. T. 2012a. Invariant visual object and face recognition: Neural and computational bases, and a model. *Frontiers in Computer Neuroscience* 6(35): 1–70.
Rolls, E. T. 2012b. *Neuroculture: On the implication of brain science*. Oxford: Oxford University Press.
Rolls, E. T. 2013. What are emotional states, and why do we have them? *Emotion Review* 5: 241–247.
Rolls, E. T. 2014a. *Emotion and decision-making explained*. Oxford: Oxford University Press.
Rolls, E. T. 2014b. Emotion and decision-making explained: Precis. *Cortex* 59: 185–193.

Rolls, E. T. 2016a. *Cerebral cortex: Principles of operation*. Oxford: Oxford University Press.

Rolls, E. T. 2016b. Pattern separation, completion, and categorisation in the hippocampus and neocortex. *Neurobiology of Learning and Memory* 129: 4–28.

Rolls, E. T. 2016c. A non-reward attractor theory of depression. *Neuroscience and Biobehavioral Reviews* 68: 47–58.

Rolls, E. T., N. C. Aggelopoulos, L. Franco, and A. Treves. 2004. Information encoding in the inferior temporal cortex: Contributions of the firing rates and correlations between the firing of neurons. *Biological Cybernetics* 90: 19–32.

Rolls, E. T., H. D. Critchley, and A. Treves. 1996. The representation of olfactory information in the primate orbitofrontal cortex. *Journal of Neurophysiology* 75: 1982–1996.

Rolls, E. T., H. D. Critchley, J. V. Verhagen, and M. Kadohisa. 2010. The representation of information about taste and odor in the orbitofrontal cortex. *Chemosensory Perception* 3: 16–33.

Rolls, E. T., and F. Grabenhorst. 2008. The orbitofrontal cortex and beyond: From affect to decision making. *Progress in Neurobiology* 86: 216–244.

Rolls, E. T., and A. Treves. 2011. The neuronal encoding of information in the brain. *Progress in Neurobiology* 95: 448–490.

Rolls, E. T., A. Treves, R. G. Robertson, P. Georges-Francois, and S. Panzeri. 1998. Information about spatial view in an ensemble of primate hippocampal cells. *Journal of Neurophysiology* 79: 1797–1813.

Rolls, E. T., A. Treves, and M. J. Tovee. 1997. The representational capacity of the distributed encoding of information provided by populations of neurons in the primate temporal visual cortex. *Experimental Brain Research* 114: 177–185.

Rosenthal, D. M. 1986. Two concepts of consciousness. *Philosophical Studies* 49: 329–359.

Rosenthal, D. M. 1990. A theory of consciousness. In *ZIF*. Bielefeld, GER: Zentrum fur Interdisziplinaire Forschung.

Rosenthal, D. M. 1993. Thinking that one thinks. In M. Davies and G. W. Humphreys (Eds.), *Consciousness*, pp. 197–223. Oxford: Blackwell.

Rosenthal, D. M. 2004. Varieties of higher-order theory. In R. J. Gennaro (Ed.), *Higher order theories of consciousness*, pp. 17–44. Amsterdam: John Benjamins.

Rosenthal, D. M. 2005. *Consciousness and mind*. Oxford: Oxford University Press.

Rushworth, M. F., M. P. Noonan, E. D. Boorman, M. E. Walton, and T. E. Behrens. 2011. Frontal cortex and reward-guided learning and decision-making. *Neuron* 70: 1054–1069.

CHAPTER 6
Moral Sedimentation

JESSE PRINZ

1. INTRODUCTION

Existentialism is often regarded as a philosophy of radical freedom. Though reconciled to the absurdity of the world, leading existentialists emphasized the human capacity for choice and self-creation. At the same time, there is a countercurrent in existentialist thought that calls freedom into question. This countercurrent draws attention to the ways in which behavior is determined by forces outside of our control. This theme is especially vivid in the moral domain. Beginning with Nietzsche's claim that Christians are self-deceived and extending through feminist and decolonial perspectives within postwar existentialism, we find key authors pointing to ways in which deeply held values get shaped by social forces. The term "sedimentation," which emerged out of phenomenology, is sometimes used to characterize this phenomenon. Here, I want to suggest that recent work in neuroscience, psychology, and other social sciences add support to the thesis that we are vulnerable to sedimentation.

I will begin by tracing the idea of sedimentation and related concepts in existentialist thought, with special emphasis on the moral domain. After that, I will turn to recent empirical work, which adds support to the thesis that morality is socially conditioned. Then, in a concluding section, I will consider various tactics against sedimentation that have been proposed. I will suggest that some of the more prominent historical tactics are problematic while also pointing to some alternatives. I will end with some reflections on ways in which empirical methods might facilitate existential revolt.

2. MORAL SEDIMENTATION

2.1. Moral Sedimentation Defined

I want to begin by examining ideas about moral sedimentation. The phrase "moral sedimentation" does not appear in existentialist literature, and it has roots in traditions that predate existentialism. The term "sedimentation" originates in the work of the phenomenologist, Edmund Husserl. Husserl is particularly concerned with the way that theories in mathematics, science, and philosophy become enshrined in language, inherited, and passively accepted (see, e.g., Husserl 1936/1970: 362). Sedimentation can lead to prejudicial thinking that closes off certain possibilities (1936/1970: 72). Husserl sees this as contributing to an unnoticed historicity of the present. He tells us that sedimented traditions, "extend enduringly through time, since all new acquisitions are in turn sedimented and become working materials. Everywhere the problems, the clarifying investigations, the insights of principle are *historical*" (1936/1970: 369; emphasis in the original).

The concept of sedimentation was imported into France by Maurice Merleau-Ponty. Merleau-Ponty is primarily concerned with perception, and he uses the term to refer to the ways in which concepts and prior experience structure ongoing perceptual episodes (1945/2012: 131f). Here, sedimentation is a usually a matter of placing our perceptual states under ideas, but it can also involve embodied habits. Whether cognitive or corporeal, prior knowledge informs present encounters with the world, shaping how we interpret things, and gives the impression of a pregiven order.

Merleau-Ponty also extends the notion of sedimentation beyond perception and uses it to encompass the way attitudes taken up in the past can come to be taken as immutable facts (e.g., 1945/2012: 416). Pushing further, Merleau-Ponty refers to the ways in which civilization can constitute a kind of sedimentation. For example, artifacts, such as a pipe, or a spoon, or a church, are physical manifestations of cultural practices, and knowledge of how to interact with them, which seems completely obvious, is the result of enculturation (1945/2012: 363; see also Husserl 1936/1970: 26). Thus, we do not just inhabit a natural world; we also inhabit cultural worlds, and our ability to travel through this world, to comprehend it and make use of the items that populate it, depends on our personal history of learning. That personal history depends, in turn, on cultural history. The world we experience strikes us as fixed and self-evident, but it really is contingent and constructed.

Merleau-Ponty sees sedimentation as an obstacle to freedom. Challenging the radical freedom promulgated in Jean-Paul Sartre's existentialism, he writes:

> [W]e must recognize a sort of sedimentation of our life: when an attitude toward the world has been confirmed often enough, it becomes privileged for us.... [For example, A]fter having built my life upon an inferiority complex, continuously reinforced for twenty years, it is not *likely* that I would change.... This means that I am committed to inferiority, that I have decided to dwell within it, that this past, if not a destiny, has at least a specific weight, and that it is not a sum of events over there, far away from me, but rather the atmosphere of my present. (1945/2012: 466–467)

Sedimentation is the phenomenon of experiencing the world and acting in it through the filter of the past, without necessarily realizing it. Merleau-Ponty accepts the existentialist credo that we can free ourselves from such sediments, but their deep inculcation and embodiment in habitual practices and material culture renders this unlikely. "We would arrive at the same result," he adds, "by examining our relations with history."

Let me turn now to the second key term in this analysis: "moral." Morality is the domain of human life where we evaluate character and action in accordance with rules that license condemnation and punishment. At its most explicit level, it consists in rules of conduct and lists of virtues, which we articulate in passing judgment on ourselves and others. Morality also has a structuring role in human society: it dictates where people belong in social hierarchies and what privileges, entitlements, and rights may be afforded to whom. In some case, morality exerts its structuring influence implicitly. Silencing, disenfranchisement, and exploitation can occur without perpetrators taking notice. This, too, is part of morality, for the privileged and powerful feel that they have right to their position, and the oppressed either acquiesce or else resist. In either case, social organization and power distributions are assessed and underwritten by attitudes about what is permissible, or even proper. There are sanctions against those who step out of place.

Bringing these threads together, we can define moral sedimentation as the past patterns of proscription that shape present attitudes and guide current behavior. It is a historically articulated set of values that we have come to accept as obvious. Such inherited moral convictions are experienced as expressions of ourselves, but they are, in fact, constraints on freedom, insofar as we obey them uncritically and they prevent us from seeing alternatives. Next, we will see how this idea has played out in existentialist thought.

2.2. Sedimentation in Existentialist Thought

Although the phrase, "moral sedimentation" does not appear in any major writings by existentialists, it is a persistent theme. Here, I will bring that out through several examples. The paradigm case is undoubtedly Nietzsche. Existentialist interest in moral sedimentation has roots in his seminal, *On the Genealogy of Morality*. There, Nietzsche takes up the issue of Christian morality. He argues that Christian understandings of core moral concepts emerged through historical happenstance. These concepts include punishment, responsibility, freedom, evil, and conscience, as well as core moral values, such as concern for the weak and the virtue of poverty. The details of Nietzsche's historical analyses vary. What matters here is that he takes Christians to be oblivious to the genealogies of their values, and, as a result, they have lost critical perspective and follow these values as if they were revealed truths. For Nietzsche's readers, Christian values had become self-evident, even though they are actually the residue of historical processes that provide no normative foundation. In other words, Christian values are a case of moral sedimentation.

Nietzsche's *Genealogy* had considerable influence on the existentialists. It informed their belief that moral values are not God-given, but are rather human inventions—subjective, relative, contingent, and culturally conditioned. On the face of it, however, Nietzsche's polemic goes against a central credo within twentieth-century existentialist thought: the idea that we are fundamentally free. In the *Genealogy* and other works, Nietzsche says that free will is an illusion (e.g., Nietzsche 1887/1994: sec. II.7; see also Nietzsche 1878/1986: sec. 18). This makes Nietzsche an unlikely hero for the existentialists. But this is only half the story. Somewhat paradoxically, Nietzsche also announces that unfreedom is a myth (Nietzsche 1886/2001: sec. 21). Both freedom and its opposite are linked to the concept of causation, which Nietzsche sees as an invention, used for communication, but lacking explanatory purchase on the real world. Nietzsche also envisions free spirits or "sovereign individuals" who are capable of self-creation (Nietzsche 1887/1994: sec. II.2). These rare individuals break free from moral sedimentation by rejecting morality altogether—any adherence to moral rules would compromise sovereignty.

The idea that cultural values are an illusion from which we might break free is a trope that runs through existentialist theorizing in the twentieth century. We see it at work in Albert Camus's *The Stranger* (1942/1946), whose protagonist commits a random murder and refuses to cry at his mother's funeral. In a similar spirit, Jean-Paul Sartre argues that moral values exist entirely because we endorse them, and we can, at any moment, relinquish them:

> Values are sown on my path as thousands of little real demands, like the signs which order us to keep off the grass. . . . But as soon as the enterprise is held at a distance from me . . . then I discover myself suddenly as the one who gives its meaning to the alarm clock, the one who by a signboard forbids himself to walk on a flower bed or on the lawn . . . the one finally who makes the values exist in

order to determine his action by their demands. I emerge alone and in anguish confronting the unique and original project which constitutes my being; all the barriers, all the guard rails collapse, nihilated by the consciousness of my freedom. (Sartre 1943/1956: 38–39)

Both Camus and Sartre differ from Nietzsche methodologically. Camus argues through allegory rather than social history, and Sartre operates at a high level of abstraction. These tendencies arguably lead them to overestimate the ease with which we can repudiate the existing moral order. That is, they underestimate the power of sedimentation.

In contrast, consider the work of Simone de Beauvoir. Like Nietzsche, and unlike Sartre and Camus, she focuses on a very concrete real-world case when discussing moral sedimentation. In her most known work, *The Second Sex* (1949/2009) she examines the subjugation of women in Western society. This is a moral issue because the treatment of women depends on an elaborate range of practices, attitudes, and values that approve and promote gender inequality. Men find it acceptable to earn more, to enjoy higher status, to opt out of childcare and domestic work, to harass and belittle, to perpetuate skewed employment ratios, and to occupy more positions of power. And women, too, take on gender roles that contribute to these harms. There is widespread acceptance, both explicit and implicit, of a world in which women are treated as "secondary beings" (Beauvoir 1949/2009: 15). Much of Beauvoir's book is dedicated to describing the social factors underlying this situation. These include a long history of male dominance, socialization, and rigid gender roles that govern every stage of life. Beauvoir describes ways in which this conditioning is embodied, both in material culture (girls given dolls and pretty dresses) and in the exploitation of female bodies in contexts of sex, pregnancy, and maternal labor. Knowledge of one's place as a woman in the West is inherent in internalized attitudes, inculcated habits, and physical realities. In these ways, women are encouraged to be passive and dependent, to cry, to worship men, to be ashamed of their bodies, and so on.

The construction of female identities is another case of moral sedimentation. Beauvoir's narrative primarily concerns relatively wealthy, white, Western women, but it illustrates some general points about moral learning. It shows how social standards can be transmitted through multiple channels and physically embodied. Here, she shares more in common with Merleau-Ponty than with Sartre. This comes out in a favorable review she wrote on Merleau-Ponty's magnum opus, where she expresses her enthusiasm for his embodied approach: "our existence never grasps itself in its nakedness but as it is expressed by our body. And this body is not enclosed in the instant but implies an entire history, and even a prehistory" (Beauvoir 1945/2009: 163). Beauvoir takes this approach further in her own work by applying it, with devastating detail, to culturally learned values. In the case of gender roles, contingent expectations of culture get intertwined with physiological facts in profound ways. Beauvoir is sometimes accused of having too sharp a sex/gender divide—her book begins with a detailed discussion of biological similarities and differences—but she emphasizes the plasticity of the body, suggesting that there can be no sharp divide between indoctrinated values and the somatic conditions of human life. Indeed, these two factors (the psychical and the psychological) become the main components of her earlier work, *The Ethics of Ambiguity* (1947/1976). The title refers to the fact that we are both subjects and things, and these two sides are inextricably connected, in contrast to the Cartesian picture, which divides mind and body and assigns bodies a secondary status.

Beauvoir's work differs from Sartre's in another respect as well. Sartre, we have seen, implies that we can easily escape moral sedimentation: a bit of philosophical reflection will help us see that values depend on us, and this gives us the freedom to comply willfully or to devise new values. Beauvoir's treatment of gender roles in the *Second Sex* exposes this as a naïve and idealistic theory. Social forces are extremely powerful. They lead us to essentialize gender, and, even when the falsity of essentialism is exposed, the person who wants to reject prevailing norms faces serious hurdles: she must retrain her body, circumvent material culture, and overcome condemnation and aggression. Sedimentation is not a benign residue that we can bring into the light and then wipe away; it is a pervasive lattice of forces that make deviation difficult or even impossible.

Kruks (1987) argues convincingly that Beauvoir influenced Sartre and that he came to see freedom as a more difficult achievement than he had originally appreciated. Webber (forthcoming) takes up this observation nicely and traces Beauvoir's influence on Sartre in his 1952 book on Jean Genet. Webber emphasizes a positive side of sedimentation; he argues that repeated pursuit of a project in life can lead to a pattern of habitual behavior that qualifies as sedimentation. Here, sedimentation is a manifestation of commitment. But I think the case of Genet can also be read in a more sinister light, along the lines suggested by Beauvoir's analysis of gender roles. Just as gender is constructed and then essentialized, so, too, is criminality. Genet, an abandoned child, is prodded into a life of crime by social forces that are beyond his control. He ultimately finds freedom through art, but his earlier years are marked by constant struggle: "thefts become external events which he is powerless to oppose" (Sartre 1952/1963: 39). Genet is condemned to a life of thievery because just society needs an object to oppose (1952/1963: 20). Sartre compares Genet's situation to that of African-Americans, who must cope with the impact of being constantly treated as inferior (1952/1963: 55).

Another case study in moral sedimentation can be found in Frantz Fanon's (1952/1967) *Black Skin, White Masks* (cf. Webber forthcoming: ch. 8). Fanon was from Martinique, in the French West Indies, but studied in France, at Lyon, where he earned a doctorate in psychiatry. *Black Skin, White Masks* was going to be his dissertation, but he published it as book instead, and it remains one of the greatest literary documents about the colonial experience. Much of it deals with his own anguish as a highly educated person of color living on an island with a small white minority. Both at home and in France, Fanon was never regarded as a person or a man, but always as a black man, a category that comes loaded with a disturbing array of negative stereotypes: dangerous, primitive, oversexed, and so on. Fanon describes how his skin color alone instilled a sense of shame and insecurity. Even in high-status professions, pigmentation is inescapable:

> We had physicians, professors, statesmen . . . [but] it was always the Negro teacher, the Negro doctor. . . . What could one expect, after all, from a Negro physician? As long as everything went well, he was praised to the skies, but look out, no nonsense, under any conditions! The black physician can never be sure how close he is to disgrace. (Fanon 1952/1967: 88–89)

Fanon's treatment of these themes falls within the existentialist tradition. He had attended lectures by Merleau-Ponty, and he engages with Sartre's work on racism and anti-Semitism in the book. His analysis is phenomenological and antiessentialist, and much of his meditation concerns the theme of alterity. But, he is more concerned than Sartre was with the impact of language (black as a metaphor for bad) and with embodiment (the inescapable reality of pigmentation). This puts him in alignment with Merleau-Ponty and Beauvoir. Fanon also discusses the impact of material culture. For example, he comments at length on the violence of comic books and on children's stories in which adventurous white men bring civilization to dark-skinned savages— or narrowly escape being served in a cannibal feast (112–114). These are all aspects and instruments of sedimentation. They represent a framework of objects and attitudes that make white supremacy seem commonsensical and dehumanize people of color. Such attitudes are a pretext for colonialism, and, as with the patriarchy, no simple act of will can eliminate the moral frameworks that undergird habitual and structural racism.

As a final example, I want to consider some of Hannah Arendt's views on atrocity. Arendt was not an official member of the existentialist circle or any other group of philosophers, but she studied with Jaspers and Heidegger and wrote important exposés on the French existentialists for American readers. In *The Origins of Totalitarianism* (1958), she also takes up the theme of moral sedimentation in the political sphere. Arendt thinks that totalitarianism is a modern and singular development, so it is not itself a sedimented approach to governance, but the early chapters of her book trace the roots of anti-Semitism in Germany and racism in colonial contexts. Bigotry becomes part of the commonsense background for some populations, creating fertile soil for exploitation or worse. Arendt also describes the impact of bureaucracy, which breaks things down into routine procedures that can be followed without any particular intentions, grand or small. Both bigotry and bureaucracy are forms of sedimentation: they become habitual and guide behavior without being questioned. This, of course, is a toxic mix. In *Eichmann in Jerusalem* (1963: 287), Arendt describes Adolf Eichmann as an obsessive bureaucrat, one who was not animated by any grandiose desire to liquidate the world's Jews: "Except for an extraordinary diligence in looking out for his personal advancement, he had no motives at all." This turns out to be factually incorrect; Eichmann was an unapologetic ideologue, as were other architects of the Final Solution.

Arendt was wrong, in this respect, to describe his evil as banal. On the other hand, she discusses the role of regular infantrymen in mass killings. Murder at this scale requires involvement of ordinary people, as well as elaborate procedures that must be broken down into steps and bureaucratically managed. Still, bureaucracy alone is not enough. Ideology is curiously underplayed in Arendt's book on Eichmann. She scoffs at the idea that his trial was also a trial against anti-Semitism (1963: 17), and she even suggests that Eichmann wasn't especially anti-Semitic (e.g., 1963: 22–39). She grants that the Germans used anti-Semitism as a unifying ideology as they conquered their neighbors in Europe, but she thinks this was a miscalculation because, though widespread, anti-Semitism took on different guises in different nations (1963: 154). Here, I favor Arendt's earlier work on totalitarianism, which assigns a bigger role to ideology. It is true that ordinary people commit atrocities, but that is not simply because execution can be routinized. It is also because hatred can become second nature. Rather than thinking of evil as banal, we might do better to describe it as sedimented, as Arendt's earlier work brings out.

As these examples illustrate, moral sedimentation is a persistent theme in existentialist thought. Leading existentialists, as well as their precursors and allies, frequently explore ways in which values are culturally inherited, internalized, and embodied in such a way that they become habitual. Values color our outlook and influence our actions. They are so deeply ingrained that they strike us as self-evident; we cannot imagine having different views. Sedimented values constrain our freedom and delimit our potential. I now want to describe empirical lines of inquiry that support these conclusions. I will then consider available strategies for clearing away moral sediments.

2.3. Sedimentation Naturalized

The cases of sedimentation described in the existentialist literature are multifaceted; they unfold in history and get enshrined in institutions, codified in laws, and physically implemented in material artifacts. To fully investigate sedimentation, one would have to move beyond philosophical reflection and engage all of the social sciences. Sedimentation is also, and perhaps most importantly, realized in the mind. Sedimented morals are part of our psychology. The notion of sedimentation, recall, derives from phenomenological theories that emphasize the concepts that we use to comprehend the world and the bodily habits that govern our responses to things. Thus, the psychological sciences, including experimental psychology and social neuroscience, are well-suited for exploring sedimentation empirically. In treating sedimentation as something that can be studied this way—something that can be "naturalized"—we open up new opportunities to confirm and expand on the hypotheses of existentialist thinkers. Is there reason to think that our values are sedimented? If so, how does this sedimentation work? And how might it be mitigated?

To establish that moral sedimentation is a real phenomenon, it is helpful to imagine what an opposing view might look like. Here, the most obvious alternative would be moral rationalism. For the rationalist, values are derived from reason, and moral opinions can be settled, for each case, by deliberation. Rationalist models tend to assume that moral precepts can be derived from universal first-principles and are thus ahistorical. They also favor models of cognition that are disembodied; moral knowledge is conceived as propositional and dispassionate. This perspective contrasts most strikingly with sentimentalist theories, which claim the primacy of passion over reason. Sentimentalists claim that moral judgments are emotional responses, derived through evolved dispositions and emotional conditioning, rather than reasoning. Sentimentalism has roots in British moral philosophy (especially Shaftesbury, Hutcheson, Hume, and Smith), but it has undergone a recent revival, fueled, in part, by developments in neuroscience and psychology.

Neuroimaging studies consistently show that moral judgment recruits brain areas associated with emotion. For example, Moll et al. (Moll, Eslinger, and Oliveira-Souza 2001) conducted a functional magnetic resonance imaging (fMRI) study in which people were asked to assess whether sentences said something morally right or wrong (as compared to factually right or wrong). They found activation in the orbital frontal context and the posterior cingulate—two classic "emotion areas." These areas also appeared in a study by Harenski et al. (2010), using a very different methodology. They showed participants photographs of moral violations. Orbital frontal cortex and posterior cingulate activations increased with judged severity of the moral wrongs. Many other studies show the same pattern. A recent meta-analysis subsuming more than thirty fMRI studies found consistent activations in these two areas as well as anterior cingulate and insula—two other areas associated with emotion (Boccia et al. forthcoming). All of these brain areas are also associated with perception of the body (Critchley et al. 2004). This is consistent with the view that emotions are perceptions of the body's preparation for action (see Prinz 2004). In this context, it supports the idea that moral values are sedimented in the form of bodily dispositions.

Moral neuroscience provides ample support for sentimentalism, but not for rationalism. Brain structures associated with reasoning are not major players in moral cognition. The Boccia et al. meta-analysis found a relationship between moral judgment and activity in the superior temporal sulcus in addition to the aforementioned emotion areas. This is a common player in tasks that involve attribution of mental states, but they found little evidence for active engagement by brain structures associated with dispassionate cognition. In fact, there is only one laboratory in the field that consistently claims to find neuroimaging evidence for cool cognition in moral judgment—Joshua Greene's group at Harvard. They claim that some moral decisions rely on emotions while others rely on reason. Their data, however, tell a different story (see Prinz 2016 for discussion). In their seminal paper, Greene et al. (2001) divide moral judgments according to whether participants usually make consequentialist (e.g., maximize lives saved) or deontological decisions (e.g., don't kill, no matter what). They claim that the former are guided by reason, the latter by emotion. But their data show emotional activation in both cases. Consequentialist judgments show a significantly higher contribution from working memory, but that is not surprising since maximizing outcomes requires holding different outcomes in mind. Brain imaging provides little support for dispassionate moral judgments. Morality seems to have an emotional basis.

This interpretation finds further support in behavioral research. Numerous studies have established that altering people's emotions alters their moral judgments as well. For example, Seidel and Prinz (2013) exposed people to irritating or uplifting music, which had significant impact on ratings of bad and good behavior. Other studies have shown that people do poorly when asked to provide reasons for their moral judgments (e.g., Cushman, Young, and Hauser 2006; Haidt 2001). Evaluations of reasons are also unstable. For example, the assessed quality of premises in a moral argument depends on order or presentation, even among professional philosophers (Schwitzgebel and Cushman 2012). Applied to sedimentation, this suggests that moral judgments are not made by reliable reflection or reasoning but rather depend on embodied feelings.

The aforementioned behavioral studies show that our moral feelings can be shifted in laboratory settings. Other work suggests, however, that moral feelings are relatively stable over time. Consider, for example, research on political party affiliation in North America. Political values are closely related to moral values since many policy issues are moralized (war, abortion, capital punishment, wealth redistribution, etc.). Political values are also relatively stable over the life span. For example, Sears and Funk (1999) found a .80 correlation for party identification over ten-year intervals, and a .65 correlation over a period of nearly four decades. This indicates deep entrenchment of values. Moreover, there is strong evidence that our political values are strongly influenced by demographic factors, such as ethnicity and geography (e.g., Green, Palmquist, and Schickler 2002). This suggests that social background, not reasoned argument, is the biggest determinant of political morality. This, together with the fixity of values over time, looks like further evidence for sedimentation.

Research using brain imaging has shown that political values, like morals more generally, are grounded in emotions. For example, Gozzi et al. (2010) found that political interest correlated with responses in emotional brain areas when reading about political issues, and Kaplan et al. (Kaplan, Freedman, and Iacoboni 2007) found emotional responses when people viewed photographs of presidential candidates. Another study, by Westen et al. (2006) presented participants with bad arguments made by political candidates; when reading bad arguments made by a preferred candidate, participants judged that the arguments were good and brain activations indicated emotional, but not cool, rational, information processing.

The foregoing findings are representative of work in social neuroscience and empirical moral psychology. They suggest a picture of morality that echoes eighteenth-century sentimentalism. Moral judgments are based on emotion, not reason. The findings also suggest that values differ across social groups and that moral debates are relatively insensitive to rational argumentation. The neuroscientific findings also link moral attitudes to the body; they implicate brain areas that are involved in bodily perception and regulation. Thus, the acquisition of values is a matter of training the body to respond. To moralize is to react physically. Each of the emotions that operates in the moral sphere has its own action tendency: anger compels us to aggress, disgust makes us withdraw, guilt moves us to seek reparations, and shame instills the urge to conceal oneself. On the positive end of the spectrum, which I have not emphasized here, we find concern, which goads us to aid others in need; gratitude, which prompts supplication and indemnification; and moral esteem, which makes us gush and fawn.

This sentimentalist picture that has reemerged in the empirical literature is more associated with British empiricism than with existentialism, but it helps confirm and explain many of the insights in the literature of moral sedimentation. Values are acquired through embodied activity, and they come to be experienced as self-evident (Merleau-Ponty). They are culturally inherited and thus have a history that may be unknown to us (Nietzsche). Because they are acquired through socialization, rather than reason, they are difficult to change by act of will (Beauvoir and Fanon). Arational social transmission makes ordinary people susceptible to the acquisition of values that are pernicious (Arendt). Just as demography can determine whether one is a Democrat or Republican, geography and creed can determine whether one falls prey to Nazi propaganda. Thus, sentimentalism sheds light on sedimentation. It is an empirically established mechanism that can account for this major obstacle to radical freedom.

The empirical literature also explores the mechanisms underlying bigotry and hate, which were matters of grave concern to the existentialists. For example, Phelps et al. (2000) examine the way that racism is neurally implemented. They measured brain activity while showing participants faces of individuals of European and African descent and then compared these results to performance on an Implicit Association Test (IAT), which measures racial bias. They found that IAT scores correlate with differences in activation of the amygdala, a brain structure associated with fear.

Other work shows that we are less empathetic toward members of racial outgroups. Neuroscientific research has shown that we experience vicarious pain when we see harm inflicted on other people; pain centers of the brain are active when watching film clips of bodily harm. In a number of studies, it has been established that this effect is moderated by group membership. For example, Azevedo et al. (2013) presented participants with films of a hand being stroked by a cotton swab or penetrated by a needle. The needle films induced the vicarious pain effect, but this was greatly mitigated when the hand in the film belonged to a different ethnic group than that of the participant.

In another study, Golby et al. (2001) presented participants with faces of different ethnicities and measured activity in the fusiform face area (FFA), which is known to play a major role in facial recognition. The FFA also registers when the scene before us contains faces and thus prompts us to engage socially. Strikingly, Golby et al. found that there was considerably more FFA activity when participants were presented with faces of their own ethnicity. This suggests that we are less likely to see faces of ethnic outgroups as opportunities for social interaction; at the extreme, we may see them as objects rather than as persons.

The Golby et al. study can be interpreted as establishing a neural component of dehumanization: our tendency to see members of outgroups as less than human. This phenomenon has also been investigated in behavioral research. In one chilling example, Goff et al. (2008) subliminally presented participants with European or African faces and then tested performance on the recognition of rapidly presented line drawings of animals. Those who had been presented with African faces where faster to recognize line drawings of gorillas. Goff et al. relate this to a long history of racist iconography that relates people of color to apes. In another study, they found that African-American criminal defendants were more frequently described using simian metaphors than were European-American defendants matched for socioeconomic factors, ethnicity of victims, and severity of crime. Use of simian metaphors in the press also correlated with death sentences for African-American.

Dehumanization is not restricted to members of other ethnic groups. In an fMRI study, Harris and Fiske (2006) presented participants with images of various individuals including members of stigmatized social groups, such as homeless people and drug addicts. Images of stigmatized groups induced brain activations in areas associated with disgust. Brain areas that are normally associated with social perception were active when perceiving nonstigmatized groups, but not stigmatized groups. Thus, social outgroups are not only viewed as objects rather than persons; they are also viewed as repugnant.

Neuroscientific methods have also been used to study the mechanisms underlying various forms of sexism. Cikara et al. (Cikara, Eberhardt, and Fiske 2011) administered a scale that measures sexist attitudes and measured brain activity while participants viewed men and women who were either dressed or sexualized and undressed. Sexism, as measured by the scale, negatively correlated with activity in brain areas associated with social cognition for male participants viewing nude women. The authors interpret this as showing that these sexist men view nude women as mere objects. (Another interpretation consistent with the data is that these man are morally offended by the presentation of male nudes but not by female nudes.)

Cikara et al. focus on negative sexism in their study: hateful attitudes toward women. There is also a phenomenon called *benevolent sexism*: acting kindly toward women in ways that are belittling or condescending. Behavioral work has shown that benevolent sexism can have a negative impact. Dumont et al. (Dumont, Sarlet, and Dardenne 2010) showed that excessive kindness to women can cause women to form negative self-construals and facilitate access to memories of their own incompetence. Dardenne et al. (Dardenne, Dumont, and Bollier 2007) show that benevolent sexism can also impact performance on a cognitive task. After being told that women are well-groomed and kind, female participants showed diminished performance when calculating routes on a map as compared to women who had been exposed to malevolent sexist remarks or neutral remarks. Dardenne et al. (2013) have also investigated the neural impact of benevolent sexism. They found that benevolently sexist remarks are associated with activity in the left anterior cingulate cortex, a brain area associated with conflict monitoring and divided attention.

These examples could easily be multiplied. The psychology and neuroscience of bias have become major areas of research. They add empirical support to existentialist work on moral sedimentation. One might express this by saying that values get sedimented in the brain. Entrenched moral attitudes, including various forms of bigotry, are physically grounded in neural circuits that get conditioned by socialization.

Still, one might wonder what these empirical results contribute to the rich literature on moral sedimentation reviewed in the previous section. What do we learn by pointing to a metabolic response in the brain of someone who is experiencing racial hatred, for example? This question allows no simple or single answer. Each study must be looked at on its own merits, and, for each, there is an open question of whether anything has been learned.

To the extent that one can speak in general terms about the value of empirical work in this area, I would suggest three possible contributions. Such work can be used to confirm philosophical hypothesis. For example, sentimentalist theories have been based, historically, on armchair argumentation and introspection. Empirical results add new lines of support. Second, empirical methods sometimes give insight into underlying mechanisms, and these may deepen or extend theories that have been developed using other approaches. For example, the reduced activation in the fusiform face area when viewing outgroup members could lead to new insights about the psychology of objectification. Third, by identifying mechanisms, empirical methods might point to new avenues for change. If morality has an emotional basis, moral change might require emotional reconditioning.

I hasten to add that empirical methods of the kind reviewed here cannot replace other approaches. Moral sedimentation is a phenomenon that involves historical events and large-scale social institutions. Experimental psychology and neuroscience tend to be individualistic and temporally local. They measure the behavior and brain processes of individuals here and now. A full investigation of moral sedimentation must use all the resources of the social sciences, working at different scales of time and space. Psychology and neuroscience also tend to provide "thin" characterizations of mental activities. Experimental design works by isolating individual variables and measuring preselected kinds of impact. They also operate from a third-person perspective, giving limited insight into what it is like for individuals to experience their values. To the extent that evaluative experience is probed, it is usually in a laboratory setting with varying degrees of applicability to ordinary life. This approach helps us break down complex processes into simpler components, but it leaves much out.

Existentialists tend to favor phenomenological methods. They use "thick" descriptions of lived experience and focus on real-world contexts. This is done from the armchair (or café chair), but it is an empirical approach. Existentialists do not engage in mere conceptual analysis; they aim to describe what it is like to engage in social activity. This work makes essential contributions to the understanding of moral sedimentation. It can also contribute to experimental work by identifying phenomena that can be studied using laboratory methods. For example, Kenneth and Maime Clark's (1947) pioneering work on children's attitudes toward the beauty of dolls with dark and light complexions can be read as an experimental investigation of Merleau-Ponty's point that values can be realized in physical artifacts. Recent work in cognitive metaphor theory takes up Fanon's observation that language sets up negative associations with darkness (Sherman and Clore 2009). One can also imagine a flow of ideas from experimental work back into phenomenology. For example, empirical work could enrich the phenomenology of objectification.

Some of the authors associated with existentialist work on moral sedimentation have been skeptical of social science (see, e.g., Arendt [1958] on the sociology of totalitarianism). But others were deeply committed to interdisciplinarity; for example, Merleau-Ponty makes extensive use of neurology, and Fanon draws on his training as a psychiatrist. Theoretical work is needed to guide social science and as a watchdog to guard against scientific essentialism, reductionism, and bias. There is, however, great value in using a plurality of methods when studying complex phenomena, and there can be mutual enrichment for both existentialist theory and neuroscience if the two are put in conversation.

Moving beyond methodology, the work reviewed here offers another avenue into the topic of moral sedimentation. It points to ways in which values get internalized, and this raises questions about the possibility of moral change. Those who subscribe to radical freedom need to grapple with the fact that values are deeply inculcated, passionately felt, and physically embodied. Given the urgency of this issue, I want to end by discussing strategies for coping with sedimentation.

3. MORAL REVOLT

Moral sedimentation is a running theme in existentialist thought. Despite a highly publicized interest in free will, many key existentialist thinkers recognize that there are strong social pressures to adopt the prevailing values of the communities in which we are reared. These values often have deep historical roots, and they are inculcated in ways that make it difficult to simply create our own moral outlook. Most of us simply adopt the values that are sedimented in our social surroundings; these then get sedimented in our minds and behavior. Empirical work supports this thesis by showing that values are culturally relative, arational, and tenacious. Brain imaging suggests that values get physically sedimented in neural circuits. There is also evidence for widespread, deeply ingrained, ingroup bias. This is true even when superficial rhetoric indicates a commitment to egalitarian values. Historically entrenched divisions and hierarchies are difficult to shed. This is a major hurdle for intergroup amity and social justice. It also threatens existentialist ideas about radical freedom. Unsurprisingly, then, many of the thinkers who draw attention to moral sedimentation also propose solutions. I want to end by briefly considering some of these and reiterating the value of empirical methods in charting a path forward.

Nietzsche's response to moral sedimentation is, perhaps, the most radical. He proposes that we move beyond morality. The sovereign individual is one who has freed himself from the dictates of ethics "because 'autonomous' and 'ethical' are mutually exclusive" (Nietzsche 1887/1994: 37). This is a seductive idea, but it is in equal measure unsettling. Moral rules are, in part, rules for getting along with others and achieving social stability. Nietzsche is right to point out that morality is also about power: moral rules are used to oppress. But the rejection of morality is not a solution. A world of immoralists would be like the Hobbesian state of nature: a war of all against all. Nietzsche might reply that very few spirits are free, and these privileged few can live outside of custom dominating the rabble who continue on in ignorance and subordination. This comes close to the insidious picture advocated by Plato in the *Republic*, a society ruled by philosopher kings who must control the population through deception and absolute power. Nietzsche can also be accused of naïve Romanticism. Freed from cultural constraints, his sovereign individuals are depicted as more natural, more active, than the rest. But the idea of human beings existing outside culture in this way is deeply confused. We are by nature cultural to the core, and there is no way to escape social influence.

Other existentialists try to find a place for freedom within a world of sedimented rules. The paradigm case is Camus, who, in the *Myth of Sisyphus* (1942/1955), recounts the tale of a man condemned to roll a rock up a hill in perpetuity. This predicament is an allegory for the absurdity of life, an absurdity we cannot escape. But rather than concluding that there is no freedom, Camus notes that Sisyphus can choose to embrace his fate. His heroism consists in his willful acquiescence to absurdity. Applied to cases of moral sedimentation, I find Camus's suggestion problematic. Taken on its face, it is deeply defeatist. In the case of sedimented sexism and racism, should we simply acquiesce? Admittedly, Camus does not always advocate complacency. In *The Rebel* (1951/1956), he acknowledges the need for social change. On the other hand, he vehemently denounces revolutionary politics and gives little indication of how to address injustice. The formula he delivers in the final chapter is a combination of moderation and love. It is hard to see how this cocktail can bring about meaningful change. The appeal to moderation sounds like plea for conservatism, and the invocation of love is reminiscent of Christian rhetoric. Far from releasing us from sedimentation, Camus seems to embrace it.

Camus's book caused a falling out with Sartre because Sartre was much more sympathetic with the Marxist experiment in the Soviet Union. This, one might even say, was Sartre's suggestion for how to deal with moral sedimentation. We can find liberation from oppressive morality in revolutionary socialism. Sartre (1957/1963) tries to reconcile Marxism with existentialism, but the coherence of that project can be called into question. Marxism advances a grand narrative according to which history is guided by teleological and materialist forces. Human will is the effect of these forces, not the cause.

In contrast, existentialism eschews grand narratives and, on Sartre's formulation, gives almost unlimited power to will. For Sartre, the greatest threat to human flourishing is bad faith; we must live according to our own values. For Marx, the greatest threat is alienation from the fruits of one's labor, and Marxists are willing to sacrifice the individualist orientation of bourgeois society in order to achieve this end.

Camus, Merleau-Ponty, and Arendt were far less enamored with Marxism than was Sartre. They saw Soviet totalitarianism as an indication that Marx's political views could not be implemented without undermining basic liberties. But this leaves us, again, with worries about complacency. These philosophers were good at diagnosing pernicious forms of moral sedimentation, but when it came to change, they deliver fewer options than one might expect.

Of all the existentialists, none was more actively engaged in fighting for social change than Fanon. During the Second World War, he became a solider for the resistance and fought against the Nazis, earning a medal for his courage. He later joined the National Liberation Front in Algeria and helped fight for that nation's independence. Thus, Fanon clearly recognized military action as a tool for fighting colonialism and imperialism. But what about moral sedimentation? In particular, what would Fanon advise for combating pervasive racism? Contemporary philosophical discussions of this issue often revolve around the question of eliminativism. Given the negative stereotypes associated with racial categories, some think sedimentation can be most effectively addressed by trying to eliminate the concept of race. Kruks (2001: ch. 3) reads Fanon this way, noting that *Black Skin, White Masks* ends with Fanon expressing the desire to be seen as a man, rather than a black man (see also Gilroy 2000). The prevailing conceptualizations of blackness have been established by white people, and Fanon surely wants to break free from these. On another reading, however, he is not hoping to eliminate the concept of race but to change it (Gines 2003). He criticizes Sartre for describing the Négritude movement as a passing phase en route to a universalizing conception of humanity (Fanon 1952/1967: 103). Négritude was a literary philosophy that promoted positive conceptions of African identity. Fanon had sympathies with the movement and shared their commitment to Pan-Africanism. He also expresses pride in African history (e.g., 1952/1967: 99) and ambivalence about intermarriage (chs. 2–3), which he sees as a symptom of white supremacy. Thus, it is hard to confidently place Fanon in the eliminitivist camp.

Rather than looking for a decisive stance on the question of eliminativism, one might read Fanon as engaged in an exercise of consciousness-raising. By describing his experience, readers can gain insight into the plight of denigrated ethnic groups and the psychological impact of colonization. Consciousness-raising may also be the main objective of Beauvoir in *The Second Sex*. For female readers with a similar background, Beauvoir's work functions as a mirror: here are our lives, and this is how we have been belittled, delimited, and constrained. For other readers, the text poses the question, "What role are you playing in oppression?"

Consciousness-raising is a powerful tool against moral sedimentation since it places a spotlight on values that have become deeply internalized. What once seemed automatic and obvious is shown to be contingent and fraught. This is a first step toward change. But consciousness-raising does not always tell us where to go from there. Once we see that our values are malleable, how do we go about changing them, and what changes do we make?

Empirical methods, including lessons from neuroscience, can help with the question of implementing change. For example, there is research that identifies effective strategies for overcoming implicit racial bias. In one recent effort, researchers examined the comparative efficacy of eighteen different proposals for reducing implicit bias (Lai et al. 2014). Using performance on the IAT as a measure, they found that half of these were ineffective. They got null results on some strategies that seem intuitively and theoretically promising: perspective taking, empathy training, and reflecting on racial injustice. These strategies are commonly deployed in consciousness-raising narratives. This raises empirical doubts about prevailing methods to reduce bias. The most effective intervention was vividly imagining counterstereotypical scenarios, in which white people are presented as vicious and people of color are presented as virtuous. These were especially effective when the participant was actively engaged in the imaginative task. Another study, by Devine et al. (2012), used a combination of effective strategies and showed that these can have long-term impact on the IAT. Empirical studies can and should look beyond the IAT as well, since this is only a superficial measure of bias. What interventions can reduce racial discrimination and sexual harassment? Is there an intervention that could allow white people to see people of color without automatically encoding race? A recent neuroimaging study indicates that white people can look beyond race when making social judgments, but it requires effortful processing (Amodio and Potanina 2008). Can this become automatic, as Fanon desired?

Empirical methods can play an invaluable role in identifying effective ways to free ourselves from sedimented values. Empirical methods can also help establish that certain values are based on factual mistakes. Scientific methods, informed by social critique, can be used to debunk scientific racism and sexism (Prinz 2012). But what guidance can science provide in selecting new values, once pernicious moral sediments are cleared away? This is a question that deserves considerable reflection. I will end with a modest suggestion: if the existentialists are right that values are a human construction with much variation and many degrees of freedom, then effort should be invested in figuring out how multiple systems of value can coexist without trespassing against each other. Empirical methods might be recruited, at multiple scales, in trying to establish workable approaches to moral pluralism. Such an effort would use philosophical theory to guide empirical inquiry.

ACKNOWLEDGMENTS

I am very grateful to Fiona Schick for invaluable editorial assistance.

REFERENCES

Amodio, D. M., and P. V. Potanina. 2008. Roles of the medial and lateral prefrontal cortex inregulating intergroup judgments. Paper presented at the annual meeting of the Social and Affective Neuroscience Society, Boston, MA.

Arendt, H. 1958. *The origins of totalitarianism*. New York: Meridian Books.

Arendt, H. 1963. *Eichmann in Jerusalem: A report on the banality of evil*. New York: Viking Press.

Azevedo, R. T., E. Macaluso, A. Avenanti, V. Santangelo, V. Cazzato, and S. M. Aglioti 2013. Their pain is not our pain: Brain and autonomic correlates of empathic resonance with the pain of same and different race individuals. *Human Brain Mapping* 34(12); 3168–3181.

Beauvoir, S. D. 1947/1976. *The ethics of ambiguity*. New York: Citadel Press.

Beauvoir, S. D. 1949/2009. *The second sex* (trans. C. Borde and S. Malovany-Chevallier). New York: Random House.

Boccia, M., C. Dacquino, L. Piccardi, P. Cordellieri, C. Guariglia, F. Ferlazzo, S. Ferracuti, and A. M. Giannini. 2017 Neural foundation of human moral reasoning: An ALE meta-analysis about the role of personal perspective. *Brain Imaging and Behavior* 11(1): 278–292.

Camus, A. 1942/1946. *The stranger* (trans. Stuart Gilbert). New York: Random House.

Camus, A. 1942/1955. *The myth of Sisyphus and other essays* (trans. Justin O'Brien). New York: Vintage-Random House.

Camus, A. 1951/1956. *The rebel* (trans. Anthony Bower). New York: Random House.

Cikara, M., J. L. Eberhardt, and S. T. Fiske. 2011. From agents to objects: Sexist attitudes and neural responses to sexualized targets. *Journal of Cognitive Neuroscience* 23: 540–551.

Clark, K. B., and M. P. Clark. 1947. Racial identification and preference among Negro children. In E. L. Hartley (Ed.), *Readings in social psychology*, pp. 169–178. New York: Holt, Rinehart, and Winston.

Critchley, H. D., S. Wiens, P. Rotshtein, A. Ohman, and R. J. Dolan. 2004. Neural systems supporting interoceptive awareness. *Nature Neuroscience* 7: 189–195.

Cushman, F. A., L. Young, and M. D. Hauser. 2006. The role of conscious reasoning and intuitions in moral judgment: Testing three principles of harm. *Psychological Science* 17: 1082–1089.

Dardenne, B., M. Dumont, and T. Bollier. 2007. Insidious dangers of benevolent sexism: consequences for women's performance. *Journal of Personality and Social Psychology* 93: 764–779.

Dardenne, B., M. Dumont, M. Sarlet, C. Phillips, E. Balteau, C. Degueldre, A. Luxen, E. Salmon, P. Maquet, and F. Collette. 2013. Benevolent sexism alters executive brain responses. *Clinical Neuroscience* 24: 572–577.

Devine, P., P. Forscher, A. Austin, and W. Cox. 2012. Long-term reduction in implicit race bias: A prejudice habit-breaking intervention. *Journal of Experimental Social Psychology* 48: 1267–1278.

Dumont, M., M. Sarlet, and B. Dardenne. 2010. Be too kind to a woman, she'll feel incompetent: Benevolent sexism shifts self-construal and autobiographical memories toward incompetence. *Sex Roles* 62: 545–553.

Fanon, F. 1952/1967. *Black skin white masks: The experiences of a black man in a white world* (trans. C. L. Markmann). New York: Grove Press.

Gilroy, P. 2000. *Against race: Imagining political culture beyond the color line.* Cambridge, MA: Harvard University Press.

Gines, K. 2003. Fanon and Sartre 50 years later: To retain or reject the concept of race. *Sartre Studies International* 9: 55–67.

Goff, P. A., J. L. Eberhardt, M. J. Williams, and M. C. Jackson. 2008. Not yet human: Implicit knowledge, historical dehumanization, and contemporary consequences. *Journal of Personality and Social Psychology* 94: 292–306.

Golby, A. J., J. D. E. Gabrieli, J. Y. Chiao, and J. L. Eberhardt. 2001. Differential responses in the fusiform region to same-race and other- race faces. *Nature Neuroscience* 4: 845–850.

Gozzi M., G. Zamboni, F. Krueger, and J. Grafman. 2010. Interest in politics modulates neural activity in the amygdala and ventral striatum. *Human Brain Mapping* 31: 1763–1771.

Green, D. P., B. Palmquist, and E. Schickler. 2002. *Partisan hearts and minds: Political parties and the social identities of voters.* New Haven, CT: Yale University Press.

Greene, J. D., R. B. Sommerville, L. E. Nystrom, J. M. Darley, and J. D. Cohen. 2001. An fMRI investigation of emotional engagement in moral judgment. *Science* 293: 2105–2108.

Haidt, J. 2001. The emotional dog and its rational tail: A social intuitionist approach to moral judgment. *Psychological Review* 108: 814–834.

Harenski, C. L., O. Antonenko, M. S. Shane, and K. A. Kiehl. 2010. A functional imaging investigation of moral deliberation and moral intuition. *NeuroImage* 49: 2707–2716.

Harris, L. T., and S. T. Fiske. 2006. Dehumanizing the lowest of the low: Neuro-imaging responses to extreme outgroups. *Psychological Science* 17: 847–853.

Husserl, E. 1936/1970. *The crisis of the European sciences* (trans. D. Carr). Evanston, IL: Northwestern University Press.

Kaplan J. T., J. Freedman, and M. Iacoboni. 2007. Us versus them: Political attitudes and party affiliation influence neural response to faces of presidential candidates. *Neuropsychologia* 45: 55–64.

Kruks, S. 1987. Simone de Beauvoir and the limits to freedom. *Social Text* 17: 111–122.

Kruks, S. 2001. *Retrieving experience: Subjectivity and recognition in feminist politics.* Ithaca, NY: Cornell University Press.

Lai, C. K., M. Marini, S. A. Lehr, C. Cerruti, J.-E. L. Shin, J. A. Joy-Gaba, A. K. Ho, B. A. Teachman, S. P. Wojcik, S. P. Koleva, R. S. Frazier, L. Heiphetz, E. E. Chen, R. N. Turner, J. Haidt, S. Kesebir, C. B. Hawkins, H. S. Schaefer, S. Rubichi, G. Sartori, C. M. Dial, N. Sriram, M. R. Banaji, and B. A. Nosek. 2014. Reducing implicit racial preferences: A comparative investigation of 17 interventions. *Journal of Experimental Psychology: General* 143: 1765–1785.

Merleau-Ponty, M. 1945/2012. *Phenomenology of perception*. New York: Routledge.
Moll, J., P. J. Eslinger, and R. Oliveira-Souza. 2001. Frontopolar and anterior temporal cortex activation in a moral judgment task: Preliminary functional MRI results in normal subjects. *Arquivos de Neuro-Psiquiatria* 59: 657–664.
Nietzsche, F. 1878/1986. *Human, all too human* (trans. R. J. Holingdale). Cambridge: Cambridge University Press.
Nietzsche, F. 1886/2001. *Beyond good and evil* (trans. J. Norman). Cambridge: Cambridge University Press.
Nietzsche, F. 1887/1994. *On the genealogy of morality* (trans. C. Diethe). Cambridge: Cambridge University Press.
Phelps, E. A., K. J. O'Connor, W. A. Cunningham, E. S. Funayama, J. C. Gatenby, J. C. Gore, and M. R. Banaji. 2000. Performance on indirect measures of race evaluation predicts amygdala activation. *Journal of Cognitive Neuroscience* 12: 729–738.
Prinz, J. J. 2004. *Gut reactions*. New York: Oxford University Press.
Prinz, J. J. 2012. *Beyond human nature*. New York: Norton.
Prinz, J. J. 2016. Sentimentalism and the moral brain. In M. Liao (Ed.), *Moral brains: The neuroscience of morality*. Oxford: Oxford University Press, pp. 45–73.
Sartre, J.-P. 1943/1956. *Being and nothingness* (trans. H. E. Barnes). New York: Washington Square Press.
Sartre, J.-P. 1952/1963. *Saint Genet: Actor and martyr* (trans. B. Fretchman). New York: George Braziller Inc.
Sartre, J.-P. 1957/1963. *Search for a method* (trans. H. Barnes). New York: Alfred A. Knopf.
Schwitzgebel, E., and F. Cushman. 2012. Expertise in moral reasoning? Order effects on moral judgment in professional philosophers and non-philosophers. *Mind and Language* 27: 135–153.
Sears, D. O., and C. Funk. 1999. Evidence of the long-term persistence of adults' political predispositions. *Journal of Politics* 61: 1–28.
Seidel, A., and J. Prinz. 2013. Mad and glad: Musically induced emotions have divergent impact on morals. *Motivation and Emotion* 37: 629–637.
Sherman, G. D., and G. L. Clore. 2009. The color of sin: White and black are perceptual symbols of moral purity and pollution. *Psychological Science* 20: 1019–1025.
Webber, J. Forthcoming. *Rethinking existentialism*. New York: Oxford University Press.
Westen D., P. S. Blagov, K. Harenski, C. Kilts, S. Hamann. 2006. Neural bases of motivated reasoning: An fMRI study of emotional constraints on partisan political judgment in the 2004 U.S. Presidential election. *Journal of Cognitive Neuroscience* 18: 1947–1958.

PART II
Autonomy, Consciousness, and the Self

CHAPTER 7

Choices Without Choosers

Toward a Neuropsychologically Plausible Existentialism

NEIL LEVY

The existentialists are often accused of having painted a bleak picture of human existence. In this chapter, I will show that, in the light of contemporary cognitive science, there are grounds for thinking that the picture is not bleak enough. For existentialists, we live in a meaningless universe, condemned to be free to choose our own values which have no justification beyond the fact that we have chosen them. But existentialists remained confident that there was *someone*, an agent, who could be the locus of the choice we each confront. Contemporary cognitive science shakes our faith even in the existence of this agent. Instead, it provides evidence that seems to indicate that there is no one to choose values; rather, each of us is a motley of different mechanisms and processes, each of which lack the intelligence to confront big existential questions and each pulling in a different direction.

While there are grounds for thinking that the picture is in some ways bleaker than the existentialists suggested, it is, however, not hopeless. The unified self that serves as the ultimate source of value in an otherwise meaningless universe may not exist, but we can each impose a degree of unity on ourselves. The existentialists were sociologically naïve in supposing a degree of distinction between agents and their cultural milieu that was never realistic. Agents are enculturated, and a realistic existentialist will recognize that. But they will also recognize that we are embodied and embedded agents: a biologically realistic picture will understand us as agents always already in process of unification but never achieving it, and always already in negotiation with values rather than choosing them. We are thrown beings: thrown into history, into culture, and into a biological and evolutionary history which we never fully understand and which we can do no more than inflect, all without foundations and lacking even the security of knowing the extent to which we choose or even what we choose. Existentialism must face up to an insecurity that is ontological and epistemological as much as it is axiological.

1. JUSTIFICATION OF EXISTENCE

The focus of the existentialists—or at least of the earlier Sartre and of those who were inspired by him—was on what we might call the *unjustifiability* of existence.[1] Existence is justified if it has a meaning or purpose independent of anyone's conferring such meaning, value, or worth on it and about which there is an opinion-independent fact of the matter. For theists, for instance, existence is justified because our lives figure in some way in God's plan for the universe; for Hegelians, our existence is justified because we play a role in Spirit's progress toward Absolute Knowledge. Existentialists reject these notions as fantasies. In fact, they go much further, rejecting the much less hyperbolic notion of moral objectivity embraced by most moral realists. There is no opinion-independent moral truths, they maintain. Our existence is unjustifiable because there is nothing that *could* justify it: no objective values or transcendent being who can confer on it meaning, value, or worth.

For the existentialists, we live in a universe that *has* no meaning, value, or worth, and we inherit this absence of meaning from it. Rather, we are brief flickers of light in an otherwise eternal darkness, destined to be snuffed out almost as soon as we come into existence and leaving behind us no trace or worth. Our birth is meaningless, from the perspective of the universe, and our death equally meaningless. For the existentialist Sartre, the only moral virtue is authenticity, which requires facing up to this lack of external meaning and value and embracing our free choice *of* values. It is up to each of us to settle on the values that our life will embody, knowing that nothing justifies this choice except that we have made it. When we face up to this choice, we live authentically. When we evade it—either by absorption in the everyday life of work, or sport, or entertainment, or by

---

1. It is a mistake to simply identify Sartre with existentialism: such an identification misses both the diversity and development of Sartre's own work (the libertarian individualist of *Being and Nothingness* [1993] became the compatibilist of *Saint Genet* [1988] and the uneasy holist of the *Critique of Dialectical Reason* [1991]) and also the diversity of the work that has been regarded as paradigmatic of existentialism. There are, for example, prominent Christian existentialists. In this chapter, however, my focus will be on that strand of existentialism which is best exemplified by early Sartre, and I will use the term "existentialist" to refer to thinkers in this school.

self-deceptively accepting some overarching metaphysical system that we take to confer justification on our lives—we are in bad faith. When we are in bad faith, we choose our values anyway—the priest chooses the truth of Christianity for himself, the businesswoman chooses money as the ultimate arbiter of value—but we refuse to recognize that our choice is free or even that we have made it.

While this picture may strike us as bleak, from some angles Sartre may be seen as reaffirming the value and worth of the human being (existentialism is, as he says [1975], a humanism). The human being is the ultimate source and foundation of value for him. The free and unjustified (from any objective standpoint) choice of the "for-itself," the human agent, is the only source of value and meaning in the universe. It, and it alone, justifies our choices. "Bad faith"—which is closely akin to self-deception—consists in denying our freedom, but when we affirm our freedom, we confer on ourselves and our world the only meaning and value it can have.

The for-itself is a kind of ultimate ground of meaning for Sartre. Sartre presents the for-itself as essentially free to choose values because, in principle, it is causally cut off from anything outside of itself. The mind, which Sartre identifies with consciousness, is the free foundation of all values, which exist only because they are chosen and persist only so long as they are affirmed. At the same time as this picture empties the universe of value and meaning, it presents a picture of the human being—of each individual, rather than of humanity as a whole—that may be seen as extremely flattering. Each of us is the unsurpassable foundation of value; each the legislator not only for ourselves but for the universe as a whole ("every one of us must choose himself; but . . . in choosing for himself he chooses for all men" [1975: 350]). We each occupy a position which has traditionally been reserved for God: uncaused causes who are the foundation of all values.

Sartre's existentialism is (as he came to recognize) subject to a number of very serious objections. To name just one, cutting the for-itself off from causal contact with the rest of the world seems to pose the problem, which plagues Cartesian views, of explaining how mind and matter can ever interact. I doubt, moreover, that the idea of radical and absolute choice which lies at the heart of Sartre's existentialism is even coherent: choice requires preexisting values for options to be available. Despite these problems, I think that a less hyperbolic existentialism, one that recognizes both the extent to which our choices are constrained (where these constraints are both limits on our freedom and enablers of it) and also that choice is a central and inescapable burden for each of us, is defensible. This is not, as Sartre would have it, our ontological predicament: rather, it is characteristic of modernity that each of us has to face multiple and sometimes conflicting live options and thereby to face the need to choose a conception of the good and thereby to impose meaning on our lives.

This realistic existentialism needs to recognize not only our enculturation, but also our embodiment. It must face up to the ways in which Sartre's optimism about agency conflicts with contemporary science. As already mentioned, Sartre identified the mind with consciousness. Consciousness was, for him, essentially free because it was separated from being—from anything that could constrain or even guide it—by nothingness. Leaving aside questions concerning the coherence of this kind of absolute freedom, we now know that the claim that the mind is essentially free and essentially independent of external (to consciousness) influences is simply false. Instead, we are each of us multiply divided minds, and much of our mind is opaque to introspection. These facts spell trouble for the claim that we choose our values freely: cognitive science threatens to dissolve the self and thereby the very agent who was supposed to do the choosing.

The implications of cognitive science for our agency and for our capacity to choose values goes much deeper than Sartre could grasp despite his recognition that the Freudian unconscious is a powerful threat to the view of the choosing self which he endorsed. The Freudian unconscious threatens the self by suggesting that authentic choice may be beyond us: when we take ourselves to be bravely affirming values that have no foundation beyond our choice, their content may in fact be a reflection of deep-seated unconscious impulses. Rather than freely choosing a foundation, our choice may be shaped by forces we do not control. Freud himself recognized the threat to our "self-love" represented by the discovery that the "I" is "not even master in its own home, but is dependent upon the most scanty information concerning all that goes on unconsciously in its psychic life" (1963a: 284–285). Sartre resisted the threat by claiming that the very idea of the unconscious was absurd. But the Freudian picture is recognizably the successor of much older views which recognize the existence of an unconscious but effectively marginalize it and thereby mitigate its threat to our self-conception.

Today, almost all educated people accept the reality of unconscious processes and states. On a very popular conception, the conscious mind functions as a kind of control center. One metaphor describes it as akin to a CEO, delegating tasks to other mechanisms. These other mechanisms may be more competent to carry out their allotted tasks than the decision center (for instance, mechanisms that control the precise trajectory of my hand when I catch a ball may be much better at the complex calculations involved in ensuring that hand and ball intersect than I, with my near-dyscalculia, would ever be). The controller delegates tasks to these mechanisms in something like the way in which the CEO may delegate a task to an accountant who is more competent to carry it out than she is, but their authority is delegated and they have neither the power nor the right to make decisions "above their pay grade." The CEO makes the decisions about company policy, about strategy and goals, leaving it to her subordinates to put them into practice. So the conscious mind selects goals for the person, in the light of the values and ends it affirms, and delegates their implementation to lower level mechanisms.

The Freudian picture retains the essential lineaments of this view. What it adds is the recognition that the nonconscious components may not always be obedient to the wishes of the conscious self. They may rebel; perhaps thereby causing bad decisions (like the decision to continue to use drugs, despite harmful consequences, or to pursue patterns of bad relationships). And they are perennially restive, putting their own stamp on our behavior and occasionally tripping us up (as might be instanced by Freudian slips—perhaps cases in which a recalcitrant desire is given expression by slipping under the guard of censor mechanisms—and by self-deception, wishful thinking, and the like). These kinds of slips and problems may arise in even the best-run organizations, after all. But the overall picture, whereby the self directs and manages the rest of the mind, is held to be at least roughly true (and, to the extent it is not, it is seen as a problem that we can and should aim to correct; "where Id was, there Ego shall be" [Freud 1963*b*: 80]).

Nothing like the CEO metaphor seems sustainable in the light of contemporary cognitive science. It is now a commonplace in cognitive science that the mind is *modular*. Minds do not consist of a central executive in addition to a multiplicity of inflexible and rather unintelligent mechanisms. Rather, the mind consists of *nothing but* such unintelligent mechanisms. There is no central executive: nothing which occupies a seat of power, and nothing which has sufficient intelligence to even understand what that power consists in, let alone use it wisely.

To say that the mind is modular is to say, at minimum, that it consists of discrete and dissociable processing units. These modules are *functionally* discrete: they each have a dedicated domain—a limited set of tasks to perform—and are sensitive only to a limited range of information. This latter feature—encapsulation—goes a long way toward explaining why these mechanisms are unintelligent. An item of information may be obviously relevant to a particular task or a particular process, but the mechanism dedicated to that task may be completely insensitive to this kind of information. Modules are (at least in very significant part) the product of evolution, and they develop sensitivity only to the kinds of cues that were reliably correlated with important information in the environment in which they developed (which is very different from the environment in which we exist today). So they are routinely insensitive to the novel. Consider the mechanisms involved in phobias. Snake detection mechanisms have a task that is (from an evolutionary perspective) important to fulfill since snakes may be deadly. Since the costs of false negatives (mistaking a snake for a stick) are on average much higher than the costs of false positives (such as mistaking a stick for a snake), these mechanisms are oversensitive to cues that suggest the presence of a snake and blind to discounting cues. So the knowledge that *this* snake is made of rubber may be powerless to prevent the person from experiencing fear, an elevated heart rate, a rush of adrenaline, and all the other somatic effects of encountering a real snake.

The evidence for the modularity of mind is very extensive and comes from all branches of cognitive science and allied disciplines (neuroscience, neurology, developmental and evolutionary biology, comparative anatomy, cognitive and social psychology, among others). One central plank consists in the raft of evidence for a variety of double dissociations between processes. Two processes, A and B, are doubly dissociable if A may persist in the absence of (or in the face of significant damage to) B, and vice-versa. Double dissociations are often evidence for neuroanatomical localization; two processes are doubly dissociable in the face of brain injury because they are realized by distinct parts of the brain. Many double dissociations have been discovered through the study of patients with brain lesions or developmental abnormalities. Damage or dysfunction in a specific area of the brain may lead to poor performance on one task without affecting another; when damage or dysfunction elsewhere produces the reverse pattern, we have a double dissociation.

Here's an example. There is an intriguing delusion called *Capgras syndrome*, sufferers of which believe that people close to them have been replaced by impostors. Capgras is widely held to be caused by damage to a pathway that causes the normal affective response to familiar people: the patient sees that the person has the features of (say) his wife, but because he fails to feel the expected affective response, he is reluctant to identify her as his wife. If emotional response and overt recognition are realized in distinct parts of the brain, however, we should expect to find that they can doubly dissociate. It appears they can: there are people who lack the ability to consciously recognize familiar faces but have preserved emotional response to these faces (as indicated by measures of their autonomic system). These people suffer from a disorder called *prosopagnosia*, or face blindness (Ellis and Lewis 2001).

Facial recognition is modular, apparently. Psychologists have identified many other mechanisms, some (like this one) perceptual and others underlying more complex cognitive capacities. An example of the latter: there appears to be a module for social contract violation detection (Stone et al. 2002). Evidence comes from the ease with which ordinary people identify violations of social contracts (e.g., taking a benefit which community rules dictate must be paid for in some way) while having great difficulty with identifying counterexamples to an arbitrary rule (e.g., everyone with pink hair rides a bicycle) despite the fact that the problems have precisely the same logical structure.

Many philosophers and psychologists argue that all there is to the mind is the collection of modules (Carruthers 2006; Pinker 1997; Sperber 1996; Tooby and Cosmides 1992). Certainly, there is little evidence for a central executive: what we are disposed to view as higher level cognitive processes are themselves subserved by specific modules, not by any equivalent of the CEO.

This picture of the mind seems to threaten to dissolve the self into nothing but a collection of dumb widgets.

The account of the mind as modular is deeply at odds with our folk psychological conception of ourselves as unified beings, delegating top-down to constitutive mechanisms. Instead, it reveals each of us as a multiplicity; more a community than a single organism. Worse, the community is fractured: our modules have different goals and different values. This fractionation may be revealed by brain injury, but it also underlies entirely everyday behavior.

Consider the kinds of disinhibition seen in dementia patients or in those who have suffered brain injuries. "Michael," one of Paul Brok's patients, lived on fish fingers and listened all day to Led Zeppelin, as though he had regressed to adolescence following damage to his right frontal lobe (Brok 2003). He says he always had liked those things; he simply had stopped pretending. Michael probably always had liked fish fingers and Led Zeppelin, but his relegation of them to an occasional treat may not have been mere pretense. Rather, his tastes for these things reflected the preferences of some of the modules constitutive of Michael but not others. Michael's overall behavior prior to brain injury might reflect competition between modules, or top-down control of some by others. We can see the conflict between modules, or between systems composed of modules, in the behavior of sufferers of *anarchic hand*. In this syndrome, the patient loses (direct) control over one hand, which engages in purposeful behavior aimed at goals she may not endorse—removing food from the plate of a fellow diner, for instance, and putting it into her mouth.

These are dramatic instances of genuine pathologies, but we may see the same conflict in ordinary weakness of will. Ainslie (2001) has modeled akratic behavior as a game theoretic conflict between two players, each of which is internal to a single agent. One player seeks immediate reward; the other delays gratification in order to achieve larger but later (sometimes much later) reward. The modeling may succeed in describing behavior because it is not entirely metaphorical: there may be a genuine case for thinking of behavior as driven by temporary or persisting coalitions of processes, internal to a single agent and competing with other such coalitions for the control of behavior. The neural basis of this kind of competition is reasonably well understood: different neural systems respond to different stimuli, encode different values, and compete with one another for access to decision-making and to the control of movement.

Where does this leave the self? It is tempting to identify the self with consciousness; those aspects of the mind to which we have some kind of introspective access. There are, I believe, grounds for thinking that consciousness (in this sense) plays a very important role in decision-making and behavior (Levy 2014), but that role is very different from the role we might naïvely assign it. Consciousness is not the executive, with the task of making decisions. It does not even have the power to set the broad direction of the organism, like a CEO making decisions about the direction of a company. It is not in virtue of its causal powers that consciousness is important; rather, it is in virtue of its role as an information clearinghouse. Information fed into it, from transducers that inform it about its environment and from internal systems that make salient their goals and values, are thereby made available to a broad range of the consuming systems which process this information, some of which have decision-making powers (again, not just in cooperation with one another, but also in competition). The consuming systems themselves are partially, and sometimes totally, inaccessible to conscious scrutiny, and the decision-making process is opaque.

We do better to identify the self with the entire collection of mechanisms and not with consciousness. It is the collection of mechanisms that encode our commitments and our values, in a persisting manner, and it is this collection that makes decisions. Identify the self with consciousness and you leave out far too much, including much of what makes each of us distinct from one another. There are grounds for seeing ourselves as a collectivity, not a unity. To be sure, there are limits on how much competition can exist between the constituents of an agent. Genes can be productively modeled, as Dawkins (1976) reminds us, as selfish replicators, but because they share a common fate when they constitute an organism, the degree to which they compete with one another is limited. Similarly, the modules that constitute me share a common fate: since they are in the same boat, they cannot defect without the risk of paying a heavy price. There is not much comfort to take from this, though, when we reflect on the gap between the ends we endorse and those which it is—or was, in the environment of evolutionary adaptiveness—adaptive to pursue. The modules that constitute us are keyed to pursue rewards that are adaptive but which we may not endorse, partly due to the enormous disparity between the environment of evolutionary adaptiveness and today (high-calorie foods, for instance) and partly due to our reflective powers and our capacity to value (sexual opportunities, for instance). Constituent modules performing their proper function threaten to defeat our projects and undermine our goals at every turn.

2. CHOICE

The existentialists urged us to face up to our absolute and terrifying freedom to choose: to choose how to live, how to be, to choose the very values which will serve as a yardstick against which to measure our success as agents. The very idea of absolute choice is probably incoherent because if choice is not to devolve into mere arbitrary plumping for one option or another, it must involve the weighing of goods (Levy 2001, 2011). In any case, it is (as Sartre came to realize) self-deceptive to think that we are capable of engaging in such a choice since we are enculturated beings who understand the world in a way that is permeated by the concepts and values which make us the agents we are. These considerations dramatically restrict the choices of ourselves and our values that we can conceivably make. We can choose only live options, and what counts as live is a function of our social and natural history. Nevertheless, the existentialist call to authenticity remains powerful. We are all tempted to confuse convention with nature, social facts for fixed regularities. The call to face up to our freedom, to recognize that we have options, and that we play a role in sustaining the values and mores of our culture is salutary and can lead us to live more reflectively and more responsibly.

However, our capacity to choose values, to whatever limited extent, is thrown into grave doubt by the considerations from cognitive science adduced earlier. We are multiplicities, pursuing disparate goals that are obscured from consciousness. We have a sense of ourselves as unified agents making choices for reasons, but the unity is fragile and the reasons we take ourselves to have may be confabulation.

Consider the agent who, convinced by Sartre, sets herself to make an authentic choice of her freedom and of values by which to live. It doesn't matter, for our purposes, what values she chooses: she might choose some kind of sub-Nietzschean nihilism or she might choose to maximize welfare or to guide her life by the teachings of the Buddha. When it comes to putting her choice into practice and making concrete choices in its light, however, there may be a large gap between the content of the action and the values to which she is committed. Her moment-to-moment decision-making is shaped by the multiplicity of modules that constitute her, and the information to which she is privy is a heavily edited subset of the information which is available to the set of modules. Under many conditions, the modules will shape her choice and cause her to have a false view of the reasons for which she chooses.

To see this in action, consider the large literature on confabulation. It is important to animals like us that we act for reasons, but the reasons we produce in response to the demand entailed are often post hoc constructions. Modules guide choice for reasons which are opaque to us and which we would reject were we to consider them, and we confabulate good rationales for the choice. Many of the most spectacular demonstrations of confabulation involve pathologies: commissurotomy or brain lesion. But confabulation can be demonstrated under experimental conditions and inferred from behavior.

The classic experimental demonstration is Nisbett and Wilson's (1977) stocking study. Consumers were asked which pair of stockings they preferred from an array. Far more often than one would expect by chance, they chose the rightmost pair. They justified this choice by reference to the properties of the stockings (its color, texture, fabric, and so on), but in fact the stockings were identical (and the order in which they were placed was randomized from subject to subject to rule out responsiveness to minor variations in what are supposed to be identical consumer goods). This kind of result has been demonstrated many times since. Recent examples are more spectacular: they involve the phenomenon that is sometimes called *choice blindness*. In separate studies, people were given the task of comparing pairs of faces in one experiment (Johansonn et al. 2005), expressing their views on important moral questions in another (Hall, Johansson, and Strandberg 2012), and on issues central to an imminent election in a third (Hall et al. 2013). In each study, participants had simply to choose which of a pair of faces or positions they preferred. In all three experiments, the experimenters were also trained magicians and used sleight of hand to place some of the dispreferred options in the preferred pile. In the second phase of the experiment, subjects were asked to justify their choices. Few noticed the switched options when they were subsequently asked to justify their choices. Even committed voters smoothly confabulated reasons for finding a policy—which they had not in fact chosen—wise.

There is direct evidence that agents make choices that are caused by representational mental states of which they are unaware, or of the influence of which they are unaware, and subsequently confabulate reasons for these choices. Unsurprisingly, given that we are multiplicities and the sense we have of our unity is fragile and partly illusory, agents often have implicit attitudes that diverge from their explicit attitudes. That is to say, roughly, we are often disposed to react in ways that conflict with our own professed values: we may sincerely claim to value racial equality but avoid eye contact with black people; sincerely claim to be opposed to sexism but find it hard to take the contributions from women as seriously as those from men, and so on. Under some circumstances, our implicit attitudes cause us both to act in ways that conflict with our values and to confabulate perfectly good reasons for our actions.

Consider the much replicated curriculum vitae (CV) studies (Dovidio and Gaertner 2000; Son Hing et al. 2008). In these studies, researchers send out CVs to potential employers. The CVs are identical except for the name of the applicant: a stereotypically white name is given for half the applications and a stereotypically black name for the other half (other studies use male and female names). The common finding is that employers are significantly more likely to express interest in interviewing the white or male "applicants" than the black or female. Together with other work which probes subjects' reasons in these kinds of studies in the lab, there are strong grounds for thinking that confabulation is at work (Uhlmann and Cohen 2005). Though some potential employers probably reject applications on grounds that are knowingly sexist or racist, many probably have an unconscious negative response to the applicant, and this response causes them unknowingly to change their criteria, such that a black or female applicant has to be more impressive than a white or female applicant to be taken seriously. What would have been seen as a relevant qualification in a white applicant, say, is seen as marginal in a black person; the years working for a charity seen as a sign of commitment to the community in the first, as a gap in the work history of the second, and so on.

The agent who chooses her values and thereby an authentic life for herself may escape bad faith (depending on how we understand it), but she does not escape self-deception. She may take herself to be bravely affirming a new path for herself, but her seeing the path as a deviation in the life plan expected of her (by parents or by society) may be a confabulation. Sartre rightly made self-knowledge central to the existentialist project, but the agent can know herself only partly, and many of her important beliefs about herself may be false.

3. THE SELF

In the previous section, I sketched a number of obstacles to existentialist authenticity which I take to be serious. Indeed, I think there is every reason to conclude that, for the kinds of animals we are, these obstacles cannot be entirely overcome. We remain prey to confabulation and we remain constituted by mechanisms that share our goals only to a limited extent—and sometimes not at all. There is nevertheless a case for thinking that something like a self constituted out of this motely, a self with goals that it may pursue and which it may choose. We are limited and constrained beings, but we can impose a degree of unity on ourselves and a purpose on our lives.

The self should be seen as an achievement and not a given. It is an achievement which is partly brought about by mechanisms that forge a fragile unity in normal development, but it is also an achievement that we may, explicitly or implicitly, pursue for ourselves. We may make ourselves. We do so, in part, by our having values (having, not choosing: having precedes, shapes, and severely constrains choosing). Affirming values is not epiphenomenal but has effects on the modules that drive behavior. Our decisions and actions are always very significantly driven by systems inaccessible to us, which respond to information of which we are sometimes unaware; but information accessible to consciousness nevertheless plays an important role in decision and action, in virtue of its being conscious. That information is made simultaneously available to a large swathe of the modules that drive behavior, which then assess that information for consistency and coherence with the representations proprietary to the module. Conflicts between such representations themselves come to be conscious or cause the inhibition or modulation of actions and decisions; this process may continue for many cycles of global broadcast of representations to the modules, each of which outputs the results of its own assessment back to the global workspace and to other modules (Levy 2014).

An important consequence of these cycles of broadcast and convergence is that modules come to operate on the same contents, given the same valence. This, too, may be an achievement: an agent may initially take a consideration to count in favor of an action, say, but come to treat it as counting against when modules signal a conflict between it and representational contents to which they have access. Such a conflict may cause the inhibition or modulation of behavior *by* altering the agents' personal-level take on their evidence. For instance, conflict between a perceptual representation—the affordances of an object say—and the agent's own values, which are represented in a distributed manner across consuming systems, may lead to a personal-level feeling of unease, which in turn biases the agent's assessment of the options she faces. Iterated cycles of convergence of information on the global workspace of consciousness and broadcast to consuming systems leads, therefore, to the functional integration of the modules: they come to constitute a single system.

We could, if we like, identify this single system with the self. The self is not a mere user illusion (as Dennett [1991] would have it), on the view I am defending. It is a system, with causal powers and the capacity to act on the world. It consists of the set of consuming mechanisms which receive representational contents from consciousness and which output their own signals to one another as well as to consciousness, plus the workspace of consciousness itself. The unification of the modules which results in the existence of a self does not eliminate multiplicity or even inconsistency. We can see the persisting conflict in valuations in ordinary weakness of the will, for example. We can also see it in the lab, under experimental conditions designed to increase the relative influence of modular processes by preventing the resource-intensive and relatively slow cycles of content integration to occur. Under these conditions—when responses must be made rapidly or when the person is under cognitive load or otherwise manipulated to lack the resources for integration—behavior is driven by subsets of modules, rather than by the entire suite, which together encode, in a distributed fashion, the agent's values and commitments. Of course, these kinds of conditions mimic those that can be expected to be found outside the lab, too: we sometimes have to respond very rapidly; we are often distracted or required to multitask; all too often, we have to make crucial decisions after too little sleep or too much alcohol. Under these conditions, too, we may act in ways that conflict with our values and commitments because our behavior is driven by a subset of components of ourselves, rather than the full ensemble.

These are important limitations on our capacity to put into practice our values or conception of the good and thereby shape our life in a manner that Sartre would regard as authentic. But they are limitations on our capacity to *enact* our values, not to *choose* them, as Sartre would have us do. For a variety of reasons, these limitations have little direct effect on our capacity to choose our values (assuming that choosing values is coherent, an issue I leave aside as irrelevant to the neuropsychological plausibility of the Sartrean view). For one thing, high-stakes choices automatically lead to heightened attention and the making available of resources (compare how reminding subjects of their values leads to the mitigation of resource depletion; Baumeister and Vohs 2007). For another, choosing values is not something we do on a particular occasion, but an extended series of actions, leading to plenty of time for deliberation, for the detection of inconsistency, and for the correction of error.

It might be thought that these claims are inconsistent with the data indicating a decisive influence of what are, very plausibly, unendorsed and probably unendorsable influences. On occasion, for instance, chance priming or unconscious inclinations may lead to momentous choices, despite our deliberating. Doesn't this show that we may be at the mercy of modules even given time to reflect and propitious circumstances? Yes, but only in a way that is less troubling than it might appear. These influences may be decisive in career choice, for instance, because, independent of their influence, the person did not lean strongly one way or another. They may tip the balance between otherwise finely balanced options. It is true that these small influences may have big effects, including effects on the values and commitments we end up endorsing. The effects of choices may snowball: having decided on law school rather than medicine due to the priming effect of a word he did not notice, say, Larry may find himself with a whole suite of values typical of a lawyer (for better or for worse). But these kinds of influences should be no more disturbing than the very many others—of environment and developmental resources like genes—which, threatening as Sartre seemed to think, we simply have to live with and accommodate.

The self we forge and maintain against the sometimes corrosive power of modules is yet dependent on these same modules for its persistence. It allows us to affirm a set of values and, for the most part, allows our big-picture values to guide our behavior. We remain vulnerable to subversion from modules, but we may limit its effects.

Does this allow for a neuropsychologically plausible existentialism (assuming we can make sense of the idea of choosing values)? The choice of our values and of our selves that it enables is, I think, not too distant from the existentialist picture to count as a realization of it. Along with other existentialists, Sartre emphasized how the recognition of our freedom is expressed in anxiety as we come to grasp the extent to which we must make our own way of being without guiderails. The picture I have presented might deepen that anxiety. To the extent to which it is accurate, we have to face up to the recognition that our self-knowledge is severely restricted, our propensity to confabulate a constant, and unity a fragile and ever-threatened achievement. We must recognize that we can never be confident that our most important choices were not influenced decisively by facts we cannot endorse or that the reasons we entertain are the reasons for which we act. That is a dizzying prospect. In our embodied existence, we find as much reason for anxiety as Sartre found in absolute freedom.

REFERENCES

Ainslie, G. 2001. *Breakdown of will*. Cambridge: Cambridge University Press.
Baumeister, R. F., and K. D. Vohs. 2007. Self-regulation, ego depletion, and motivation. *Social and Personality Psychology Compass* 1: 1–14.
Brok, P. 2003. *Into the silent land: Travels in neuropsychology*. New York: Atlantic Monthly Press.
Carruthers, P. 2006. *The architecture of mind*. Oxford: Oxford University Press.
Dawkins, R. 1976. *The selfish gene*. Oxford: Oxford University Press.
Dennett, D. 1991. *Consciousness explained*. London: Penguin Books.
Dovidio, J. F., and S. L. Gaertner. 2000. Aversive racism and selection decisions: 1989 and 1999. *Psychological Science* 11: 319–323.
Ellis, H. D., and M. B. Lewis. 2001. Capgras delusion: A window on face recognition. *Trends in Cognitive Science* 5: 149–156.
Freud, S. 1963a. Introductory lectures on psychoanalysis (Part III). In *Works–Standard Edition* (vol. 16). London: Hogarth Press and the Institute of Psychoanalysis.
Freud, S. 1963b. New introductory lectures on psychoanalysis. In *Works–Standard Edition* (vol. 22). London: Hogarth Press and the Institute of Psychoanalysis.
Hall, L., P. Johansson, and T. Strandberg. 2012. Lifting the veil of morality: choice blindness and attitude reversals on a self-transforming survey. *PloS One* 7(9): e45457.
Hall, L., T. Strandberg, P. Pärnamets, A. Lind, B. Tärning, and P. Johansson. 2013. How the polls can be both spot on and dead wrong: Using choice blindness to shift political attitudes and voter intentions. *PLoS One* 8(4): e60554.
Johansson, P., L. Hall, S. Sikström, and A. Olsson. 2005. Failure to detect mismatches between intention and outcome in a simple decision task. *Science* 310(5745): 116–119.
Levy, N. 2001. *Being up-to-date: Foucault, Sartre and postmodernity*. New York: Peter Lang.
Levy, N. 2011. *Hard luck: How luck undermines free will and moral responsibility*. Oxford: Oxford University Press.
Levy, N. 2014. *Consciousness and moral responsibility*. Oxford: Oxford University Press.
Nisbett, R., and T. Wilson. 1977. Telling more than we can know: Verbal reports on mental processes. *Psychological Review* 84: 231–259.
Pinker, S. 1997. *How the mind works*. London: Penguin Press.
Sartre, J-P. 1975. Existentialism is a mumanism (trans. P. Mairet). In W. Kaufmann (Ed.), *Existentialism from Dostoevsky to Sartre*. New York: New American Library.
Sartre, J-P. 1988. *Saint Genet: Actor and martyr* (trans. B. Fretchman). London: Heinemann.
Sartre, J-P. 1991. *Critique of dialectical reason* (trans. Alan Sheridan-Smith). London: Verso.
Sartre, J-P. 1993. *Being and nothingness* (trans. Hazel E. Barnes). London: Routledge.
Son Hing, L. S., G. A. Chung-Yan, L. K. Hamilton, and M. P. Zanna. 2008. A two-dimensional model that employs explicit and implicit attitudes to characterize prejudice. *Journal of Personality and Social Psychology* 94: 971–987.
Sperber, D. 1996. *Explaining culture: A naturalistic approach*. Oxford: Blackwell.
Stone, V. E., L. Cosmides, J. Tooby, N. E. A. Kroll, and R. T. Knight. 2002. Selective impairment of reasoning about social exchange in a patient with bilateral limbic system damage. *Proceedings of the National Academy of Sciences* 99: 11531–11536.

Tooby, J., and L. Cosmides. 1992. The psychological foundations of culture. In J. Barkow, L. Cosmides, and J. Tooby (Eds.). *The adapted mind: Evolutionary psychology and the generation of culture* pp. 19–136. Oxford: Oxford University Press.

Uhlmann, E. L., and G. L. Cohen. 2005. Constructed criteria: Redefining merit to justify discrimination. *Psychological Science* 16: 474–480.

CHAPTER 8
Relational Authenticity

SHAUN GALLAGHER, BEN MORGAN,
AND NAOMI ROKOTNITZ

How should we think of existential authenticity given the dominant view of human existence in science and in contemporary materialist philosophical approaches, where discussions frequently come down to talk of neural processes? More generally, can the 4Ms—mind, meaning, morals, and modality (Price 2004)—translate without loss into neurovocabulary? In this chapter, we argue that to understand existential authenticity it will not do to return to the individuality celebrated by classical existentialism. Nor is it right to look for a reductionist explanation in terms of neuronal patterns or mental representations that would simply opt for a more severe methodological individualism and a conception of authenticity confined to proper brain processes. Rather, we propose to look for a fuller picture of authenticity in what has been termed the "4Es"—the embodied, embedded, enactive, and extended conception of mind (Menary 2010; Rowlands 2015). One requires the 4Es to maintain the 4Ms in the face of reductionistic tendencies in neurophilosophy. The 4E approach gives due consideration to the importance of the brain, taken as part of the brain-body-environment system. It incorporates neuroscience in its explanations, but it also integrates important phenomenological-existentialist conceptions that emphasize embodiment (especially following the work of Merleau-Ponty) and the social environment. Specifically, phenomenological conceptions of intersubjectivity, or, in existentialist terms, being-with (*Mitsein*) and being-for-others, play significant roles in our rethinking of authenticity.

Heidegger's terms *Uneigentlichkeit* and *Eigentlichkeit* are usually translated as "inauthenticity" and "authenticity." The German etymology does not suggest resonances of being fake but, rather, an attempt to make things one's own (*eigen*). Indeed, John Haugeland has suggested that "owned" would be a preferable rendition of *eigentlich* in English (Haugeland 2013: 152). For thinkers in the existentialist tradition, the question is not so much how or why human life is embedded in natural and social contexts but how much we own and own up to what we are doing in those contexts. Returning to and partly adopting Heidegger's articulation of authenticity in his early philosophy, we also attempt to rescue aspects of *Being and Time* from overemphasis on the individual. In this manner, we sketch an alternative model of authenticity which focuses on shared activities as a means to learning and doing things together.

1. EXISTENTIAL AUTHENTICITY AND RELATIONS WITH OTHERS

Although there are significant analyses of being-with and being-for-others in Heidegger and in Sartre, respectively, and indeed, analyses that link an understanding of authenticity to these concepts of intersubjectivity, as we shall show, the very basic analysis of authenticity is framed in individualist terms of facing-up to one's own solitary possibility for not being: that is, one's own death (as in Heidegger's being-toward-death) or in terms of facing up to the nothingness of one's existence (understood by Sartre to be a recognition of one's own freedom). If there is a sense of relation in the concept of *Eigentlichkeit*, it is assumed to involve a form of self-relation (Varga and Guignon 2014). From this perspective, one's relations with others tend to be understood as the occasions of inauthenticity as one loses oneself in the *Das Man* (the anonymous "they") or in bad-faith relations with others.

For this reason, the primary conception of authenticity offered by Heidegger seems at first nonrelational or, at best, self-relational: that is, not an intersubjective phenomenon but a being-unto-one's (very own individual) death, revealed primarily in experiences of anxiety. "Anxiety individualizes Dasein and thus discloses it as '*solus ipse.*' . . . Dasein *is authentically itself* in [its] primordial individualization" (Heidegger 1962: 233/188, 369/22).[1] There seems to be no room in this account for the possibility of recognizing that being-toward-death is a condition that we share with others (see Marx 1987). If social relations are characterized as the occasions of inauthenticity, then

---

1. Most editions of *Being and Time*, including the Macquarrie/Robinson translation and the Heidegger *Gesamtausgabe*, give the equivalent page numbers of the Niemeyer edition of *Sein und Zeit* (1979) in the margins, so, where we quote directly, we will be citing the Macquarrie/Robinson translation and giving pages references to the translation and the German text in the form (Heidegger 1962: 233 = trans/188 = Niemeyer).

authenticity consists of being able to withdraw from being lost in the inauthentic crowd and to confront one's ownmost possibilities.

Heidegger does place some emphasis on the idea of being with others (*Mitsein*) and indeed proposes that it is part of the very existential structure of human existence. Yet, in most cases (we'll return to an interestingly anomalous moment later), the details of his analysis actually make *Mitsein* a secondary phenomenon. Despite his claims that being-with has equal primordial status with being-in-the-world (1962: 149/114; also 153/117; 1985: 238), his analysis tends to privilege the individuality of action by making our encounters with others contingent on our already established involvement with ready-to-hand instruments and worldly projects. One's encounter with others is "by way of the world" (1985: 239, 242). Thus, we encounter others primarily within the context of the pragmatic affairs of everyday life. One comes upon others as unavoidably involved in the same way that one is involved in pragmatic contexts: "Here it should be noted that the closest kind of encounter with another lies in the direction of the very world in which concern is absorbed" (1985: 241). Indeed, even if we did not encounter others, Mitsein has to be considered part of the very nature of the individual's human existence. One might take that as a strong statement of the essential nature of *Mitsein* (Blattner 2006: 67); at the same time, one could take it to indicate that others are actually not essential or necessary for *Mitsein*. Being-with as such does not depend on there being other people around at all; Dasein (human existence) "is far from becoming being-with because an other turns up in fact" (1985: 239):

> This being-with-one-another is not an additive result of the occurrence of several such others, not an epiphenomenon of a multiplicity of Daseins, something supplementary which might come about only on the strength of a certain number. On the contrary, it is because Dasein as being-in-the-world is of itself being-with that there is something like a being-with-one-another. (1985: 239)

Heidegger thus gives primacy to Dasein's own existence per se, even in its being-with, over the actual or possible relations that Dasein could have with others.[2] These relations feature mostly negatively. In asking "who" Dasein

---

2. Some commentators defend Heidegger from this charge, pointing out the equiprimordial nature of being-with (e.g., Wheeler 2005: 149). There are many commentators, however, who do see a problem. For example, Heidegger's student, Karl Löwith (1928), a year after the publication of *Being and Time*, suggested that Heidegger ignored the role of direct interpersonal contact in his account. Binswanger (1962) made similar criticisms and claimed that the idea that Dasein is being-with left him with "a knot of unresolved questions" (1962: 6). Gadamer states: "*Mitsein*, for Heidegger, was a concession that he had to make, but one that he never really got behind.... [It] is, in truth, a very weak idea of the other" (Gadamer 2004: 23). Also see Pöggeler (1989: 251);

is, Heidegger shifts the answer away from the traditional solutions of "I," self, mind, and soul. "It could be that the 'who' of everyday Dasein just is not the 'I myself'" (1962: 150/115). Rather, he suggests, the "they" (das Man) constitutes an important part of Dasein's identity. Dasein is so taken up by the social dimension and by the dominance of others that it gets lost in a social inauthenticity in which it understands itself as being the same as everyone else. "We are inauthentic because our self-relations are mediated by others" (Varga 2011: 92). Dasein, as inauthentic is "not itself," it loses itself (Selbstverlorenheit); it becomes self-alienated (Heidegger 1962: 152/116).

The road to authenticity, on this account, is in some way to escape getting lost in the others. In the following section, we will discuss brief passages where Heidegger breaks this pattern. But before turning to the pointers that Heidegger gives beyond the limits of his own assumptions, we want to give a brief account of the model of authenticity that the early Sartre, based on his reading of Heidegger, developed in dialogue with *Being and Time*.

For Sartre, authenticity goes hand in hand with autonomy. To whatever extent autonomy, as a realization of one's freedom, disappears, so, too, does existential authenticity. Thus Sartre's (1956) inauthentic waiter has given himself up as an individual in order to take up his professional role. To the extent that one considers oneself or treats oneself as a thing that "is outside, in the world . . . [one is] a being of the world, like the ego of another" (Sartre 1957: 31). On this account, authenticity is not to get lost in the world or be defined by the other in the world, but to act on the radical freedom of choosing ourselves, which involves a lonely "forlornness" (see Varga 2011: 86). "For human reality to be is to choose oneself, *without any help whatsoever*, it is entirely abandoned to the intolerable necessity of making itself be—down to the slightest detail" (Sartre 1956: 440–441, emphasis added).

Sartre's interpretation of Heidegger carries forward the individualist model of self, according to which our relations with others lead us astray from our fundamental project—our unique projection of possibilities upon which we need to act. Relations with others tend to be in bad faith. Nonetheless, Sartre finds in Hegel an important realization—that "in my essential being I depend on the essential being of the Other, and instead of holding that my being-for-myself is opposed to my being-for-others, I find that being-for-others appears as a necessary condition for my being-for-myself" (1956: 238). This strikes Sartre as insightful and important, but he immediately gives it up as a model of self–other relations because, in attempting to work out what it means, it leads directly to the realization

Theunissen (1984); Tugendhat (1986); Frie (1997); Gallagher and Jacobson (2012); and Morgan (2013: 37–45) for further discussion.

that social relations are always inauthentic. For Hegel, if there is in truth a Me for whom the Other is an object, this is because there is an Other for whom the Me is object. Knowledge here is still the measure of being, and Hegel does not even conceive of the possibility of a being-for-others which is not finally reducible to a "being-as-object" (Sartre 1956: 238). The problem with Hegel's analysis, for Sartre, is that it requires a "mutual recognition of consciousnesses brought face to face which appeared in the world and which confronted each other" (1956: 238) resulting in objectification. Sartre therefore turns to Heidegger to find a better model.

As Sartre interprets Heidegger, "The Other is not originally bound to me as an ontic reality appearing in the midst of the world among 'instruments' as a type of particular object; in that case he would be already degraded, and the relation uniting him to me could never take on reciprocity" (1956: 245). Yet this is precisely where Heidegger does find the other—ontically. Even if *Mitsein* is an ontological dimension of Dasein, this is not in the same framework of our everyday interactions with others, whom we find out there in the world, precisely among the instruments and pragmatic contexts, even if we treat them as different kinds of objects. Indeed, Sartre is pulled into Heidegger's analysis of inauthenticity at this point:

> [I]f I am asked how my "being-with" can exist for-myself, I must reply that through the world I make known to myself what I am. In particular when I am in the unauthentic mode of the "they," the world refers to me a sort of impersonal reflection of my unauthentic possibilities in the form of instruments and complexes of instruments which belong to "everybody" and which belong to me in so far as I am "everybody" . . . I shall be my own authenticity only if under the influence of the call of conscience (*Ruf des Gewissens*) I launch out toward death with a resolute-decision (*Entschlossenheit*) as toward my own most peculiar possibility. (1956: 246)

This interpretation suggests yet again that authenticity leads to solitude (Sartre 1956: 247). "I emerge alone and in anguish confronting the unique and original project which constitutes my being" (1956: 39).

We take the intersecting accounts of Heidegger and Sartre to be the classical (and perhaps paradigmatic) existential analysis of the concepts of authenticity and inauthenticity. Before we proceed to an alternative interpretation, let us first summarize this classical analysis in three points.

(1) Authenticity in Heidegger and Sartre is primarily linked with one's self, understood as one's own individuality—one's lonely anxiety in the face of the possibility of death or self-demise; or, more positively, one's facing up to the possibilities that open from one's freedom.

(2) Inauthenticity involves a denial or running away from one's ownmost possibilities, one's freedom.
(3) Being with others, or being in relations with others, is the common and perhaps unavoidable occasion for inauthenticity. If authenticity is possible, it is not clear how an intersubjective authenticity is possible.

2. HEIDEGGER'S THREE HESITATIONS: A BRIDGE TO AN ALTERNATIVE MODEL OF AUTHENTICITY

At this point, we wish to draw attention to three instances in *Being and Time* which invite reconsideration and which point the way toward an alternative relational model of authenticity. There are two main aspects to Heidegger's understanding of authenticity in *Being and Time*: being-in-the-world and the take that we develop on our being-in-the-world. "Being-in-the-world" already presupposed an embeddedness which feeds into contemporary theories regarding embodied, embedded, enactive, and extended cognition. Human beings, following this view, cannot be conceived as separate from the world in which they come to a sense of themselves and in which they realize their projects. "I am," for Heidegger, means "I am in a world" (Heidegger 1962: 368/21). At the same time, the environment is disclosed through human purposive action (Heidegger 1962: 96–100/68–70). Objects or paths of action emerge as salient as human beings go about the business of getting things done. Human life is a complex collection of situated courses of action which we cannot decompose into constituent parts, like "human mind" and "things," "subjects" and "objects" but, rather, grasp as a multifaceted whole which transforms the standard philosophical "problem of knowledge" and the "problem of other minds" (Heidegger 1962: 133/00, 60–63/23–25).

Hubert Dreyfus has been particularly influential in emphasizing this aspect of Heidegger's thought. He sets out the context of shared activities into which human beings are born and through which they come to a sense of their own identity, glossing Heidegger's term of art for human life, *Dasein*, as "the skilful ways we are accustomed to comport ourselves" (Dreyfus 1991: 75). Dreyfus's upbeat approach to shared, social practices draws attention, by its contrast, to the more problematic, second aspect of Heidegger's arguments about authenticity. The search for a point of purchase (a "take" on our being-in-the-world) is made difficult by two characteristics of Heidegger's thought. One is the level of his arguments: because he aspires to give what he calls an "ontological" account of human life, the element he is searching for must qualify as a deep structure of existence. He cannot, for instance, look to the sort of specific, everyday activities to which his contemporary Walter Benjamin turned to understand how individuals might be surprised into a new relation with their lives; for instance, an advertisement, going to the cinema, particular forms of poetry or narrative (Duttlinger, Morgan, and Phelan 2012: 105–117).

The second characteristic is an underlying assumption that he shares with Kierkegaard and the early Nietzsche: a disapproval of contemporary culture that colors his argument even as he protests that the shared activity into which Dasein "falls" is not meant to be portrayed in a moralizing way (Heidegger 1962: 211/167) and that any "authentic" form of life can arise only as a modification of the shared structures into which we are born, not as a magical removal from them (Heidegger 1962: 168/30, 224/179). To construct an argument within these strict limits, as we have seen, he takes two phenomena that remove individuals from their daily habits and throws them back on themselves—the experience of anxious alienation from received practices and the process of facing mortality. These, he argues, enable individuals, as individuals, to recommit to, or modify, their habits and activities: to choose their choices (Heidegger 1962: 437/385).

Nevertheless, despite his preference for primordial individualization, Heidegger's arguments include moments when interaction with others seems important. By returning to these moments, we can begin to get a clearer sense of how his productive and influential account of being-in-the-world might, in ways that Heidegger himself did not fully exploit, point to an alternative conception of authenticity that acknowledges the constitutive role of attunement to others, imitation, and shared activity. Specifically, three hesitations in Heidegger's arguments about authenticity show the way toward this alternative approach.

The first hesitation consists in a short comment on a form of cooperative behavior he terms *Vorausspringen* or "jumping ahead" (Heidegger 1962: 158–159/22). In our solicitude for others, we might be tempted to take over a job for them, to jump in and do it ourselves. But, in Heidegger's view, this can easily be a form of imposition. The preferred way of helping is to nurture someone's ability to take responsibility for themselves: "This kind of solicitude pertains essentially to authentic care—that is, to the existence of the Other, not to a 'what' with which he is concerned; it helps the Other to become transparent to himself in his care and to *become free for* it" (Heidegger 1962: 159/22). This is a striking passage, given the general direction of Heidegger's argument. Authenticity here emerges through interaction: not only are facing up to anxiety or mortality transformative, but also the well-judged attunement of another person to our predicament.[3]

---

3. Like Heidegger, the early Sartre does not rule out good-faith relations with others, but neither does he develop an analysis of such possibilities. Everything he has to say about this is as a disclaimer or remains an ambiguous promissory note (see, e.g., 1956: 70 n. 9).

The second hesitation in Heidegger's argument is his discussion of role models. We come to a true sense of ourselves, he argues, through creative imitation. Consistent with his view of a human identity which arises through the shared practices of a specific environment, Heidegger suggests that authenticity will necessarily involve an engagement with past achievements, not simply in order to affirm old ways of doing things or blindly break with the past, but as a means of fostering interactive encounters that Heidegger explicitly says involve choosing a "hero" (Heidegger 1962: 437/385). Via such encounters, we come to a sense of ourselves through an active and, if need be, critical exchange with this role model. The repetition of, or break from, an example "makes a *reciprocative rejoinder* to the possibility of that existence which has-been-there" (Heidegger 1962: 438/386). Authenticity, for Heidegger therefore, can be fostered by the considerate interventions of others and by figures who inspire us to imitate or rebel. Thus, despite his intellectual commitment to an individualist model, Heidegger nevertheless incorporates interaction into his theory at the most fundamental level.

Heidegger's third hesitation further acknowledges this by noting the emotional or affective attunement that is the precondition of human communication. Since, for Heidegger, we grow into the way things are done in our social milieu, and only in a second step make these habits our own, "the world is always the one that I share with Others. The world of Dasein is a *with-world [Mitwelt]*. Being-in is *Being-with* Others. Their Being-in-themselves within-the-world is *Dasein-with [Mit-Dasein]*" (Heidegger 1962: 155/18). The examples he gives, as we have seen, make this coexistence appear primarily to be one of social interdependence: a book we are reading was bought somewhere or given by someone; a boat that we see belongs to someone (Heidegger 1962: 153–154/18). But at one point in the argument, it becomes clear that our coexistence with others goes much deeper: our own emotional attunement to our predicament—the primary affective response that Heidegger calls *Befindlichkeit* or disposition (Heidegger 1962: 172–177/34–38)—arises out of something like emotional contagion or affective resonance (*Mitbefindlichkeit*), and this resonance is what makes possible more explicit, verbal forms of communication between people. Indeed, in Heidegger's view, affective communion is itself premised on an even deeper structural connectedness, which can be glossed as the fact that the process of being connected and attuned to each other has already begun before individuals are born into the stream of human interaction:

> "Communication" in which one makes assertions—giving information, for instance—is a special case of that communication which is grasped in principle existentially. In this more general kind of communication, the framework of Being-with-one-another understandingly is constituted. Through it a co-disposition [*Mitbefindlichkeit*] gets "shared," and so does the understanding of Being-with. Communication is never anything like a conveying of experiences,

such as opinions or wishes, from the interior of one subject into the interior of another. Dasein-with is already essentially manifest in a co-disposition and a co-understanding. (Heidegger 1962: 205/162) (translation revised)

In Heidegger's account, then, a space is already kept open by our shared practices in which we can be disclosed to each other at the level of mood and emotions, and it is on this visceral attunement that more explicit and conscious forms of communication depend.

If we integrate these three hesitations into Heidegger's account of authenticity, the beginnings of a modified model emerges. The shared practices through which we come to a sense of identity need no longer appear predominantly to promote conformity. On the contrary, the attunement that underpins communication can also allow us solicitously to help others develop and grow. At the same time, shared culture offers us patterns of behavior for us to adapt and adopt. Conspicuous by its absence from this revised model is the isolating moment of realization. Once we have admitted forms of enabling interaction into the argument, we no longer need the isolating experiences to initiate authenticity. On the contrary, authentic existence may be part of the shared activities.

What is needed, then, if we are to develop a richer, more relational model of authenticity, is a detailed, naturalistic, and continually revisable account of embodied human life. The following sections will draw together some of the current research which contributes to this working hypothesis.

3. NEUROSCIENCE, METHODOLOGICAL INDIVIDUALISM, AND THE QUESTION OF AUTHENTICITY

In this section, we offer some speculative and admittedly unsystematic thoughts about what has been called *existential neuroscience* (Iacoboni 2007; Quirin et al. 2012), a field that so far remains relatively undeveloped. On the one hand, a standard neuroscientific conception of human existence could easily rule out the possibility of authenticity, as it is construed in the first of the three points of Heidegger and Sartre's classical account. On certain interpretations of neuroscientific evidence, there is no such thing as a self (see Metzinger 2004); on some interpretations, there is no such thing as freedom (see interpretations of the Libet experiments; Pockett 2004; Prinz 2006). Such interpretations could directly challenge the concept of autonomy and the existential conception of authenticity.[4] On the other hand, an existential

---

4. Although this, too, is open to interpretation. The early Sartre (1957) also held that there was no self in the sense of a transcendental ego and that the production of an empirical ego (what Metzinger might call a self-model) would reflect an inauthentic investment that would take the self as an object.

analysis could equally conclude that neuroscience is itself a form of human inauthenticity (as captured in the second point) precisely because (on some interpretations) it denies human freedom. We can immediately see, however, that these are overly simplistic generalizations—on both sides.

Does neuroscience actually address the question of authenticity? Neuroscientists have studied a number of related phenomena—such as self, freedom, anxiety, and even mortality salience—that seemingly underpin the notion of authenticity. The main thrust of the research on mortality salience develops out of *terror management theory*, which, working with behavioral indices, has established that people cope with the threat of death by reaffirming aspects of their cultural world view ("cultural worldview defense"; see Greenberg et al. 1995). Recent work uses functional magnetic resonance imaging (fMRI) studies of neural activation to understand in more detail the brain's responses to the threat of death. For example, Quirin et al. (2012) show increased activation in right amygdala, left rostral anterior cingulate cortex, and right caudate nucleus (CN) under these conditions. However, the priming stimuli in this experiment were not actual existential threats so much as simple questions pertaining to death on a "fear of death scale" (Boyar 1964). Quirin et al. (2012: 196) suggest that such activation is the neural correlate of "existential fear" or anxiety. Quirin's team point out that Heidegger's existential philosophy interprets cultural worldview defense—that is, identifying with values of the larger social group (the "they")—as an attempt to relieve mortality threat. They speculate that activation of the CN actually suggests this type of response since the CN is associated with stereotypical, habitual behavior (see Packard and McGaugh 1992) and the experience of love (Fisher, Aron, and Brown 2005) and thus a kind of existential security.

The experimenters are careful to say that the connections they make are tenuous and highly speculative. Even if these connections were more clear, it would not necessarily confirm Heidegger's analysis (in a non–question-begging way) since Heidegger starts with the thought that the social group phenomenon of *Das Man* makes us forget death—distracts us from experiencing being-toward-death—and prevents us from becoming anxious, leading to inauthenticity. The experimental results could, instead, be read as suggesting the opposite direction—namely, that death-related information and possibly an accompanying anxiety motivates us to turn to others. To think of that as a form of inauthenticity, one would have to simply assume that being-with-others and even love-related experience just are inauthentic forms of existence.

This brings us back to the third point in our summary of the classical existential conception of authenticity and to a point that neuroscience in some way shares with this analysis. It concerns methodological individualism. Neuroscientists, of course, study the brain, and, as one might expect, their experiments focus on what is happening inside the individual brain.

Even when neuroscientists are doing social neuroscience—when they are trying to understand how we relate to others—they are looking inside individual brains. The supposition is that that is where they will find the mechanisms that will explain social cognition and interaction. This fits nicely with Heidegger's conception of being-with—specifically that being-with is part of the very structure of Dasein.[5] If Heidegger were a neuroscientist, he might say that being-with is hard-wired into the brain. And as we saw, it does not depend on whether there are in fact others in the world. This, too, reflects a type of methodological individualism—one need not look outside of the individual (e.g., to social interactions themselves) to understand the mechanisms (or the existentials) that explain intersubjective existence.[6]

4. 4E EXISTENCE

Moving away from the individualism (methodological or otherwise) in both neuroscience and existentialism, we suggest that there is a completely different way to think about existential authenticity, roughly hinted at in the previous sections. To articulate an ecologically valid theory of relational existential authenticity, we introduce the wider scope of the 4E (embodied, embedded, enactive, and extended) conception of human existence. To be clear, discussion of the 4Es is usually confined to studies of cognition or mind (in cognitive science and philosophy of mind). Here, we suggest that we can think more widely about human existence in these terms and, in doing so (in the next section), come to a different conception of existential authenticity.[7]

Marco Iacoboni, a well-known neuroscientist, provides a welcome shortcut that will help us make our case. What Iacoboni describes as "existential neuroscience" is in fact the 4E approach inspired in large part by existential phenomenology, including Heidegger and Sartre, but also the work of Merleau-Ponty.

---

5. We are in disagreement with Hubert Dreyfus's (1991: 7) interpretation of Heidegger on this point. He argues "Heidegger rejects the methodological individualism that extends from Descartes to Husserl to existentialists such as the pre-Marxist Sartre and many contemporary American social philosophers." Indeed, the later Heidegger criticizes himself on just this point. See Gallagher and Jacobson (2012) for further discussion.

6. Methodological individualism characterizes not only neuroscientific studies of social cognition, but also psychological and philosophical explanations—theory of mind and simulation approaches (see De Jaegher, Di Paolo, and Gallagher 2010). In recent social neuroscience and philosophy, there has been some critical movement against methodological individualism (see, e.g., Schilbach et al. 2013).

7. Although originating in cognitive science and philosophy of mind, discussion of the 4Es is now becoming fruitfully deployed and extended by scholars in the humanities, yielding new insights. Studies of authentic being that arise from this hybrid "cognitive humanities" include, so far, two proposals specifically aimed at reformulating existentialist notions of authenticity (see Rokotnitz 2014, forthcoming 2016).

The neuroscience that Iacoboni points to is not mainstream, but nevertheless is an approach that has a growing influence in this area. Here, we reframe Iacoboni's analysis more explicitly in terms of the 4Es.

First, the human agent is *embodied*. The subject is not just a brain but an embodied perceiver and agent. As Iacoboni puts it, "a human brain ... needs a body to exist in a world of shared social norms in which meaning originates from being-in-the-world" (Iacoboni 2007: 440). The brain functions the way it functions because it evolved along with the body. Not only does the brain regulate the body; the body also regulates the brain (Sternberg 1997; Sternberg et al. 2011).

Second, the human agent is *embedded* in an environment that is both physical and social, and the unit of explanation is not just the brain, not just the body, but the brain-body-environment, which is characterized by dynamical coupling. This goes against what Iacoboni calls the notion of subject–world dichotomy and includes the notion of lived space—that is, an understanding of space as pragmatic rather than geometrical. Iacoboni cites well-known experiments by Iriki et al. (Iriki, Tanaka, and Iwamura 1996) showing changes in experiential peripersonal space by tool use and studies by Graziano and colleagues (e.g., Graziano et al. 2002) showing how cross-modal neural processes integrate with complex embodied responses to rearrange peripersonal space. These studies support Merleau-Ponty's claim that "my body's spatiality is not, like the spatiality of external objects or of 'spatial sensations,' a positional spatiality; rather, it is a situational spatiality" (2012: 102). As Iacoboni puts it, these studies suggest that, rather than implementing an abstract representation of objective space, "these neural ensembles are relevant to a direct sense of being-amid, of being-in-the-world" (Iacoboni 2007: 444). Importantly, being-in-the-world is not just a physical relation but also social, and our responses to the environment are shaped by the presence of others. Research on imitation in infants (Meltzoff and Moore 1977), cited by Iacoboni, indicates that imitation is an important factor in social cognition that shows an interdependence of self and other, "rather than the expression of a separation between the perceiver (the imitator) and the perceived (the model)" (Iacoboni 2007: 442). One might say that imitation is an ontic mode of being-with.

Third, the human agent is enactive—that is, action-oriented. For enactivists (e.g., Noë 2004; Varela, Thompson, and Rosch 1991), following Merleau-Ponty, an agent perceives in terms of action-affordances. According to the enactivist view, the brain does not create internal representations of the world; it does not act on some inner model of the surrounding environment; it is instead involved in responding to the environment in an ongoing dynamic adjustment, as part of and along with the larger organism, finding the right kind of attunement with the physical and social environment (Gallagher et al. 2013). Iacoboni cites neuroscience experiments by Fadiga et al. (1995) showing that action observation facilitates corticospinal excitability—that is, something more like a whole-body response, rather than something akin to a mental representation. It is not just activation of mirror neurons, but activation that includes central and peripheral aspects of the motor system down to the level of selective muscle facilitation relevant to the perceived action. Moreover, tradeoff activation between canonical neurons and mirror neurons depends on differences in intentional pragmatic context, thus undermining the idea that there is an atomism of input and reinforcing Merleau-Ponty's idea of the intentional arc, "more readily explained by a holistic stance, which assumes that the nature of any given element is determined only by the 'whole' and that this 'whole' is never simply determined by the sum of its elements" (Iacoboni 2007: 447).

Fourth, cognition is extended. The "extended mind hypothesis" (EMH; Clark and Chalmers 1998) is typically applied to cognition, but it can also reflect enactivist and pragmatic (rather than strictly functionalist) perspectives on action and intentionality (see, e.g., Gallagher and Miyahara 2012). Some aspects of human cognition are not only causally supported or scaffolded by the surrounding environment, but are actually constituted by our use of and interaction with the environment. Iacoboni does not mention the concept of extended mind and perhaps one might think that this is where neuroscience meets its limits since the claim of the EMH is that, in some cases, the vehicles of cognition are not neurons in the head but instruments, artifacts (Malafouris 2012), and perhaps even social practices and institutions (Gallagher 2013a). In fact, however, Clark's most recent work (2016) involves a large-scale argument that EMH is entirely consistent with the most recent work on predictive processing in neuroscience. The details of this argument go beyond what we can discuss here, but we can take the point as a reinforcement of the basic idea that the human being is not just a brain, but is rather an extended system-in-the-world, in ways that are enactively relational.

One of the important implications of the 4Es is that, as we have already noted and as Iacoboni puts it, we are social by default—we are inextricably embedded in social environments and enactively engaged with others, even at levels of behavior that are considered to be automatic and nonconscious. Iacoboni understands this in neuroscientific terms and quite literally in terms of what is called the *default network* in the brain (Iacoboni 2007: 448–449). This idea brings us back to the question of authenticity, specifically because it brings us back to the question of self and our relations to others.

The default network, which comprises a set of midline cortical structures, has been something of a puzzle to neuroscientists, eliciting a number of different interpretations (see Andrews-Hanna et al. 2010; Buckner and Carroll 2007; Raichle et al. 2001). There is gathering consensus, however, that processes in these midline structures integrate two important (and one might say existential) features of human existence. The processes are said to be self-related and other-related. Northoff and Bermpohl (2004; see also Qin and Northoff 2011) have argued that the cortical midline structures are involved in every function that relates to self, including self–other differentiation and reflective aspects of self-experience (see Gallagher 2013*b*). A number of other researchers have pointed to the fact that what is usually identified as "theory-of-mind" areas (i.e., social cognition areas; Iacoboni mentions the dorsomedial prefrontal cortex and the medial parietal cortex and their increased activation when confronted with others and with social relations) are part of this default network of the cortical midline structure (see, e.g., Spreng, Mar, and Kim 2009). Theory of mind and default network areas may also be activated in moral decision-making (Reniers et al. 2012). Some have hypothesized that, in resting states, the default processes of the brain are activated in reflections on social affordances and our relations with others (Schilbach et al. 2008).

To put this more clearly into the 4E context, we can refer to developmental studies that make the same connections between embodied self and others. Ever since Colwyn Trevarthen first published his description of "primary intersubjectivity" in 1979, a body of research has been devoted to forms of attunement, reciprocal interaction, mirroring, and attention sharing that enable and shape the development of infant identity (Carpendale and Lewis 2006; Hobson 2002; Reddy 2008; Seemann 2011; Trevarthen and Aitken 2001) and that are also the basis on which linguistic communication and conscious forms of mindreading and analysis are founded (Bruner 1983; Hutto 2008; Tomasello 2008).

The notion of primary intersubjectivity (Trevarthen 1979) refers to basic, embodied sensory–motor processes of self–other interaction that are present from birth and that persist throughout the life span. Neonatal imitation, mentioned by Iacoboni, is an initial expression of primary intersubjective processes (Gallagher and Meltzoff 1996). Such processes are primarily processes of dyadic interactions between self and other that nonetheless depend on the autonomy of the participants (De Jaegher et al. 2010). Secondary intersubjectivity, which begins with joint attention and joint action in the first year of life and also continues throughout the life span, involves relations with others in pragmatic and social contexts and the generation of worldly meaning (sometimes referred to as "participatory sense making"; De Jaegher et al. 2010). All of these intersubjective processes are primarily interactive in nature and not reducible to internal neural representations, or even to processes that are purely individual-based—they are embodied, embedded, enactive, and extended as indicated earlier, and they clearly push us toward an understanding of human existence that is not defined by methodological individualism.

5. RELATIONAL AUTHENTICITY

Where a reductionist form of neuroscience represents a problem for the 4Ms—mind, meaning, morals, and modality (Price 2004)—neuroscience informed by the 4Es (Iacoboni's "existential neuroscience") underwrites a new conception of the mind (as embodied, embedded, enactive, and extended), provides an account of how meaning is generated (in intersubjective contexts), and suggests that moral decisions are grounded in socially rich action-oriented processes and that all of these phenomena are directed to human possibilities, including the social affordances that define the brain-body-environment system. If humans are (ontologically, existentially) embodied, embedded (physically and socially), enactive, and cognitively extended—if this in fact describes their facticity and the way they are, and if being-with is not just hard-wired, but entirely dependent on interaction with others, understood now in terms of primary and secondary intersubjective relations so that we understand self as relational or socially situated—it would be strange to suggest that the authentic life is nonintersubjective, nonrelational, a retreat to our ownmost existence, as if that existence were not being-in-the-world.

A concept of authenticity that is informed by a 4E version of existential neuroscience will differ precisely on the three points that summarize the Heideggerian-Sartrean view presented in the first section:

(1) Rather than taking authenticity as essentially linked with one's self, understood as one's own individuality characterized by a lonely anxiety in the face of the possibility of death or self-demise, authenticity should be understood as relational since the self is relational. That is, as the 4E approach and recent neuroscience suggest, the self is, by default, and in continuing development, an intersubjective accomplishment, dynamically intertwined with others from the beginning of life. Autonomy is itself relational (Mackenzie and Stoljar 2000). The classical existential conception of autonomy is akin to traditional conceptions that focus on self-sufficiency, self-legislation, or self-determination, although they do not put "undue emphasis on the ideal/assumption of rational self-awareness and cognitive mental operations," as one finds in Kant, for example (Christman 2004: 13). Still, such conceptions remain individualistic, ignoring the importance of embodied space as given in socially embedded forms of action. If autonomy and the self are relational, so must authenticity be.

(2) It may still be right to think of inauthenticity as a denial or running away from one's ownmost possibilities, but it would be a mistake to think that these ownmost possibilities are strictly "ownmost." Authenticity is not facing up to the nothingness of one's existence (as Sartre would have it),

but facing up to the richness and complexity of our situated existence that comes from being in-the-world-with-others. If it is an *Eigentlichkeit*, it is important to understand that being one's own (*eigen*), even one's own embodiment, already involves an otherness. This is a point made by Ricoeur, who criticizes Heidegger for suppressing embodiment and failing to consider it as a topic (Ricoeur 1992: 327). Ricouer points out that Heidegger failed to develop the notion of Leib (lived body) "as a distinct existentiale" (1992: 327).[8] Ricoeur himself conceives of the lived body as the "primary otherness" required for our openness to other people, and he thinks that there is an overly strong association between embodied spatiality and inauthenticity in Heidegger. "If the theme of embodiment appears to be stifled, if not repressed, in *Being and Time*, this is doubtless because it must have appeared too dependent on the inauthentic forms of care" (1992: 328). The message we can take from this is that an impoverished conception of the body leads to an impoverished conception of being-with (1992: 341ff). This suggests that inauthenticity is more properly a denial or a running away from one's embodied intersubjective relations.

(3) Thus, being with others, or being in relations with others, rather than the common and perhaps unavoidable occasion for inauthenticity, should be seen as the unavoidable occasion for authenticity. It strikes us as much more existential to think of an engaged authenticity, open to possibilities, risking failures as well as successes. Indeed, it is not clear how a non-intersubjective authenticity is possible. Would that not be a running away from what makes our existence existential in the most intense way? As Heidegger's student Werner Marx (1987) pointed out, shared mortality is precisely the thing that calls for a responsibility toward one another and moves us toward a very different conception of authentic relations with others (as speculated in the studies by Quirin et al. 2012 and Fisher et al. 2005, mentioned earlier). As we have suggested, Heidegger's concept of the solicitude of "leaping ahead," which targets the existence of the other as such; his observation of the importance of role models; and his brief comments on fundamental, shared attunement point the way to a new paradigm. We suggest, however, that this is possible only on the condition that we are capable of relations that are best described in terms of embodiment and primary intersubjectivity.

We suggest that the speculations of existential neuroscience can be cashed out more explicitly in 4E approaches to existence. We find in such approaches indications that lead, not to the classical conception of existential authenticity,

---

8. Heidegger does return to the idea of *"Leib"* in the *Zollikon Seminars* during the 1960s (Heidegger 2006: 105–115).

but to an alternative view that understands, as Heidegger may have done, hesitantly, in his clearer moments, that there is no authenticity in social isolation. Our view of the enterprise of life and of authenticity is social, pluralistic, dynamic. Transformation emerges not just through arduous self-directed and lonely invention but through participation in shared, enriching, enlivening, and often joyous social situations.

REFERENCES

Andrews-Hanna, J. R., J. S. Reidler, J. Sepulcre, R. Poulin, and R. L. Buckner. 2010. Functional-anatomic fractionation of the brain's default network. *Neuron* 65(4): 550–562.

Blattner, W. 2006. *Heidegger's being and time*. London: Bloomsbury.

Binswanger, L. 1962. *Grundformen und Erkenntnis Menschlichen Daseins*. Ernst Reinhardt Verlag.

Boyar, J. I. 1964. The construction and partial validation of a scale for the measurement of the fear of death. *Dissertation Abstracts International* 25: 20–21.

Bruner, J. S. 1983. *Child's talk: Learning to use language*. Oxford: Oxford University Press.

Buckner, R. L., and D. C. Carroll. 2007. Self-projection and the brain. *Trends in Cognitive Sciences* 11(2): 49–57.

Carpendale, J. I. M., and C. Lewis. 2006. *How children develop social understanding*. Oxford: Basil Blackwell.

Christman, J. 2004. Relational autonomy, liberal individualism, and the social constitution of selves. *Philosophical Studies* 117(1): 143–164.

Clark, A. 2016. *Surfing uncertainty: Prediction, action, and the embodied mind*. New York: Oxford University Press.

Clark, A., and D. Chalmers. 1998. The extended mind. *Analysis* 58(1): 7–19.

De Jaegher, H., E. Di Paolo, and S. Gallagher. 2010. Can social interaction constitute social cognition? *Trends in Cognitive Sciences* 14(10): 441–447.

Dreyfus, H. L. 1991. *Being-in-the-world: A commentary on Heidegger's Being and Time, Division I*. Cambridge, MA: MIT Press.

Duttlinger, C., B. Morgan, and A. Phelan (Eds.). 2012. *Walter Benjamins anthropologisches denken*. Freiburg im Breisgau: Rombach.

Fadiga, L., L. Fogassi, G. Pavesi, and G. Rizzolatti. 1995. Motor facilitation during action observation: A magnetic stimulation study. *Journal of Neurophysiology* 73: 2608–2611.

Fisher, H., A. Aron, and L. L. Brown. 2005. Romantic love: An fMRI study of a neural mechanism for mate choice. *Journal of Comparative Neurology* 493: 58–62.

Frie, R. 1997. *Subjectivity and intersubjectivity in modern philosophy and psychoanalysis*. New York: Rowman & Littlefield.

Gadamer, H-G. 2004. *A century in philosophy: Hans-Georg Gadamer in conversation with Riccardo Dottori*. New York: Continuum.

Gallagher, S. 2013a. The socially extended mind. *Cognitive Systems Research* 25–26: 4–12.

Gallagher, S. 2013b. A pattern theory of self. *Frontiers in Human Neuroscience* 7(443): 1–7. doi: 10.3389/fnhum.2013.00443

Gallagher, S., D. Hutto, J. Slaby, and J. Cole. 2013. The brain as part of an enactive system. *Behavioral and Brain Sciences* 36(4): 421–422.

Gallagher, S., and R. Jacobson. 2012. Heidegger and social cognition. In J. Kiverstein and M. Wheeler (Eds.), *Heidegger and cognitive science*, pp. 213–245. London: Palgrave-Macmillan.

Gallagher, S., and A. Meltzoff. 1996. The earliest sense of self and others: Merleau-Ponty and recent developmental studies. *Philosophical Psychology* 9: 213–236.

Gallagher, S., and K. Miyahara. 2012. Neo-pragmatism and enactive intentionality. In J. Schulkin (Ed.), *Action, perception and the brain*, pp. 117–146. Basingstoke, UK: Palgrave-Macmillan.

Graziano M. S., C. S. Taylor, T. Moore, and D. F. Cooke. 2002. The cortical control of movement revisited. *Neuron* 36(3): 349–362.

Greenberg, J., L. Simon, E. Harmon-Jones, S. Solomon, T. Pyszczynski, and D. Lyon. 1995. Testing alternative explanations for mortality salience effects: Terror management, value accessibility, or worrisome thoughts? *European Journal of Social Psychology* 25: 417–433.

Haugeland, J. 2013. *Dasein disclosed: John Haugeland's Heidegger*. Cambridge, MA: Harvard University Press.

Heidegger, M. 1962. *Being and time* (trans. J. Macquarrie & E. Robinson). New York: Harper & Row.

Heidegger, M. 1979. *Sein und Zeit*. Max Niemeyer: Tübingen.

Heidegger, M. 1985. *The basic problems of phenomenology* (trans. A. Hofstadter). Bloomington: Indiana University Press.

Heidegger, M. 2006. *Zollikoner seminare*. Frankfurt am Main: Klostermann.

Hobson, R. P. 2002. *The cradle of thought*. London: Macmillan.

Hutto, D. D. 2008. *Folk psychological narratives: The sociocultural basis of understanding reasons*. Cambridge, MA: MIT Press.

Iacoboni, M. 2007. The quiet revolution of existential neuroscience. In E. Harmon-Jones and P. Winkielman (Eds.), *Social neuroscience: Integrating biological and psychological explanations of social behavior*, pp. 39–453. New York: Guilford Press.

Iriki A., M. Tanaka, and Y. Iwamura. 1996. Coding of modified body schema during tool use by macaque postcentral neurones. *NeuroReport* 7: 2325–2330.

Löwith, K. 1928. Das individuum in der rolle des mitmenschen. In K. Stichweh (Ed.), *Sämtliche Schriften*, vol. 1, pp. 9–197. Stuttgart: J. B. Metzler, 1981.

Mackenzie, C., and N. Stoljar. 2000. Introduction: Autonomy refigured. In Mackenzie, C., and Stoljar, N. (Eds.), *Relational autonomy: Feminist perspectives on automony, agency, and the social self*, pp. 3–31. Oxford: Oxford University Press.

Malafouris, L. 2012. *How things shape the mind*. Oxford: Oxford University Press.

Marx, W. 1987. *Is there a measure on Earth? Foundations for a nonmetaphysical ethics* (trans. T. J. Nenon and R. Lilly). Chicago: University of Chicago Press.

Meltzoff, A. N., and M. K. Moore. 1977. Imitation of facial and manual gestures by human neonates. *Science* 198(4312): 75–78.

Menary, R. 2010. Introduction to special issue on 4E cognition. *Phenomenology and the Cognitive Sciences* 9: 459–463.

Merleau-Ponty, M. 2012. *Phenomenology of perception* (trans. D. Landes). London: Routledge.

Metzinger, T. 2004. *Being no one: The self-model theory of subjectivity*. Cambridge, MA: MIT Press.

Morgan, B. 2013. *On becoming God: Late medieval mysticism and the modern western self*. New York: Fordham University Press

Noë, A. 2004. *Action in perception*. Cambridge, MA: MIT Press.

Northoff, G., and F. Bermpohl. 2004. Cortical midline structures and the self. *Trends in Cognitive Sciences* 8(3): 102–107.

Packard, M. G., and J. L. McGaugh. 1992. Double dissociation of fornix and caudate nucleus lesions on acquisition of two water maze tasks: further evidence for multiple memory systems. *Behavioral Neuroscience* 106: 439–446.

Pockett, S. 2004. Does consciousness cause behaviour? *Journal of Consciousness Studies* 11(2): 23–40.

Pöggeler, O. 1989. *Martin Heidegger's path of being* (trans. D. Magurshak and S. Barber). Atlantic Highlands, NJ: Humanities Press.

Price, H. 2004. Naturalism without representationalism. In M. DeCaro and D. Macarthur (Eds.), *Naturalism in question*, pp. 71–88. Cambridge, MA: Harvard University Press.

Prinz, W. 2006. Free will as a social institution. In S. Pockett, W. P. Banks, and S. Gallagher (Eds.), *Does consciousness cause behavior?* pp. 257–276. Cambridge, MA: MIT Press.

Quirin, M., A. Loktyushin, J. Arnd, E. Küstermann, Y. Y. Lo, J. Kuhl, and L. Eggert. 2012. Existential neuroscience: A functional magnetic resonance imaging investigation of neural responses to reminders of one's mortality. *Social Cognitive and Affective Neuroscience* 7(2): 193–198.

Qin, P., and G. Northoff. 2011. How is our self related to midline regions and the default-mode network? *Neuroimage* 57(3): 1221–1233.

Raichle, M. E., A. M. MacLeod, A. Z. Snyder, W. J. Powers, D. A. Gusnard, and G. L. Shulman. 2001. A default mode of brain function. *Proceedings of the National Academy of Sciences* 98(2): 676–682.

Reddy, V. 2008. *How infants know minds*. Cambridge, MA: Harvard University Press.

Reniers, R., B. Rhiannon Corcoran, A. Völlm, A. Mashru, R. Howard, and P. F. Liddle. 2012. Moral decision-making, ToM, empathy and the default mode network. *Biological psychology* 90(3): 202–210.

Ricoeur, P. 1992. *Oneself as another*. Chicago: University of Chicago Press.

Rokotnitz, N. 2014. "Passionate reciprocity": Love, existentialism, and bodily knowledge in *The French Lieutenant's Woman*. *Partial Answers: Journal of Literature and the History of Ideas* 12: 331–354.

Rokotnitz, N. Forthcoming. True self: From despair to authenticity in *Either/Or* and *Passion Play*.

Rowlands, M. 2015. Bringing philosophy back: 4e cognition and the argument from phenomenology. In D. O. Dahlstrom, A. Elpidorou, and W. Hopp (Eds.), *Philosophy of mind and phenomenology: Conceptual and empirical approaches*, pp. 310–325. New York: Routledge.

Sartre, J-P. 1956. *Being and nothingness* (trans. Hazel Barnes). London: Routledge.

Sartre, J-P. 1957. *The transcendence of the ego* (trans. Forrest Williams & Robert Kirkpatrick). New York: Noonday Press.

Schilbach, L., S. B. Eickhoff, A. Rotarska-Jagiela, G. R. Fink, and K. Vogeley. 2008. Minds at rest? Social cognition as the default mode of cognizing and its putative relationship to the "default system" of the brain. *Consciousness and Cognition* 17(2), 457–467.

Schilbach, L., B. Timmermans, V. Reddy, A. Costall, G. Bente, T. Schlicht, and K. Vogeley. 2013. Toward a second-person neuroscience. *Behavioral and Brain Sciences* 36(4): 393–414.

Seemann, A. (Ed.). 2011. *Joint attention: New developments in psychology, philosophy of mind, and social neuroscience*. Cambridge, MA: MIT Press.

Spreng, R. N., R. A. Mar, and A. S. Kim. 2009. The common neural basis of autobiographical memory, prospection, navigation, theory of mind, and the default mode: A quantitative meta-analysis. *Journal of Cognitive Neuroscience* 21(3): 489–510.

Sternberg, E. M. 1997. Neural-immune interactions in health and disease. *Journal of Clinical Investigation* 100(11): 2641.

Sternberg, E., S. Critchley, S. Gallagher, and V. V. Raman. 2011. A self-fulfilling prophecy: Linking belief to behavior. *Annals of the New York Academy of Sciences* 1234: 83–97.

Theunissen, M. 1984. *The other* (trans. C. Macann). Cambridge, MA: MIT Press.

Tomasello, M. 2008. *Origins of human communication.* Cambridge, MA: MIT Press.

Trevarthen, C. 1979. Communication and cooperation in early infancy: A description of primary intersubjectivity. In M. Bullowa (Ed.), *Before speech: The beginning of interpersonal communication,* pp. 321–348. Cambridge: Cambridge Univeristy Press.

Trevarthen, C., and Aitken, K. J. 2001. Infant intersubjectivity: Research, theory, and clinical applications. *Journal of Child Psychology and Psychiatry* 42: 3–48.

Tugendhat, E. 1986. *Self-consciousness and self-determination* (trans. P. Stern). Cambridge, MA: MIT Press.

Varela, F., E. Thompson, and E. Rosch. 1991. *The embodied mind.* Cambridge, MA: MIT Press.

Varga, S. 2011. *Authenticity as an ethical ideal.* New York: Routledge.

Varga, S., and C. Guignon. 2014. Authenticity. In E. N. Zalta (Ed.), *The Stanford encyclopedia of philosophy* (Fall 2014 edition). http://plato.stanford.edu/archives/fall2014/entries/authenticity/.

Wheeler, M. 2005. *Reconstructing the cognitive world: The next step.* Cambridge, MA: MIT Press.

CHAPTER 9
Behavior Control, Meaning, and Neuroscience

WALTER GLANNON

In his essay, *Existentialism Is a Humanism*, Jean-Paul Sartre claims that "man is not only that which he conceives himself to be, but that which he wills himself to be, and since he conceives of himself only after he exists, and wills himself to be after being thrown into existence, man is nothing other than what he makes of himself" (Sartre 2007: 18). For Sartre, this is the First Principle of Existentialism. In a godless universe, there are no objective models to guide our behavior. Existence precedes essence. A person makes herself through her own actions as a being in the world. Meaning in one's life is entirely the product of individual choice grounded in subjectivity.

Advances in neuroscience have increased our understanding of the neurobiological underpinning of the conscious mental states that figure in our explanations of agency, free will, and responsibility. Yet, if all of our thoughts and actions can be completely explained in terms of neurons, neural networks, and neurotransmitters operating outside of our awareness, then subjectivity seems to have no place in our behavior. A neuroscientific account of behavior obviates the need to appeal to any aspect of consciousness to explain how and why we think and act. If we are nothing more than our brains, then a person is not what she makes of herself but what her brain makes of her. Being free *to* create ourselves from our actions in a world devoid of meaning has evolved into the problem of establishing how we can be free *from* unconscious processes in the brain and mind that drive our behavior and seem to preclude meaningful action. The challenge that reductive accounts of the mind–brain relation present to us is to show that our conscious mental states have some causal role in our behavior and some impact on events in the world. The existential angst of neuroscience is not the result of having to choose in the absence of religious or cultural models. Rather, the angst results from the idea that the subjectivity and conscious choice that presumably define us as persons can be completely explained—if not explained away—by neural and psychological factors to which we have no access.

But the fact that unconscious processes drive many of our actions does not imply that conscious mental states have no causal role in our behavior and that we have no control over it. Some degree of unconscious neural constraint on our conscious mental states is necessary to modulate thought and action and promote flexible behavior and adaptability to the demands of the environment. Too much conscious reflection can impair behavior control, and some degree of unconscious neural processing is necessary for this control. A nonreductive materialist account of the mind–brain relation makes it plausible to claim that mental states can cause changes in physical states of the brain. Examination of some psychiatric and neurological disorders shows how the conscious mind can have a causal role in the etiology of these disorders as well as in therapies to control them and behavior more generally. Lower level unconscious neural functions and higher level conscious mental functions complement each other in a constant process of bottom-up and top-down circular causal feedback that enables interaction between the organism and the external world. The motivational states behind our actions and the meaning we attribute to them cannot be explained entirely by appeal to neural mechanisms. Although the brain generates and sustains our mental states, it does not determine them, and this leaves enough room for individuals to "will themselves to be" through their choices and actions.

1. CONSCIOUS AND UNCONSCIOUS PROCESSES

Conscious mental states are necessary for the deliberation and decision-making in rational and moral agency. The capacity to consciously recognize and respond to reasons for or against certain actions and act in accord with these reasons provides us with the requisite sort of control that makes us free and responsible agents. Yet some cognitive psychologists have questioned the extent to which consciousness is involved in agency. John Bargh and Ezequiel Morsella, for example, claim that unconscious perceptual, evaluative, and motivational systems grounded in priming and automaticity largely guide behavior. "Actions of an unconscious mind precede the arrival of a conscious mind—these actions precede reflection" (Bargh and Morsella 2008: 73). Consciousness is more of a passive conduit of information than an active force of control (Morsella et al. 2015; Wegner 2002). Similarly, other psychologists claim that our behavior is driven largely by "System 1," which is fast, instinctive, and emotional rather than by "System 2," which is slow, rational, and deliberative (Kahneman 2011). Some interpret these claims to mean that we lack free will and control of our behavior since most of us assume that free will and behavior control are functions of the conscious mind. Although psychologists do not always explain which regions of the brain correspond to conscious and unconscious mental processes, there is a rough correlation between their description of conscious processes and cortical brain regions, on the one hand, and unconscious processes and subcortical regions, on the other. The general claim by these psychologists is not that consciousness has no role in behavior but that it has a much more limited role than what philosophers and others have assumed.

This is an oversimplified and misleading dichotomous conception of human behavior. Conscious and unconscious processes in the brain and mind do not operate independently or in just one direction—from unconscious to conscious—but interdependently and bidirectionally. Both types of processes and the interacting cortical and subcortical neural pathways that mediate them are necessary for behavior control. Instead of the "either . . . or" model that some psychologists and cognitive neuroscientists appear to endorse, theoretical and clinical neuroscience suggest that a "both . . . and" model is more accurate in explaining the causal role of mental and neurobiological processes in our decisions and actions. This should dispel concern that our thoughts and actions are not our own and that we have no control of our behavior.

Claims by psychologists about the reduced role of consciousness in agency are consistent with the results of experiments conducted by neuroscientist Benjamin Libet in the 1980s (Libet 1985). Libet used electroencephalography (EEG) to record activity in neural signals, or readiness potentials, in motor, premotor, and prefrontal areas of the brain when subjects were asked to flex their fingers or wrists. The experiments indicated that neural activity preceded the subjects' awareness of their intention to perform these actions by several hundred milliseconds. The results of Libet's experiments, and replications of them, suggest that if neural activity initiates actions, then conscious human agents are not the originators of their actions. On the assumption that having free will implies that we are the originators or authors of our actions, these experiments suggest that we do not have free will. Psychiatrist Sean Spence spells out the apparent upshot of Libet's studies thus: "If this is the case, then what space is left for freedom?" Spence adds that, on this view, "freedom, if it exists, may be *unconscious*. And what kind of freedom is that?" (Spence 2009: 12).

Libet acknowledged a negative conception of free will, whereby a subject could consciously intervene in a neurally initiated process and "veto" an intention to act. Not all behavior is beyond willful control. A recent study has confirmed that subjects can consciously intervene in the decision-making process and interrupt a movement initiated at the unconscious neural level

(Schultze-Kraft et al. 2016). Nevertheless, these studies at best only contribute to a negative account of agency in confirming that we can cancel intentions. They do not contribute to a positive account of agency in confirming whether or how we form and execute conscious intentions in desired actions. In this respect, they do not provide a sufficiently causally robust account of decision-making and action. The idea that actions are initiated by unconscious neural events may be a threat to agent-causal libertarian theories of free will, which hold that no events antecedent to the conscious action itself have a causal role in the performance of the action. However, most compatibilist and event-causal libertarian versions of free will acknowledge that events other than the agent's decision at the time of action can have a causal role in action without diminishing the conscious causal role of the agent.

Even if they allow a positive role for conscious decision-making in agency, Libet-type experiments are limited in what they imply for how much control human agents have of their behavior. Finger- and wrist-flexing are not representative of the actions we value because they lack normative content. They are not the types of actions for which we can be praised or blamed and held morally and legally responsible. Also, the artificial experimental setting in which subjects are instructed to perform certain actions does not accurately reflect real-life situations in which we act in response to social cues and other environmental stimuli. The role of unconscious neural events in initiating actions is at most necessary but not sufficient for a satisfactory account of whether or how actions are performed. The process extending from the formation of an intention to its execution in an action typically involves a temporal framework that is much broader than what Libet-type experiments encompass. Alfred Mele distinguishes proximal from distal intentions (Mele 2009, 2014). While the first type of intention may correspond to what Libet and other neuroscientists describe as the initiating neural event in the performance of an action, the second type extends beyond the narrow time frame extending from a neural event in motor cortices to a bodily movement. Distal intentions are long-range conscious plans that may precede the performance of an action by days, weeks, months, or even years. Actions performed at a particular time may have a physical and psychological history that extends into the past (Bratman 1987, 2007: ch. 10). Insofar as these plans prepare the organism for future activity, and the organism engages both conscious mental and unconscious neural processes to do this, it is possible that distal intentions could influence the activity of readiness potentials in motor cortices at specific times. Unconscious neural events alone cannot provide a satisfactory account of action. Just because neural events may initiate an action does not imply that conscious mental states have no causal role in that action.

Intuitively, any form of constraint on our motivational states would impair or undermine behavior control. Constraint seems at odds with the idea of free will. Indeed, most compatibilist defenses of free will claim that constraint (together with coercion and compulsion) is one of the conditions that can interfere with control. Despite sounding counterintuitive, some degree of constraint from unconscious processes in subcortical regions of the brain on conscious thought and motivation is not only compatible with but necessary for behavior control. Spence states that "the human capacity for volition, for voluntary control, or the apparent expression of 'willed' actions, is subject to multiple constraints" (2009: 363) in the neural networks mediating this capacity. These constraints operate in both top-down and bottom-up pathways in the brain. For example, prefrontal regions of the brain mediating the cognitive capacity for reasoning inhibit impulsive behavior associated with what otherwise could be an unchecked reward system in subcortical regions. Subcortical regions mediating proceduralized behavior constrain prefrontal regions mediating conscious reflection and deliberation in preventing these processes from becoming overactive in increasing the subject's cognitive load as an impediment to action. Constraint at different neural levels corresponds to constraint at different mental levels in promoting optimal levels of brain and mental function that in turn promote flexible behavior.

The neural networks responsible for constraint in the brain consist of a series of reentrant loops extending from anterior and posterior regions of the cerebral cortex associated with planning and decision-making, to the basal ganglia associated with motor functions, and the thalamus, the brain's main information relay station. They then project from the thalamus back to the cortex in generating and sustaining consciousness. These loops ensure that neural functions and the mental functions they mediate are neither underactive nor overactive. Spence adds that "if the human agent possesses any freedom at all, then it is a freedom that is expressed under optimal conditions, be they structural, neurochemical, interpersonal or situational" (2009: 378). In the remainder of this chapter, I first consider structural and neurochemical factors inside the brain and then interpersonal and situational factors outside the brain and how all of them influence our mental states and actions.

The idea that unconscious neural and mental processes regulate much of our thought and action does not undermine the conviction that we have some conscious control of our behavior. On the contrary, unconscious processes are necessary to maintain optimal levels of the cognitive, affective, and volitional states that lead to action. We would not want these processes to be conscious because they could increase the perceived need to process large amounts of information in deliberating and deciding which actions to perform. Many motor and cognitive tasks are performed automatically as a matter of course. This reduces cognitive load and allows the conscious mind to attend to more demanding tasks. A division of neural and mental labor ensures that the agent is not overwhelmed by information entering and remaining in consciousness.

Most normal brain processes are not transparent to us and operate outside of our awareness. We have no direct access to the afferent neurons carrying sensory information from the peripheral to the central nervous system or to the efferent neurons carrying sensory information in the other direction. We only experience the sensorimotor consequences of these inputs and outputs in the way we form and carry out action plans. The fact that much of this process is unconscious does not eliminate the subject's control of her behavior, but instead promotes it by not overloading the conscious mind with excess information. Genetic transcription factors in the brain such as cyclic adenosine monophosphate (cAMP)-response element binding protein (CREB) are good examples of neurochemical constraint. Binding is a neural selection mechanism that prevents information overload in the brain. It does this by limiting the amount of sensory information available to us in conscious awareness. Without this and other unconscious mechanisms, we would be constantly aware of too much information, which could overwhelm our cognitive capacity to process it in deliberating and making decisions. Not only do unconscious mechanisms as such not interfere with agency, but they are necessary for balanced levels of the cognitive, affective, and volitional capacities necessary for effective agency.

Absence of unconscious neural or mental constraints on behavior could impair or undermine free will and free agency. In impulse control disorders, dysfunctional inhibitory mechanisms in the prefrontal cortex can allow the reward circuit in the limbic system to become hyperactive. In addition, studies of obsessive-compulsive disorder (OCD) have shown that a dysfunctional pathway connecting the prefrontal cortex with the subcortical thalamus and striatum disrupts normal cognitive, affective, and motor functions (Figee et al. 2013; Melloni et al. 2012). The disrupted functional connectivity in this pathway causes hyperreflective behavior that impairs the ability to perform basic actions without having to think about performing them. It generates the obsessions and compulsions that characterize the disorder. Obsessions include thoughts, images, or impulses about contamination, exactness, or other concerns that occur repeatedly and feel outside of the person's control. Compulsions are repetitive thoughts or behaviors such as handwashing or checking door locks that the affected person uses to neutralize or rid himself of the obsessions. People with OCD tend to be aware of these thoughts. Like most types of major depression and generalized anxiety, OCD can be described as an ego-dystonic disorder, where the patient retains insight into the fact that he has the disorder. Unlike ego-syntonic disorders such as schizophrenia with hallucinations and delusions, where patients lack this insight, the OCD patient realizes that the obsessions and compulsions are incongruent with the mental states he wants to have and wants to be rid of them. The insight one has into one's condition can motivate one to seek therapy and is an example of how the person takes some control of his thoughts and actions (Meynen 2012).

One group of psychiatry researchers studying patients with OCD reported that people without the disorder "are able to process procedural strategies outside of awareness, whereas in patients with OCD there instead appears to be an intrusion of information and emotion into consciousness" (Stein, Goodman, and Rauch 2000: 343). The problem is that there is an intrusion of too much information into consciousness, a substantial amount of which ordinarily remains unconscious. The dysregulation of the neural pathways causes the hyperreflection preventing them from performing basic actions without much cognitive or motor effort. Commenting on the pathology of OCD patients and what it implies for an optimal level of awareness for behavior control, another group of psychiatrists argues as follows: "To feel free, one needs to find a proper balance between deliberation and conscious control, on the one hand, and spontaneous unreflective actions, on the other. In other words, being free also requires losing control" (de Haan, Rietveld, and Denys 2015: 90). This last point needs qualification, however. Contrary to the claim that free will requires losing control, a more accurate claim is that being free requires less awareness of our thought and action than what our intuitions might suggest and that too much awareness can impede control. The significance of OCD as a psychiatric disorder is that it illustrates that it is possible for a person's agency or will to be impaired or disabled to a significant degree if one is too conscious of the actions one performs. The more general upshot is that, contrary to what many psychologists, cognitive scientists, and some philosophers claim, the idea that unconscious processes drive our behavior to a considerable extent does not mean that the person is not in control of that behavior. Claims that unconscious neural and mental processes undermine control and free will are not supported by the evidence from psychiatry in particular and clinical neuroscience in general.

Behavior control implies a balance between complementary rather than competing reflective conscious and unreflective unconscious processes. This balance is made possible through the constraining activity of the brain at cortical and subcortical levels in a circular causal series of reentrant loops operating in both bottom-up and top-down directions. Too much of the latter can result in behavior that is overly automatic and at odds with our intuitive notion of behavior control. It conflicts with the model of conscious deliberation and reflection we associate with the normative aspect of agency making us candidates for responsibility, praise, and blame. Yet, as the example of OCD illustrates, too much conscious thought and too little automatic unreflective behavior can result in psychopathology that can have equally debilitating effects on agency. Interacting unconscious and conscious processes ensure balanced cognitive, affective, and volitional functions necessary for agents to form and execute a range of action plans.

2. MIND–BRAIN INTERACTION AND THE CAUSAL ROLE OF MENTAL STATES

There is another challenge to our conviction of having at least some conscious control of our actions and another source of existential angst about whether we really have any control. Some neuroscientists claim that we are identical to our brains (Churchland 2013; Crick 1993). This claim appears to imply either ontological or explanatory reductionism. It may imply that conscious mental states that presumably define us as persons are nothing more than physical states in the brain. Alternatively, it may imply a form of reductive materialism whereby mental processes can be completely explained in terms of physical processes in the brain. On this second view, mental processes can be fully accounted for by their component neurobiological and neurochemical parts and organizational properties between and among these parts (Craver 2007). There are significant differences between reductive and mechanistic explanations of human behavior. But both effectively rule out any causal role for conscious mental states in upregulating or downregulating unconscious brain activity. If reductive materialism is true, then neural processes obviate the need for psychological explanations of our behavior. This can be described as the "causal exclusion problem" (Bayne 2011). Mental properties are excluded from having a causal role in agency because the causal efficacy of the neural properties underlying them provides a complete explanation of agency. The conscious desires, beliefs, reasons, and intentions that we think drive our actions and give normative significance to them are causally inert.

According to reductive materialism, phenomena at one level can be completely explained in terms of more basic elements at a different level. Applied to the mind–brain relation, this theory says that normal and abnormal mental states can be completely explained in terms of brain function and dysfunction. According to nonreductive materialism, the brain necessarily generates and sustains mental states but cannot account for all of their properties. Persons are constituted by but not identical to their brains. Genetics, neuroendocrine and neuroimmune interactions, and the environment all influence brain structure and function and how the conscious and unconscious mental states that define persons emerge from the brain. These states are shaped by factors both inside and outside the brain. Describing or explaining them at a brain-systems level is necessary but not sufficient to account for the content and quality of our mental states and how they issue in decisions and actions. As psychiatrist and philosopher Thomas Fuchs puts it:

> [T]he brain is not a creator, but a relational organ; it is embedded in the meaningful interactions of a living being with its environment. It mediates and enables these interactive processes, but it is in turn also continuously formed and restructured by them. The mind may be regarded as a continuous process

of relating to the environment which is constantly transformed into the more stable structures of neural networks and dispositions. This reciprocal relationship ... in the joint development of mind and brain ... strongly contradicts any reductionist notions of the brain as the creator of the mind. (Fuchs 2011: 198)

Neuropsychiatrist Todd Feinberg's concept of the mind as a process emerging from the brain in a nested hierarchy also supports nonreductionism about the mind–brain relation. Higher level processes associated with conscious and unconscious mental states are compositionally dependent on, or nested within, lower level processes associated with circuits in the brain (Feinberg 2001: 129–131). Similar to Spence's interpretation of the concept, Feinberg's understanding of "constraint" can explain how interacting neural and mental processes promote homeostasis within an organism and its adaptability to the external world. Constraint refers to the control that one level of a system exerts over another level of the same system. The "system" at issue is a human organism, and the relevant "levels" correspond to neural and mental processes constituting the brain and mind. Neural circuits constrain mental states so that they accurately interpret information from the environment and respond appropriately to stimuli. Mental states constrain neural circuits to ensure that they are neither overactive nor underactive. Beliefs with heightened negative emotional content due to psychosocial stress can overactivate the limbic fear system, disable cortical constraint on this system, and lead to depression, anxiety, or panic disorders. Disabled constraints on belief content from dysfunctional prefrontal, parietal, and auditory cortices can result in the hallucinations and delusions in the positive subtype of schizophrenia. Lack of constraint can cause neural and psychological imbalance that can impair agency. Bidirectional constraint as I have described it is necessary for behavior control.

Proponents of nonreductive materialism about the mind–brain relation claim that mental properties are part of the material world. They also claim that mental properties can be causally efficacious in inducing changes in material properties of the brain (Baker 2009; Northoff 2014: part II). Critics of this position argue that if mental events are not reducible to physical or material events, then they are epiphenomenal (Kim 1998, 2005). They argue that mental events are the effects of material causes but cannot cause any material events to occur. If mental processes associated with beliefs, desires, and emotions are not reducible to their neural correlates and are epiphenomenal, then they can be excluded from any explanation of human behavior. But there are many examples in neuropsychiatry of how mental states have a causal role in disrupting and modulating neural pathways. This phenomenon is not limited to OCD. The content of a person's beliefs and emotions can influence the etiology of abnormal behavior and the restoration of normal, healthy behavior through their effects on physical processes in the brain. They can either impair agency and behavior control or restore them to normal levels.

Chronic psychosocial stress can cause a cascade of adverse biochemical events in the brain and body, including hyperactivation of the amygdala fear system and dysregulation of the hypothalamic-pituitary-adrenal (HPA) axis and sympathetic nervous system. This can disrupt frontal-limbic connectivity mediating cognitive and affective processing and result in impaired cognition and mood (Gold and Charney 2002). Persistent stress can also cause high levels of the hormone cortisol to circulate in the body and brain, which can result in neuronal degeneration in the prefrontal cortex and increase the risk of cognitive impairment and dementia. The etiology of these conditions shows how mental states can induce adverse events and processes in neural networks. These in turn can result in further impairment of cognitive, emotional, and volitional capacities necessary for agency and behavior control.

Mental states can play an equally important positive role in therapies to treat these disorders and restore some degree of control of thought and action. In cognitive behavioral therapy (CBT), patients with major depression or generalized anxiety disorder can be trained to reframe their beliefs and emotions and rewire some regions of the brain, resulting in significant improvement in depressive symptoms (Goldapple et al. 2004). Studies using brain imaging have shown that CBT can modulate function in specific sites in limbic and cortical regions mediating mood and cognition. Bottom-up effects of dysregulated neural circuits on disordered mental states can be reversed to some extent by top-down modulating effects of the mind. Similarly, a recent study of mindfulness meditation showed positive effects in frontal-limbic pathways regulating stress-related reactions to stimuli (Cresswell et al. 2016). Neurofeedback (NFB) is another technique whereby mental states can modulate brain activity (de Charms et al. 2005). With this technique, a person can be trained to downregulate brain hyperactivity causing chronic pain or attention deficit-hyperactivity disorder (ADHD) through their cognitive and emotional responses to the visual feedback of neural function they receive from EEG or functional magnetic resonance imaging (fMRI). This type of self-regulation can restore some degree of behavior control and relieve symptoms associated with pain and ADHD as well as with anxiety and posttraumatic stress disorder (PTSD). It may also be possible for depressed patients with anhedonia (inability to experience pleasure from previously pleasurable activities) and avolition (inability to motivate oneself to act) associated with an underactive nucleus accumbens in the reward system to upregulate activity in this region and improve motivation and mood (Linden et al. 2012). NFB demonstrates that persons can induce changes in their brains through their own mental states without having to rely on psychoactive drugs or devices implanted and stimulated in specific neural circuits.

More importantly, the fact that the cognitive and emotional responses that induce these changes depend on indices of brain activity fed back to the patient show that mind and brain are not independent but interdependent and interacting processes. NFB and other techniques highlight the erroneous assumption that nonreductive materialism implies dualism between mental and neural properties and that the first cannot influence the second. It is because of mind–brain interaction in NFB and CBT that these techniques can produce their therapeutic effects. Some investigators describe NFB as "a holistic approach that overcomes bio-psychological dualisms" (Linden et al. 2012: 8). The causal exclusion thesis and reductionism rest largely on the tendency of many cognitive neuroscientists to dissociate the mental from the physical. This thesis reflects the idea of a clear separation between a first-person subjective perspective and a third-person objective perspective and the idea that causation pertains only to the latter. Mind–brain interaction in the examples I have presented shows that these perspectives are not separable.

Some cognitive neuroscientists might insist that the purported effects of mental states on brain states are illusory. These phenomena can be explained in materialist terms as nothing more than higher level neural functions. But it is unclear how a brain-level explanation alone could completely account for the modulating effects of mental states on underactive or overactive brain regions and their therapeutic effects in symptom relief. In NFB, for example, salutary changes in the brain are caused by the person's subjective response to the visual feedback she is receiving from the EEG or fMRI about her brain. It is persons, not their brains, that experience pain or mood disturbances, and persons, not their brains, that experience relief from these states through their cognitive and affective responses to brain activity. The phenomenology of what it is like to visualize one's own brain activity and modulate it and associated symptoms cannot be explained entirely in materialist terms. Techniques such as NFB and CBT demonstrate that behavior control occurs at both neural and mental levels. They show that the mind can induce changes in the brain and have a causal role in this control.

3. MEANING AND THE WILL ARE NOT IN THE BRAIN

Despite the extent to which unconscious neural processes regulate our behavior, they allow enough room for us to be the authors of at least some of our actions and have some impact on events in the world. As Spence puts it, "an action creates a new event in the world. Michelangelo's hands created images that had not existed previously. Charlie Parker's fingers improvised new melodies. Before these movements were made, there were 'gaps,' potential spaces for action" (2009: 60). These are examples of what Spence calls "the (potential) response space, the space for responses" (2009: 367). The brain has largely laid out the chords that guide our behavior. But the neurobiological structures that enable the mind allow room for individual improvisation in how we respond cognitively and emotionally to situations in the world. These responses may be sufficiently causally robust to retain the conviction that we are the authors of at least some of our actions. In these respects, "the contents of consciousness *influence* our behaviors, even if they do not initiate them" (2009: 394). Consciousness "provides the venue, the medium for the goals that we pursue" (2009: 383).

The influence of contextual factors on choice also underscores the limitations of neuroreductionist accounts of behavior. According to the biological disease model of addiction, disabled prefrontal inhibitory mechanisms allow overactivation of the reward system and the pattern of relapse and remission (Koob and Volkow 2010). Social cues can trigger the desire to have the substance to which one is addicted. Changing the environment so that the cues and desire are no longer present is one way the addict can prevent relapse (Hyman 2007). A different environment with different stimuli can gradually modulate activity in the brain's reward system. Historical and social context can also influence the meaning an individual attributes to an action over and above its neural substrate. The Nazi salute of Hitler and the Black Power salute of Tommie Smith and John Carlos at the 1968 Summer Olympic Games in Mexico City involved the same nerve and muscle fibers and the same neural circuits in motor and other regions of the cerebral cortex. But they had very different meanings for these agents because of the different times, places, and social contexts in which they were performed. The contextual factors behind these meanings suggest that the actions were not literally initiated by particular neural events at discrete times but by complex distal intentions extending into the past. Many actions are shaped by beliefs and values that develop over a lifetime and are uniquely one's own. The content and significance of our motivational states and actions are not confined to the brain but expand into the natural and social environment. Again, Spence: "In the right circumstances, and in the right company, conscious awareness is potentially redemptive; it tells us about ourselves, and it may tell us where we are going. It is not the instant of the act but its context that seems to matter" (2009: 395).

There are cases where the exercise of a person's will also resists purely neuroscientific explanation. One such case is the autobiography of the Russian soldier Lyova Zazetsky, written in collaboration with his doctor, the neuropsychologist Alexander Luria (Luria 1987). During the Battle of Smolensk in the Second World War, Zazetsky sustained a severe head injury that caused extensive damage to the left occipital-parietal region of his brain. This caused his memory, visual field, and bodily perception to become fragmented. It left him with the experience of an unstable self in what for him had become a shattered world. The injury did not destroy his agency, because enough of his brain was intact for him to try to make sense of his experience. He retained his capacity for imagination, fantasy, and empathy. To cope with his experience, he kept a journal for twenty years, recording his thoughts and memories as they occurred on a daily basis. While this required excruciating effort, he wrote thousands of pages. Zazetsky's determination to reshape his identity and construct a meaningful narrative of his life after his injury illustrates how one can retain and exercise the will in spite of significant neurological and psychological impairment.

Others with neurological disorders may devise psychological strategies to maintain some control of their behavior. With the onset of an aura, some people with epilepsy can avoid seizures by focusing their attention in listening intently to music or to other people's conversations. Neurologist Adam Zeman points out that "while specific experiences sometimes engender seizures, attacks can also be resisted by an act of will" (Zeman 2004: 121). Another example of a psychological control strategy is habit-reversal training. Some people with Tourette syndrome can consciously anticipate and suppress seemingly uncontrollable motor and vocal tics. These mental acts are preceded by the event or process of trying to perform them, which has a phenomenological aspect that cannot be captured by appeal to brain processes alone. Although their effects may be limited, through these acts of will, some people can expand the space of agency beyond what normal or abnormal neurobiology has given them.

4. CONCLUSION

Our brains shape much of our behavior by providing the neurobiological underpinning of the mental capacities that lead to our decisions and actions. The brain enables the mind. We are more than our brains, though, and neural function does not determine that we perform some actions rather than others. Our behavior is constrained by brain structure and function in important respects. The idea of constraint and associated neuroscientific explanations of behavior may raise worries about whether they leave any room for individual choice based on the causal power of our mental states. These worries are unfounded. Brain and mind are not distinct entities but interdependent processes of human organisms living and acting in the world. Neural and mental functions interact in a constant process of bottom-up and top-down circular causation involving reentrant loops running to and from subcortical and cortical levels of the brain and unconscious and conscious levels of the mind. There is both brain–mind and mind–brain causal interaction. These loops in turn are shaped by the interaction between the central nervous and other bodily systems and between the organism and the natural and social environment in which she is embedded. The constraints in these relations do not undermine but are necessary for behavior control by ensuring that neural and mental functions operate at optimal levels. This promotes flexible behavior and adaptability to the demands of the environment. Depending on the extent to which an ordered or disordered brain enables or disables the mind, there is enough room for us to be the authors of our actions and to be praised, blamed, and held responsible for them. The normative content of these actions is what makes them meaningful. We do not "find" meaning in the brain, any more than an existentialist "finds" meaning in the world. Rather, we construct it from the actions we perform on the basis of our brain-enabled mental capacities.

This meaning often depends on the social context in which one thinks and acts. Contextual factors may limit the extent to which one can translate one's mental and physical capacities into actions. They also can provide the motivation for actions and their normative content. Two bodily movements involving the same cortical and subcortical activity and the same efferent projections from the central to the peripheral nervous system can have different meanings for agents performing them in different circumstances and at different times. In some cases, individuals can devise psychological strategies to prevent adverse neural events or exert the volitional capacity to make sense of living in a neurologically compromised state. All of these considerations suggest that, while neuroscience can explain much of who we are and what we do, it cannot provide a complete explanation of our behavior. There is more to persons than can be dreamed of in our neuroscience.

REFERENCES

Baker, L. R. 2009. Non-reductive materialism. In A. Beckermann, B. McLaughlin, and S. Waller (Eds.), *The Oxford handbook of philosophy of mind*, pp. 109–120. Oxford: Oxford University Press.

Bargh, J., and E. Morsella. 2008. The unconscious mind. *Perspectives on Psychological Science* 3(1): 73–79.

Bayne, T. 2011. Libet and the case for free will skepticism. In R. Swinburne (Ed.), *Free will and modern science*, pp. 25–46. Oxford: Oxford University Press.

Bratman, M. 1987. *Intention, plans and practical reason*. Cambridge, MA: Harvard University Press.

Bratman, M. 2007. *Structures of agency: Essays*. New York: Oxford University Press.

Churchland, P. S. 2013. *Touching a nerve: The self as brain*. New York: W. W. Norton.

Craver, C. 2007. *Explaining the brain: Mechanisms and the mosaic unity of experience*. Oxford: Clarendon Press.

Cresswell, J. D., A. Taren, E. Lindsay, C. Greco, D. Gianaros, H. Fairgrieve, et al. 2016. Alterations in resting-state functional connectivity in mindfulness meditation reduced interleukin-6: a randomized controlled trial. *Biological Psychiatry* 79. doi: 10.1016/j.biopsych. 2016.01.008

Crick, F. 1993. *The Astonishing hypothesis: The scientific search for the soul*. New York: Scribners.

De Charms, R., F. Maeda, G. Glover, D. Ludlow, J. Pauly, D. Someji, et al. 2005. Control over brain activation and pain learned by using real-time functional MRI. *Proceedings of the National Academy of Sciences* 102: 18626–18631.

De Haan, S, E. Rietveld, and D. Denys. 2015. Being free by losing control. In W. Glannon (Ed.), *Free will and the brain: Neuroscientific, philosophical and legal perspectives*, pp. 83–102. Cambridge: Cambridge University Press.

Feinberg, T. 2001. *Altered egos: How the brain creates the self*. New York: Oxford University Press.

Figee, M., J. Luigies, R. Smolders, C. Valencia-Alfonso, G. Van Wingen, B. de Kwaasteniet, et al. 2013. Regaining control: deep brain stimulation restores frontal-striatal network activity in obsessive-compulsive disorder. *Nature Neuroscience* 16: 366–387.

Fuchs, T. 2011. The brain—a mediating organ. *Journal of Consciousness Studies* 18: 196–221.

Gold, P., and D. Charney. 2002. Depression: a disease of the mind, brain and body. *American Journal of Psychiatry* 159: 1826.

Goldapple, K., Z. Segal, C. Garson, M. Lau, P. Bieling, S. Kennedy, et al. 2004. Modulation of cortical-limbic pathways in major depression: Treatment-specific effects of cognitive behavior therapy. *Archives of General Psychiatry* 61: 34–41.

Hyman, S. 2007. The neurobiology of addiction: implication for voluntary control of behavior. *American Journal of Bioethics* 7(1): 8–11.

Kahneman, D. 2011. *Thinking, fast and slow*. New York: Farrar, Straus and Giroux.

Kim, J. 1998. *Mind in a physical world: An essay on the mind-body problem and mental causation*. Cambridge, MA: MIT Press.

Kim, J. 2005. *Physicalism, or something near enough*. Princeton, NJ: Princeton University Press.

Koob, G., and N. Volkow. 2010. Neurocircuitry of addiction. *Neuropsychopharmacology* 35: 217–238.

Libet, B. 1985. Unconscious cerebral initiative and the role of conscious will in voluntary action. *Behavioral and Brain Sciences* 8: 529–566.

Linden, D., I. Habes, S. Johnston, S. Linden, R. Tatineni, L. Subramanian, et al. 2012. Real-time self-regulation of emotion networks in patients with depression. *PLoS ONE* 7: e38115.

Luria, A. 1987. *The man with a shattered world: The history of a brain wound* (trans. L. Solotaroff). Cambridge, MA: Harvard University Press.

Mele, A. 2009. *Effective intentions: The power of conscious will*. New York: Oxford University Press.

Mele, A. 2014. *Free: Why science hasn't disproved free will*. New York: Oxford University Press.

Melloni, M., C. Urbistanda, L. Sedeno, C. Gelormini, R. Kichic, and A. Ibanez. 2012. The extended frontal-striatal model of obsessive-compulsive disorder: convergence from event-related potentials, neurophysiology and neuroimaging. *Frontiers in Human Neuroscience* 6: 259. doi: 10.3389/fnhum.2012.00259

Meynen, G. 2012. Obsessive-compulsive disorder, free will and control. *Philosophy, Psychiatry & Psychology* 19: 323–332.

Morsella, E., C. Godwin, T. Jantz, S. Krieger, and A. Gazzaley. 2015. Homing in on consciousness in the nervous system: an action-based synthesis. *Behavioral and Brain Sciences* 38: 1–106.

Northoff, G. 2014. *Minding the brain: A guide to philosophy and neuroscience*. New York: Palgrave Macmillan.

Sartre, J-P. 2007. *Existentialism is a humanism*. New Haven, CT: Yale University Press.

Schultze-Kraft, M., D. Birman, M. Rusconi, C. Allefeld, K. Gorgen, S. Dahne, et al. 2016. Point of no return in vetoing self-initiated movements. *Proceedings of the National Academy of Sciences* 113. doi: 10.1073/pnas.1513569112

Spence, S. 2009. *The actor's brain: Exploring the cognitive neuroscience of free will.* Oxford: Oxford University Press.

Stein, D., W. Goodman, and S. Rauch. 2000. The cognitive-affective neuroscience of obsessive-compulsive disorder. *Current Psychiatry Reports* 2: 341–346.

Wegner, D. 2002. *The illusion of conscious will.* Cambridge, MA: MIT Press.

Zeman, A. 2004. *Consciousness: A user's guide.* New Haven, CT: Yale University Press.

CHAPTER 10

Two Types of Libertarian Free Will Are Realized in the Human Brain

PETER U. TSE

In my book *The Neural Basis of Free Will* (2013), I described various developments in neuroscience that reveal how volitional mental events can be causal within a physicalist paradigm. I began by (1) attacking the logic of Jaegwon Kim's (1993) exclusion argument (EA). According to the EA, information in general, and mental information in particular, cannot be causal and must be epiphenomenal because particle-level physical-on-physical causation is sufficient to account for apparent causation at all higher levels. If this is true, then mind cannot be causal in the universe. It would follow that there cannot be any free will or morality that made any difference to physical outcomes because quark-level descriptions (or whatever is operative at the rootmost level of physical causation), where there is no need for informational, mental, or moral descriptors, would be sufficient to account for the causal unfolding of events. The first task of anyone interested in free will or mental causation, therefore, must be to show where the EA breaks down. I will summarize here my past arguments that the EA falls apart if indeterminism is the case. If I am right, I must still build an account of how mental events are causal in the brain. To that end, I take as my foundation (Section 2) a new understanding of the neural code that emphasizes rapid synaptic resetting over the traditional emphasis on neural spiking. Such a neural code is an instance of "criterial causation," which requires modifying standard interventionist conceptions of causation such as those favored by Judea Pearl (2000) and John Woodward (2003) (see Section 3). A synaptic reweighting neural code provides (discussed in Section 4) a physical mechanism that accomplishes downward informational causation, a middle path between determinism and randomness (Section 5), and a way for mind–brain events to turn out otherwise (Section 6). This "synaptic neural code" allows a constrained form of randomness parameterized by information realized in and set in synaptic weights, which in turn allows physical/informational criteria to be met in multiple possible ways when combined with an account of how randomness in the synapse is amplified to the level of randomness in spike timing. This new view of the neural code also provides a way out of self-causation arguments against the possibility of mental causation (see Section 7). It leads to an emphasis on imaginative deliberation and voluntary attentional manipulation as the core of volitional mental causation rather than, say, the correlates of the unconscious premotor computations seen in Libet's readiness potentials (discussed in Section 8). And this new view of the neural code leads to a new theory of the neural correlates of qualia as the "precompiled" informational format that can be manipulated by voluntary attention, which gives qualia a causal role within a physicalist paradigm (Section 9).

Finally, it is not enough to simply have the "first-order free will" afforded by the preceding kind of nervous system that can choose actions freely. Only if present choices can ultimately lead to a chooser becoming a new kind of chooser—that is, only if there is second-order free will or meta-free will—do brains have the capacity to both have chosen otherwise and to have meta-chosen otherwise. Only such a meta-free will allows a brain to not only choose among options available now, but to cultivate and create new types of options for itself in the future that are not presently open to it. Only then can there be responsibility for having chosen to become a certain kind of person who chooses from among actions consistent with being that kind of person. In Section 10, I will discuss how the brain can choose to become a new kind of brain in the future, with new choices open to it than are open to it now. In Section 11, I will argue that criterial causation gets around luck arguments against self-forming actions. I elaborate each of these ten points in the eleven sections below.[1]

1. OVERTURNING KIM'S EXCLUSION ARGUMENT

It is necessary to challenge this philosophical claim because if the EA is correct, a central assumption of neuroscience and psychology, namely that mental information can be causal of subsequent brain events, falls apart. A philosopher might object that Kim's argument is only an argument against antireductionism or nonidentity theories, and that, if one adopted a type identity

---

1. This chapter has been adapted from an excerpt from my upcoming book *Imagining Brains: The neural sources of human freedom and creativity*.

theory, then mental information would indeed be causal, but only by virtue of being physical. But that would still make mental events, like pain, not causal in the universe by virtue of their informational characteristics, such as hurting, but only causal via their physical instantiations having physical causal efficacy. If mental events cannot be causal by virtue of being informational—and they would not be if the EA is correct—then it would follow that there can be no free will (i.e., free mental events) that makes a difference to physical outcomes. It would also follow that there can be no morality or immorality; mental decisions cannot be held accountable for their having made a difference to physical outcomes, when, as would be the case if the EA holds, mental decisions can make no difference in a universe where quark-on-quark interactions (or whatever units of energy are at the very lowest level) are sufficient to account for all causal chains. So confronting the EA is a first step to any argument for mental causation or free will since informational or mental causation is a necessary (though not sufficient for) form of free will and moral agency that makes a difference to physical outcomes.

Let us examine Kim's (1993) EA more closely. If true, Kim's EA would logically rule out that mental information can be causal. (Note that this argument could also be used to argue that genetic information is epiphenomenal, although no one argues that—probably because we understand the genetic code quite well now, whereas we do not yet fully understand the neural code). The argument rests on a premise of the causal closure of the physical. "Causal closure" means that causality at the level of particles is *sufficient* to account for all outcomes and interactions at the level of particles.[2] Kim (2005: 17), applying Occam's razor, advocates the "exclusion of over-determination" when modeling physical causation. In his words: "If event *e* has a sufficient cause *c* at *t*, no event at *t* distinct from *c* can be a cause of *e*." Note that without the sufficiency of *c*, Kim cannot apply the "exclusion of over-determination" principle, so cannot rule out mental causation. The sufficiency of *c* is crucial if the EA is to succeed at excluding mental causation. If particle-level causality is

---

2. Here is how Kim (1996: 147) defines the causal closure principle: "Pick any physical event, say, the decay of a uranium atom or the collision of two stars in distant space, and trace its causal ancestry or posterity as far as you would like; the principle of causal closure of the physical domain says that this will never take you outside the physical domain. Thus, no causal chain involving a physical event will ever cross the boundary of the physical into the nonphysical." Caruso (2012: 13) defines causal closure as follows: "If x is a physical event and y is a cause or effect of x, then y, too, must be a physical event." Note, however, that these "physical from physical" definitions of causal closure ignore the sufficiency of a cause. They hold under determinism or indeterminism as we walk causal chains backward into the past. But as we walk causal chains into the future, which, barring backward causation in time, is the way they in fact go, under indeterminism, but not under determinism, the majority of physically possible causal chains do not happen and therefore never become actually physical. They do not happen because, under indeterminism, a cause is not sufficient to specify which one of many possible outcomes will happen.

sufficient to account for particle behavior, and neurons are made of particles, then mental events, assuming that they supervene on neuronal events, can play no causal role in neuronal behavior. In other words, mental events qua mental events cannot cause fundamental particles to behave differently than they otherwise would have if they had only interacted according to the laws obeyed by particles.

Put succinctly (Kim 1993: 206–210): if (i) the "realization thesis" is the case, then each mental state is synchronically determined by underlying microphysical states, and if (ii) "the causal or dynamical closure of the physical thesis" is the case, then all microphysical states are completely diachronically necessitated by antecedent microphysical states, then it follows that (iii) there is no causal work left for mental states as such to do. If the logic here is valid, then only if either (i) or (ii) is incorrect is there potentially room to develop a theory of mental causation. So any theory of mental causation that attempts to meet "Kim's challenge" must explicitly state which premise, (i) and/or (ii), is incorrect.

If quantum domain indeterminism is correct, then (ii) is incorrect because any particular present microphysical state is not necessitated by its antecedent microphysical state or states. In other words, the traditional definition of causal closure that "every physical event has an immediately antecedent sufficient physical cause" is not satisfied because when a cause c can be indeterministically followed by any number of possible effects e_i, then c is not a sufficient cause of any of the possible e_i because they might not happen if they have not yet happened, and they might not have happened even after they have happened. Papineau (2009) tries to handle the problem of causal nonsufficiency of c introduced by indeterminism by appending a qualifier to the more traditional definition of causal closure as follows: "Every physical effect has an immediate sufficient physical cause, in so far as it has a sufficient physical cause at all." A similar attempt to make—in this case Davidson's—definition of causal closure consistent with indeterminism is to say that "every physical event *that has an explanation* has a physical explanation." But neither of these attempts to dodge the nonsufficiency of c imposed by indeterminism gives existing physical explanations enough credit. Quantum-domain effects are not unexplained. It is not the case that just anything can happen inexplicably. Rather, the set of possible outcomes and their likelihoods of occurrence are very precisely defined by quantum theory, arguably the most accurately predictive theory in the history of science.

Classical deterministic laws are laws that hold among sufficiently causal actualia, where both c at $t1$ and e at $t2$ are actual events. Quantum mechanical laws are deterministic at the level of possibilia, but indeterministic at the level of actualia, because which possible outcome will occur upon measurement is only probabilistically specifiable. Nonetheless, under quantum mechanics, c is sufficiently causal of its entire set of possible outcomes e_i with their associated probabilities of occurring. It is just that c is not a sufficient cause of any particular one of its many possible effects that happens to happen when measured. Classical deterministic and modern quantum mechanical laws both operate deterministically, and causation is sufficient, but over different types of physical entities: actualia and possibilia, respectively. Actualia and possibilia, while both physical, have mutually exclusive properties. Actualia are real and exist now or in some past moment; they have a probability equal to 1 of happening or having happened. Possibilia are not yet real and may never become real; they exist in the future relative to some c and have a probability of happening between 0 and 1. A given event cannot be both actual and possible at the same time.

Closure, therefore, applies to different types of physical events under ontological determinism and indeterminism. "Closure" entails that the set of physical events is closed; any particular effect will be a member of the same set to which a sufficient cause itself belongs. Determinism is closed at the level of actualia; any particular cause or effect will be a member of the set of all actual events in the universe across all time. Indeterminism, in contrast, is not closed at the level of actualia. This is because a nonsufficient actual cause and one of its possible outcomes that may never happen are not both members of the set of actualia. Rather, quantum theory is closed (and deterministic!) at the level of possibilia: any particular outcome or event will be a member of the set of all possible outcomes or events in the universe across all time, and any possible cause is sufficient to account for the set of all of its possible effects. Under indeterminism physical explanations are of a different type than under determinism, though both actualia and possibilia are physical, and theories of either actualia or possibilia are physical explanations.

An indeterministic causal closure thesis could be restated as follows: "(ii*) the set of all possible microphysical states is completely diachronically necessitated by antecedent possible microphysical states." The realization thesis for the indeterministic case might be: "(i*) all mental states are synchronically determined by underlying sets of possible microphysical states." But claim (i*) is contrary to the definition of supervenience. Mental events do not supervene on sets of possible physical states, they supervene on specific, actually occurring physical states. Since it is absurd to maintain that mental events synchronically supervene on sets of possibilia, we can rule (i*) out. It remains to be shown whether (i) (i.e., supervenience on actualia) can be combined with (ii*) (i.e., causal sufficiency and closure among possibilia) to yield (iii). We will see later that this combination fails to deliver causal closure.

An actual microphysical state and the set of all possible microphysical states are different kinds with mutually exclusive properties (e.g., real/~real; present/~present). The essentially syllogistic structure of the EA requires staying within a logical kind. It is logically valid to draw from the major premise (ii) "all physical events are caused by preceding sufficient physical causes" and the minor premise (i) "mental events are realized in physical events" the conclusion (iii) that "the physical events that realize mental events have preceding sufficient physical causes." But now we are splitting "physical" into two types with mutually exclusive properties, possibilia and actualia. The conclusion (iii) of the syllogism holds only if both the major and minor premises hold and are both about actualia as in (ii) and (i), or both are about possibilia as in (ii*) and (i*). If one premise is about possibilia and the other about actualia, the conclusion does not follow because the premises are about exclusive entities. For example, (ii) and (i*) would read "all actual physical events are caused by preceding sufficient actual physical causes" and "mental events are realized in sets of possible physical events," which violates syllogistic logic as much as "all men are mortal" and "Socrates is a robot." Conversely, (ii*) and (i) would read "the set of possible physical events are caused by preceding sufficient possible physical causes" and "mental events are realized in actual physical events," which similarly violates syllogistic logical form. Thus, assuming indeterminism, mental causation is not logically ruled out by Kim's argument.[3]

I wrote a version of the preceding argument first in *The Neural Basis of Free Will*, and then on the philosophy blog *Flickers of Freedom*, where I battled with free will denier Neil Levy. He wrote that the preceding argument ". . . is badly confused. It rests on a misunderstanding regarding the causal closure

---

3. As an aside, there is another argument that (i) with (ii*) cannot logically entail (iii). Obviously, causes must precede effects. The usual EA is that (ii) diachronic actual elementary particle interactions *preceding* the moment t of (i) synchronic mental supervenience on actual particle configuration p leave no room for mental events qua mental events to have any causal effect since those preceding physical interactions are sufficient to cause p. However, if (ii*) is taken to refer to a diachronic set of possible events preceding (i) mental supervenience on p, then there is a problem because possibilia do not exist in the past of p; only actual events, such as those described in (ii), do. Once we have reached time t and p is not a possibility but an actuality, then all events prior to t must also be actual; events in the past are actual events that happened and are no longer possible. If they were possible, they would lie in the future. Possibilia only exist in the future relative to some actual or possible event. But p we agree is actual since supervenience makes no sense for possibilia, as in (i*), which we have rejected. Alternatively, if we want to think of the possibilia in (ii*) "collapsing" into p, where p was one among many possibilities, much like the quantum mechanical collapse of the wave function, we are again left with the problem that the set of possibilia is not sufficient to cause p per se because p might not have happened at all, and some other possible outcome might instead have happened. However, if the possibilia in (ii*) are taken to temporally follow (i) the actual p at t, well, that is certainly consistent with the idea that possibilia can exist in the future of p. But then possibilia in the future of p would be seen as being sufficiently causal of p, which would entail impossible backward causation in time. Thus, the possibilia described in (ii*) can neither precede nor follow the actualia described in (i) and be sufficiently causal of them. In sum, (i) and (ii*) do not together entail (iii), whether on logical (syllogistic) grounds or on the grounds that possibilities can only exist in the future and not in the past of actual events such as those on which mental events supervene. Again, assuming indeterminism, mental causation is not logically ruled out by Kim's argument.

principle. Tse understands the principle to claim that physical causes are sufficient for the occurrence of physical effects. If indeterminism is true, then physical causes sometimes or often are not sufficient for the occurrence of later events. Tse therefore concludes that the closure principle is false for indeterministic systems, so it is no obstacle to mental causation. But the causal closure principle is, roughly, the principle that physical events can be accounted for by physical causes, or (equivalently) that physics is causally complete. It is silent on whether physics is deterministic or not. The brain may be indeterministic; causal closure remains an obstacle to mental causation."[4]

In response to Levy, I did not invent the definition of causal closure as "every physical effect having an immediately antecedent *sufficient* physical cause"; many philosophers have written variants of just such a definition, including Papineau and Kim, cited earlier. If we eliminate the requirement that c be sufficient to cause its physical effects, we lose Kim's elegant "exclusion of over-determination" argument against any possible causal role of the mental qua mental and can no longer rule that out. In the absence of sufficient physical causation, we could at most argue that an action or outcome would be overdetermined if it has both a physical cause (whether deterministic or indeterministic) and a mental cause. Under that move, all the causal work is still done by, presumably, particle-on-particle interactions, not by mental events qua mental events (e.g., when pain, by virtue of consciously hurting, causes a trip to the dentist). As a physicalist, I agree that "physical events can be accounted for by physical causes." But there is an ambiguity in Levy's phrase "accounted for" here. Deterministic physical laws account nonprobabalistically (or rather, with a probability of 1) for a deterministic succession among actualia, whereas indeterministic physical laws account probabilistically for an indeterministic succession among actualia; or, as is the case with the evolution of the wave function in quantum mechanics, physical laws account deterministically for a changing probability distribution of possible outcomes of measurements. If we are to take the idea of closure of the physical seriously, then a physical cause c and its physical effect(s) must belong to the same closed set. We agree that this closed set includes only physical events, whether determinism or indeterminism is the case. But, under determinism, that closed set of physical events includes physical actualia across time, whereas, under indeterminism, it includes physical possibilia across time. In principle, classical physics is a causally complete and deterministic account of the sequence of

4. I find the views of Levy and others who deny mental causation and free will to be wrong, nihilistic, and impoverished; wrong for reasons covered in my book and here, nihilistic because there can be no moral responsibility or self forming acts under such views, and impoverished because they fail to recognize the astounding elaboration of modes of top-down informational causation that have evolved in biological systems, including, principally, the causal roles of our minds in realizing our own envisioned futures.

actualia over time, and quantum physics is a causally complete and deterministic account of the sequence of possibilia over time. But standard versions of quantum physics do not give a complete account that can explain why one possible outcome becomes actual upon measurement or observation rather than other possible outcomes that did not occur. It just happens, with no reason given beyond chance. If c does not provide sufficient grounds for why one possible outcome occurs over another, exclusion of overdetermination cannot be used to rule out the possibility that the physical realization of present mental events might bias which particle possibilia will become actualia in the imminent future. Note that this does not require positing any bizarre notions like consciousness collapsing the wave packet. It just requires that present physically realized informational criteria placed on inputs can be met in the future in multiple possible ways.

In sum, Kim's EA amounts to saying that the physical substrate does all the causal work that the supervenient mental state is supposed to do, so mental or informational events can play no causal role in material events. One might say that this does not hold if the mental and physical are identical, but, even then, it is the physical side of the equation where causal efficacy resides. On Kim's reductionistic view, all causation "seeps away," as Ned Block put it, to the rootmost physical level (i.e., particles or strings or whatever physicists next model the most basic level to be like). Add to that an assumption of determinism and the laws of physics applicable at the rootmost level are sufficient to account for event outcomes at that level and every level that might supervene on that level. So informational causation, including voluntary mental causation or any type of libertarian free will that relies on information being causal in this universe, is ruled out. I argue that indeterminism undermines this sufficiency, so provides an opening whereby physically realized mental events could be downwardly causal.

Exploiting this opening, biological physical systems evolved to emphasize a new kind of physical causation, one based on triggering physical actions when detected spatiotemporal patterns in energy meet the criteria for triggering. This is a very different kind of causation than traditional Newtonian conceptions of the causal attributes of energy, such as mass, momentum, frequency, or position, which seem to underlie deterministic and exclusionary intuitions. But patterns, unlike amounts of energy, lack mass and momentum and can be created and destroyed. They only become causal if there are physically realized pattern detectors that respond to some pattern in their energetic inputs. Basing causal chains on successions of detected patterns in energy, rather than the transfer of energy among particles, opens the door not only to informational downward causation but to causal chains (such as mental causal chains or causal chains that might underlie a game of baseball or poker) that are not describable by or solely explainable by the laws of physics applicable at the rootmost level. Yes, a succession of patterns must be realized in a physical causal chain that is consistent with the laws of physics, but many other possible causal chains that are also consistent with physical laws are ruled out by informational criteria imposed on indeterministic particle outcomes. Physically realized informational criteria set in synaptic weights effectively sculpt informational causal chains out of the "substrate" of possible physical causal chains. Information is not causal as a force. Rather, it is causal by allowing only those possible physical causal chains that are *also* informational causal chains (i.e., that meet particular preset informational criteria) to become real (i.e., to switch ontological status from possibilia to actualia).

The old argument that there is no middle ground between utter randomness and determinism is wrong. If indeterminism is ontologically the case, then parameters placed on possible outcomes can select from among possible particle paths just that subset that also satisfies specified informational parameters. Causation via informational reparameterization would not be possible if the neural code were based on spikes ballistically triggering spikes like Newtonian billiard balls deterministically triggering the motions of the billiard balls that they collide with. But if presynaptic neural spikes reparameterize the informational criteria that will make postsynaptic neurons spike, given possible future presynaptic neural spike inputs, then many neural causal chains are possible that would be consistent with those reset informational parameters or criteria for firing. By itself, neural causation via informational reparametrization does not get us the control or the ability to settle outcomes that is needed for free will and moral responsibility. Much more is needed and has indeed evolved to be present in our brains, which I will return to later. But an informational criterial neural code is a necessary condition for having such control. For example, if I say to you "Name a female politician with red hair" your response will likely not be utterly random because you will state a name that meets these three criteria of being a woman, a politician, and having red hair. But your response is also not determined because your answer might have turned out otherwise. For example, if you responded "Angela Merkel," had I been able to rerun the universe again from the moment of my question, this time you might have said "Margaret Thatcher." This is because the brain has in fact evolved to amplify quantum domain randomness (as I explore in depth in *The Neural Basis of Free Will*), up to a level of neural spike timing randomness. And since neurons are effectively spike coincidence detectors, this randomness affords the possibility of other solutions to any given finite set of informational criteria. This kind of criterial neural code in turn affords the possibility that events might turn out otherwise, yet not be utterly random, because they will have to meet the informational criteria that were preset. Information, then, is not causal as a force again, but more as a filter that allows possibilia (at the particle level) that are consistent with informational parameters to become actualia, and those that are not consistent with informational parameters to get weeded away. The brain will need more causal powers to get to a full-blown libertarian free will, and these have also evolved, as I will argue later. But a criterial or parametric neural code is necessary (even when not sufficient) for free will and moral responsibility because informational reparameterization via synaptic weight resetting is the core engine whereby information can be causal of subsequent events in the brain. Thus all possible or actual informational causal chains are also possible or actual physical causal chains, whereas the vast majority of possible physical causal chains are not informational causal chains. Only those who have yet to appreciate that causation in the brain can proceed via informational criterial or parameter resetting via rapid synaptic weight changes can continue to bring out the tired Humean argument that there can be no libertarian free will realized in the brain because there is nothing between determinism (where events could not have turned out otherwise) and utter randomness (where an agent plays no role in the chance events that happen next).

Accounts of free will that require supernatural or contra-causal interventions have given libertarianism a bad name and violate basic assumptions of physicalism and science. For any naturalistic variant of libertarian free will to exist, several necessary conditions must be met. First, indeterminism must be ontologically real, rather than just a matter of epistemic uncertainty. Under indeterminism, I argued earlier, Kim's EA fails to rule out mental causation, leaving us with an opening to develop a believable account of mental or informational causation that is not epiphenomenal. To get there, the following facts must in turn be true of neural processing: (1) quantum domain randomness must be amplified up to the level of randomness in macroscopic neural information processing, which (2) would have to be able to harness this randomness to fulfill information processing aims, and (3) there would have to be a role for the subclass of information processing that we call "mental," particularly the subclass of the mental that we regard as consciously volitional, in the specification of the ends to which such harnessing will apply, if conscious willing is to be agentic or causal of the realization of such aims. In *The Neural Basis of Free Will*, I laid out a detailed case that these conditions are met, permitting the physical realization of a "type-1 libertarian free will." In order to have a "type-2 libertarian free will," however, an additional condition would have to be met: namely, the nervous system would have to (5) be able to make decisions about how it would like to change itself and then have the means to change itself, over time, into the intended type of decider.

Libertarian free will requires nonillusory downward mental causation. "Downward" here means that events at a supervening level can influence outcomes at the rootmost level. In this context, it would mean that information can bias which possible particle paths are realized. There is no wiggling out of this. If we want mental causation, and a free will and moral responsibility rooted in mental events that cause real consequences, we must defend the position that an informational entity, such as an intention or plan developed or held in working memory, can bias what possible particle paths open at the rootmost level can and do become real. But an entity at a supervening level cannot, logically, change its own physical basis because there can be no *causa sui*. This is where criterial causation via informational reparameterization comes in because this allows what supervenes now to place informational constraints on what can supervene in the future.

How might such constraining work in the brain? The key pattern to which neurons respond is temporal coincidence. A neuron will only fire if it receives a certain number of coincident inputs from other neurons. Criterial causation occurs where physical criteria imposed by synaptic weights on coincident inputs in turn realize informational criteria for firing. This permits information to be downwardly causal regarding which indeterministic events at the rootmost level will be realized; only those possible rootmost physical causal chains that meet physically realized informational criteria can drive a postsynaptic neuron to fire and thus become causal at the level of information processing. Typically, the only thing that the set of all possible rootmost physical causal chains that meet those criteria have in common is that they meet the informational criteria set. To try to cut information out of the causal picture here is a mistake; the only way to understand why it is that just this subset of possible physical causal chains—namely, those that are also informational causal chains—can occur is to understand that *informational* criteria delimit that class of possible outcomes.

As Eddy Nachmias put it on my October 2013 thread on the blog *Flickers of Freedom*: "the fact that informational state S1 could be realized by a range of physical states P1–PN and that informational state S2 counterfactually depends on S1 but *not* any one of the specific physical states, including the one that actually realizes S1 on this occasion (e.g., P3) suggests that S1 is what makes a difference to S2 in a way that P3 does not. If we want to causally manipulate S2, manipulating P3 may not do it (e.g., if we alter it to P1 or P4, or one of the other S1 realizers); rather, we need to manipulate S1 (yes, by altering its realizers in the right way, but the right way will involve considerations of the S-level, not the P-level). S2 *rather than* S7 occurs *because* S1 *rather than* S1' occurred, and not because P1 rather than P4 occurred."

Information only comes into existence by virtue of a decoder receiving input that matches its conditions (typically placed on the phase relationships or patterns in incoming energy) for the release of some effect—say, an action potential sent to other such decoders. But a decoder also serves as a "filter" on the set of all potentially causal inputs since it will only change the system of decoders in which it is embedded, namely, by firing, if its physically realized informational criteria are met.

Information cannot be anything like an energy that imposes forces because it is not material even when it is realized in the material substrate. Information's causal power consists in "filtering" informational causal chains out of the set of all possible physical causal chains by constraining which sets of possible physical causal chains can occur. Although every informational causal chain is also a physical causal chain, most physical causal chains are not informational causal chains. Information is downwardly causal not as a material force, but as constraints that only allow the realization of sets of possible physical causal chains at the rootmost level that also comprise informational causal chains. Physical laws are not violated by this. Every possible physical causal chain conserves energy and momentum and so forth. But only those possibilities allowed by physical laws which *also* meet informational criteria pass the physically realized informational filter and become informationally causal, either by reparameterizing the criteria by which other neurons will assess future input: namely, by changing their synaptic weights or by triggering other neural firing.

Information is multiply realizable because which particular set of spike inputs—and thus what particular information—will make the neuron fire is unforeseeable so long as the physical/informational criteria for firing are met. If neural causal chains are also informational causal chains, and informationally equivalent informational causal chains are realizable in multiple different neural or particle causal chains, then the parsimonious model is one of information causing information. Yes, there must always be some physical realization of information, but, under physical/informational criterial causation, which one it happens to be is irrelevant so long as informational criteria are met. Chains of successive informational criterial satisfactions and criterial resettings afford the physical realization of downward mental causation.

On the same *Flickers of Freedom* blog, Derk Pereboom said, "if on some proposal, a dualist or a nonreductivist one, M and P are distinct causes of E, the threat posed by exclusionary reasoning will be neutralized by any response on which the number of causes is reduced to just one. There are two ways to achieve this: a first is by eliminating all but one of the causes, and the second is by identifying the causes." If mental events are a type of information, and information is identical to acts of decoding immaterial (i.e., not made of mass) relationships or patterns among physical inputs, then mental events are identical to some class of acts of decoding. But note, this identity does not make mental events have physical properties like mass or momentum because the identity is not with physical events at some instant, but with a process that is realized in physical events. Moreover, acts of decoding patterns cannot be reduced to a level where the patterns are not explicit—say, the rootmost level—because the decoder only responds to a pattern at a level where it is explicit. And, at that level, potentially countless configurations of rootmost events are equivalent in that they each realize the same pattern as far as the decoder is concerned. Thus the identification is not with events at the microscopic level or even the neuronal level, but at the level of decoding the nonphysical patterns to which the decoder is sensitive. Under determinism, supervening informational criteria cannot filter out possible but noninformational causal chains at the rootmost level because there is only one possible causal chain. But if indeterminism is the case, supervening informational criteria can make a difference regarding which possibilities at the rootmost level happen. That is, under indeterminism but not determinism, there is nonredundant causal work for informational criteria to do.

But how does this give the brain the capacity to freely will? It is not enough for neurons to filter out noninformational possible physical causal chains. It must be the case that some neural activity that we associate with volition can control the parameters that neurons will apply in the future to enact such acts of filtering. Control comes from executive circuits that can plan, imagine, deliberate, and make decisions in light of highest level demands and needs and that can "rewire" circuits and reparameterize neurons by changing synaptic weights to embody new informational criteria for firing that will fulfill current executive ends. The downward causation afforded by the informational filtering of possible rootmost causal chains becomes agentic downward causation when executive circuits can rewire lower level circuits to fulfill whatever criteria they demand.

2. A NEW VIEW OF THE NEURAL CODE

How might informational reparameterization and causation work at a neural level? In *The Neural Basis of Free Will*, I developed a new understanding of the neural code that emphasizes rapid and dynamic synaptic weight resetting over neural firing as the core engine of information processing in the brain. The neural code on this view is not solely a spike code, but a code whereby information is transmitted and transformed by flexibly and temporarily changing synaptic weights on a millisecond timescale. One metaphor is the rapid reshaping of the mouth (analogous to rapid, temporary synaptic weight resetting) that must take place just before vibrating air (analogous to spike trains) passes through if information is to be realized and communicated. What rapid synaptic resetting allows is a moment-by-moment changing of the physical and informational parameters or criteria that have to be met before a neuron will fire. This dictates what information neurons will be responsive to and what they will "say" to one another from moment to moment. Thus the heart of criterial causation in the brain is the resetting, by other neural inputs, of the synaptic weights that realize informational parameters that have to be met by a neuron's subsequent inputs in order for that neuron to fire, which in turn will reset the parameters that will make other neurons subsequently fire.

3. RETHINKING INTERVENTIONIST MODELS OF CAUSATION

Interventionist/manipulationist models of causation (e.g., Pearl 2000; Woodward 2003) are rooted in the intuition that if some event A causes some event B, then one should be able to manipulate A in some way and see corresponding changes in B after changing A. If A is modeled as causing B, then there should be an intervention on A (in at least some state of the model) that results in B changing its value. These kinds of models of causation basically describe what scientists already do to determine causal relationships among variables. Scientists have for centuries tried to control for all independent variables (Woodward calls this "screening off" the other variables in addition to A that likely partially cause B by holding their values constant) save one, A, which they vary, in order to see the consequences or changes expressed by some outcome or dependent variable B. If B changes with an intervention on A, it is concluded that A, among perhaps other causal variables, in part causes B.

A counterfactual formulation of interventionism would be: "If A had not occurred, with all screened off variables that may cause B held constant, then B would not have occurred." A core point of criterial causation is that we need to enhance the interventionist account by saying: "If A had not occurred, with all screened off variables that may cause B held constant, *and with the parameters by which B evaluates its inputs also held constant*, then B would not have occurred."

Standard interventionist models of causation carry out some intervention on A to determine what effects, if any, there might be on B (and other variables). If, instead of manipulating A, or A's output to B, however, we instead manipulate the criteria, parameters, or conditions that B places on its input (including on input from A), which must be satisfied before B changes or acts, then changes in B do not follow passively from changes in A as they would if A and B were, say, billiard balls. Inputs from A can be identical, but in one case B changes in response to A, and in another case it does not, depending on B's criteria for responding. This reparameterization of B is what neurons do when they change each other's synaptic weights, such that a neuron now responds optimally to different inputs than prior to the act of physical and informational reparameterization or criterial resetting. Criterial causation emphasizes that what can vary is either outputs from A to other nodes (the traditional and, I would say, incomplete view of causation) or how inputs are decoded by receiving nodes B, B', B'' and so on. On this view, standard interventionist models (hearkening all the way back to "Newtonian" models of causation that emphasize energy transfer and conservation; e.g., P. Dowe's views [1992]) are a special case where B places no particular conditions on input from A that have to be met before B changes state. But the brain, if anything, emphasizes causation via reparameterization of B, by, for example, rapidly changing synaptic weights on postsynaptic neurons. Let me emphasize that I do not think that Woodward or Pearl are wrong. But they also make no mention, as far as I can tell, that causation might partly depend on reparameterizations of B. Thus their views of causation were incomplete and need to be amended by emphasizing the role of informational reparameterization of the response characteristics of B.

Changing the code or parameters or criteria that B uses to decode, interpret, or respond to input is a manipulation that might make no apparent changes to A or any other variable in the system for long and uncertain spans of time, until just the right pattern of inputs comes along. This is quite different from the unamended traditional manipulationist view, where a manipulation of A is expected to alter B within a short duration that is dictated by whatever physical laws are thought to apply, and certainly not after unspecifiably long durations, as is the case under my amended view. For example, the Mossad might program a cell phone to explode only when a particular phone number, known "only" to their target, is dialed. Manipulating "A" here appears to have no effect on any dependent variable "B" and might not, in principle, for as long as you like (think of a booby trap in a king's tomb set by the pharaoh's builders that only kills archaeologists millennia later). It might take years to work this phone up the ranks and into the hands of their targeted Hamas leader. But when the bomb maker dials his "secret" number to call his uncle in Paris, his head is blown off. This kind of reparameterization of B need not have immediate noticeable or measurable effects within the system, so seems to violate the assumption of the traditional view that causation is transferred at some fast speed (say the speed of light). But reparameterization of B is a causal intervention nonetheless, even though this subclass of causation has been relatively ignored by philosophers so far. It is at the heart of what I mean by "criterial causation." Other names for this might be "reparameterization causation," "pattern causation," or "phase causation."

I believe, however, that it was the "discovery" of this class of causation by evolution that led to the explosion of physical systems that we now call biological systems. Once causation by reparameterization came not only to involve conditions placed on physical parameters (e.g., molecular shape of a neurotransmitter before an ion channel would open in a cell membrane), but also conditions placed on informational parameters (fire above or below baseline firing rate if and only if the criteria for a face are met in the input) that were realized in physical parameters (fire if and only if the criteria on the simultaneity of spike inputs are met), we witnessed a further revolution in natural causation. This was the revolutionary emergence of mind and informational causation in the universe, as far as we know, uniquely on Earth, and perhaps for the first time in the history of the universe.

4. HOW DOWNWARD CAUSATION WORKS

Downward causation means that events at a supervening level can influence outcomes at the rootmost level. In this context, it would mean that information could influence which possible particle paths are actualized. While it would be impossible self-causation if a supervening event changed its own present physical basis, it is not impossible that supervening events, such as mental information, could bias future particle paths. How might this work in the brain? The key pattern in the brain to which neurons respond is temporal coincidence of arriving action potentials from other neurons. A neuron will only fire if it receives a certain number of coincident inputs from other neurons. Criterial causation occurs where physical criteria imposed by synaptic weights on coincident inputs in turn realize informational criteria for firing. This permits information to be downwardly causal regarding which indeterministic events at the rootmost level will be realized; only those rootmost physical causal chains that meet physically realized informational criteria can drive a postsynaptic neuron to fire and thus become causal at the level of information processing. Typically, the only thing that the set of all possible rootmost physical causal chains that meet those criteria have in common is that they meet the informational criteria set. To try to cut information out of the causal picture here is a common but serious mistake; the only way to understand why it is that just this subset of possible physical causal chains—namely, those that are also informational causal chains—can occur is to understand that it is informational criteria that dictate that class of possible outcomes.

The information that will be realized when a neuron's criteria for firing have been met is already implicit in the set of synaptic weights that impose physical criteria for firing that in turn realize informational criteria for firing. That is, the information is already implicit in these weights before any inputs arrive, just as what sound your mouth will make is implicit in its shape before vibrating air is passed through it. Assuming indeterminism, many combinations of possible particle paths can satisfy given physical criteria, and many more cannot. The subset that can satisfy the physical criteria needed to make a neuron fire is also the subset that can satisfy the informational criteria for firing (such as "is a face") that those synaptic weights realize. So sets of possible paths that are open to indeterministic elementary particles which do not also realize an informational causal chain are in essence "deselected" by synaptic settings by virtue of the failure of those sets of paths to meet physical/informational criteria for the release of a neural spike. A neural code based on informational reparameterization of subsequent neural firing affords the possibility of top-down causation because an informational command such as "think of a woman politician with red hair," whether externally heard or internally generated by executive processes, can reparameterize subsequent physical neural activity such that the result is, randomly within those parameters, Angela Merkel or, equivalently, Margaret Thatcher.

5. BETWEEN DETERMINISM AND RANDOMNESS

Let us return to Hume. Way back in 1739, he wrote "'tis impossible to admit of any medium betwixt chance and an absolute necessity." Many other philosophers have seen no middle path to free will between the equally "unfree" extremes of determinism and randomness. They have either concluded that free will does not exist or tried to argue that a weak version of free will, namely, "freedom from coercion," is compatible with determinism. A weak free will, where events could not have turned out otherwise than they were destined to turn out since the beginning of a deterministic universe, is by definition compatible with determinism in that our determined decisions are uncoerced while certainly playing a causal role in our subsequent actions. But a strong free will, where events could have turned out otherwise, is incompatible with determinism because only given an ontological indeterminism can events really have turned out otherwise than they did. Indeterminism is a necessary condition of a strong free will, such as a type-1 libertarian free will, or a stronger free will, such as a type-2 libertarian free will or meta-free will. Thus compatibilism holds regarding weak free will while incompatibilism holds regarding strong free will and meta-free will. A lot of confusion occurs because compatibilists and incompatibilists are talking past each other, assuming conflicting notions of free will.

A Humean freedom from coercion offers a "weak" conception of free will that is by definition compatible with determinism since no mention is made of a need for outcomes to have the possibility of having turned out otherwise. A libertarian conception of free will, according to which events really might have turned out otherwise, however, is not compatible with either determined or random choices because, in the determined case there are no alternative outcomes so events could never turn out otherwise, while in the random case what happens does not happen because it was willed. A libertarian free will requires meeting four high demands: beings with strong free will (1) must have information processing circuits that have multiple courses of physical or mental activity open to them; (2) they must really be able to choose among them; (3) they must be or must have been able to have chosen otherwise once they have chosen; and (4) the choice must not be dictated by randomness alone, but by the informational parameters realized in those circuits. (Although an agent is not needed at the stage that settles which option will happen, one is needed at the stage that settles that this set of criteria will be set rather than others). This is a tough bill to fill since it seems to require that acts of free will involve acts of self-causation. I argue that these conditions for a libertarian free will are realized in the nervous system. We have no choice but to have a libertarian free will because evolution fashioned our nervous systems to have it. Those animals that had a nervous system that realized a libertarian free will survived to the point of procreation better than those that did not.

Criterial causation offers a middle path between the two extremes of determinism and randomness that Hume was not in a position to see; namely, that physically realized informational criteria parameterize what class of neural activity can be causal of subsequent neural events. The information that meets preset physical/informational criteria may be random to a degree, but it must meet those preset informational criteria if it is to lead to neural firing, so it is not utterly random. Preceding brain activity specifies the range of possible random outcomes to include only those that meet preset informational criteria for firing. Thus volitionally present informational parameterization of future firing is causal in the universe because it is a special subclass of that information that is causal in the universe. Such information is causal in the universe, and not epiphenomenal, by virtue of allowing only that subset of possible futures open at the particle level to become real which also realize informational causal chains. These are those that are consistent with the informational parameters or criteria that were preset by prior neural firing in present neural synaptic weights that realize informational constraints on allowable triggers of future neural firing.

6. HOW BRAIN/MIND EVENTS CAN TURN OUT OTHERWISE

The key mechanism, I argue, whereby atomic level indeterminism has its effects on macroscopic neural behavior is that it introduces randomness in spike timing. There is no need for bizarre notions such as consciousness collapsing wave packets or any other strange quantum effects beyond this. For example, as described in detail in *The Neural Basis of Free Will*, quantum-level noise expressed at the level of individual atoms, such as the single magnesium atoms that block N-methyl-D-aspartic acid (NMDA) receptors, is amplified to the level of randomness and near chaos (criticality domain) in neural and neural circuit spiking behavior. A single photon can even trigger neural firing in the retina in a stunning example of amplification from the quantum to macroscopic domains. The brain evolved to harness such "noise" for information processing ends. Since the system is organized around coincidence detection, where spike coincidences (simultaneous arrival of spikes) are key triggers of informational realization (i.e., making neurons fire that are tuned to particular informational criteria), randomizing which incoming spike coincidences might meet a neuron's criteria for firing means informational parameters can be met in multiple ways just by chance.

7. SKIRTING SELF-CAUSATION

A synaptic account of the neural code also gets around some thorny problems of self-causation that have been used to argue against the possibility of mental causation. The traditional argument is that a mental event realized in neural event x cannot change x because this would entail impossible self-causation. Criterial causation gets around this "no *causa sui* argument" by granting that present self-causation is impossible. But it allows neurons to alter the physical realization of possible *future* mental events in a way that escapes the problem of self-causation of the mental upon the physical. Mental causation is crucially about setting synaptic weights. These serve as the physical grounds for the informational parameters that must be met by unpredictable future mental events realized in unpredictable future spike inputs to a neuron that will fire or not depending on whether those physically realized informational parameters were met or not.

8. VOLUNTARY ATTENTION AND FREE WILL

I argue that the core circuits underlying free choice involve frontoparietal and default mode circuits that facilitate deliberation among options that are represented and manipulated in executive working memory areas. Playing out scenarios internally as virtual experience allows a superthreshold option to be chosen before specific motoric actions are planned. The chosen option can best meet criteria held in working memory, constrained by conditions of various evaluative circuits, including reward, emotional, and cognitive circuits. This process also harnesses synaptic- and, ultimately, atomic-level randomness to foster the generation of novel and unforeseeable satisfactions of those criteria. Once criteria are met, executive circuits can alter synaptic weights on other circuits that will implement a planned operation or action. For example, someone—say, one of the Wright brothers—can imagine different flying machines and then, after much deliberation, go and build an airplane that will transform the physical universe.

9. A NEW THEORY OF QUALIA

Paradigmatic cases of volitional mental control of behavior include voluntary attentional manipulation of representations in working memory and the voluntary attentional tracking of one or a few objects among numerous otherwise identical perceived objects. If there is a flock of indistinguishable birds, for example, there is nothing about any individual bird that makes it more salient. But with volitional attention, any bird can be marked and kept track of. This salience is not driven by anything in the stimulus. It is voluntarily imposed on bottom-up information and can lead to eventual motoric acts, such as shooting or pointing at the tracked bird. This leads to viewing the neural basis of attention and consciousness as not only realized in part in rapid synaptic reweighting, but also in particular patterns of spikes that serve as higher level units that traverse neural circuits (aside for neuroscientists: what I call the "NMDA channel of communication," commonly associated with gamma and high-gamma power in EEG data). Qualia are necessary for volitional mental causation because they are the only informational format available to volitional attentional operations. Actions that follow volitional attentional operations, such as volitional tracking, cannot happen without consciousness. Qualia on this account are a "precompiled" informational format made available to attentional selection and operations by earlier, unconscious information processing. The relationships between qualia, free will, working memory, and volitional attention are therefore very intimate. The domain of qualia is the domain of representations that either are now being volitionally attended or that could be so attended in the next moment in light of current criteria held in working memory.

10. THE HUMAN BRAIN REALIZES META-FREE WILL

Even a tiger would have the kind of "first-order or type-1 libertarian free will" afforded by the kind of nervous system summarized in the preceding points in that a tiger can choose among considered actions freely, and those choices/actions could have turned out otherwise. But only if present choices can ultimately lead to a chooser turning into a new kind of chooser—that is, only if there is second-order or type-2 libertarian free will or a meta-free will—do brains have the capacity to both have chosen otherwise and to have meta-chosen otherwise. Only such a meta-free will allows a brain to not only choose among options available to it now, but to cultivate and create new types of options for itself in the future that are not presently open to it. Only then can there be responsibility for having chosen to become a certain kind of person who chooses from among actions consistent with being that kind of person. Thus, in addition to meeting the four conditions that must be met by a strong free will or first-order libertarian free will, discussed earlier, a second-order or type-2 libertarian free will, or meta-free will, must meet an additional fifth condition: (5) present choices must trigger actions that, after perhaps long durations of training or practice, ultimately lead to the reconformation of the nervous system such that a brain or chooser can decide to train itself to become a new kind of brain or chooser in the future. The human brain can choose to become a new kind of brain in the future, with new choices open to it then that may not be open to it now. This is possible because of a slower kind of neural plasticity, rooted in long-term potentiation of synaptic weights that lead to the reconformation of neural circuits. For example, one can choose to learn Chinese, and then, within a year, become a brain that can adequately process and produce Chinese inputs. A tiger might have type-1 libertarian free will, but it lacks type-2. No tiger thinks to itself, "next year I would like to be a different kind of tiger." This is why animals are amoral despite having type-1 freedom of choice and action. Unlike humans, although they can choose, they cannot choose to become a new kind of chooser. Humans, in contrast, bear a degree of responsibility for having chosen to become the kind of chooser who they now are.

Incompatibilists like Kane write about two different kinds of freedom of will, both of which are incompatible with determinism by definition. The first type of libertarian free will requires that one could have done otherwise. This is, by definition, not possible under determinism barring exotic philosophical moves like granting agents the capacity to change the past or the laws of physics, which most scientists, I suspect, would argue no one could actually accomplish. Choosers can only have chosen otherwise if some physical events themselves could have turned out otherwise. This makes indeterminism a necessary condition for the truth of incompatibilism. But the reality of indeterminism is not sufficient because indeterminism might turn out to be true, yet an agent could lack the freedom to shape future events; unless the agent plays some role in defining which chance events can happen, chance just happens to the agent randomly. The freedom afforded by a nervous system that met the four conditions of a libertarian free will, such as might be realized in the brain of a tiger or other nonhuman vertebrate, would be what Kane might call "freedom of action." But this would not be what Kane would mean by a truly free libertarian free will. For that, an agent would need type-2 incompatibilist free will or meta-free will: the capacity to choose to become a new kind of chooser in the future. According to this notion of meta-free will, the agent must have freedom of action that allows the agentic shaping of the nature of future volitional decisions. In effect, the agent must be able to shape the basis or grounds of future volitional decisions. The agent must be able to volitionally choose the kind of volitional being she will become in the future. This seems like a tall order, but really the seeds of this idea were already discussed in Aristotle's *Nicomachean Ethics*. According to his virtue ethics, a person's decisions and the moral consequences of those decisions flow from the fact that a person has made past choices that have cultivated them (or not) into being the kind of person they are now, in particular, by having reached a point where certain kinds of—ideally virtuous—decisions have been automatized because of habit formation.

To reiterate then, there are at least two kinds of free will demanded by incompatibilists. One is what Kane regards as freedom of action, and the other is what might be regarded as the freedom to choose one's capacity to choose, not right now, but in the perhaps distant future after much cultivation or training. This type-2 incompatibilist free will or meta-free will requires the capacity to develop one's nervous system, or the character realized in it, in an intended way. This is the freedom to choose what kind of chooser one will become weeks, months, or years down the line.

According to Kane, ultimate responsibility for an action and its immediate consequences requires responsibility for anything that is a sufficient cause or motive of that action. Thus, if an action follows necessarily from having a certain character (which we can use as a shorthand notion for the conglomeration of desires, values, tendencies, capacities, and principles that one has), then to be ultimately responsible, even if in part, one must be in part responsible for the character that one has. This hearkens back to Aristotle's insight, in the *Nicomachean Ethics*, that, in order to be in part responsible for the wicked acts that follow from having a wicked character, one must be in part responsible for having the wicked character that one has.

Kane recognizes that there would be a logical regress if we chose our present character based on the character we had in the past, which we in turn chose based on the character we had in the more distant pass, and so on, until we came into existence. To escape this regress, Kane argues that there must be a break in the chain of sufficient causes. This happens when a character-forming decision is made that is not sufficiently caused by one's present character or state. This, Kane argues, requires that this "self-forming" decision is not determined because if it were, it would be sufficiently caused by the preceding state of affairs. Only under indeterminism is an outcome not sufficiently caused because identical causes can lead to various different effects or outcomes just by chance. But if a self-forming act or decision were utterly random, then our resulting new character would not have been one that we willed. It would instead be one that just happened by chance. We might make a random self-forming decision and find that suddenly we are a murderer for no reason at all.

To get around this, Kane, in *Four Views on Free Will* (2007), and Balaguer, in *Free will as an Open Scientific Question* (2009), zoom in on a very special class of decisions where people are torn between two options, both of which they have willed. These "torn decisions" might happen only rarely in a person's life, but unless at least one such self-forming act happened in their life, they could not be even partly ultimately responsible for anything they decide or do on the basis of their present character because of the preceding regress. To break sufficiency, at least these kinds of decisions have to be undetermined, even if every other act or decision in a person's life is determined.

Kane writes: "undetermined self-forming actions . . . occur at those difficult times in life when we are torn between competing visions of what we should do or become. . . . yet the outcome can be willed (and hence rational and voluntary) either way [that we decide] owing to the fact that in such self-formation,

[our] prior wills are divided by conflicting motives. . . . When we . . . decide in such circumstances, and the indeterminate efforts we are making become determinate choices, we make one set of competing reasons or motives prevail over the others then and there by deciding" (2007: 26).

Even though which way we will decide is not determined, since either choice is one we want and have reasons for, Kane argues that it is willed because it reflects our purposes and intentions and is not utterly random. Indeterminism in essence selects among a class of options, each of which some part of us has agentically specified as one "we" (note the need for a divided self here) want and intend to do. No matter how we decide by chance, we succeed in doing what at least one part of us knowingly and willingly was trying to do. Note that the reasons and motives for the two options have to be different for this to work because otherwise we are like Buridan's ass choosing between options that are equivalent to us (see Balaguer 2009: 74–75; it is unlikely that Buridan's ass kinds of decisions will transform our character, so reduce to type-1 libertarian free will choices). Kane considers, as a prototypical example of a torn decision, a businesswoman who wants to help someone out of empathy, but who also wants to get to an important meeting because of ambition, and she cannot do both because the clock is ticking. She must choose, either/or.

I think Kane and Balaguer do not need to try to rest the possibility of the existence of a type-2 libertarian free will or a meta-free will on such rare events as torn decisions or the notion of a divided self with conflicting subagendas or subdesires, even if this is a valid way of grounding them. All that is required to ground both type-1 and type-2 libertarian free will is that options reflect our reasons and motives and that the option selected is undetermined.

This is where criterial causation can help. To ground a type-1 libertarian free will, it is enough to specify some criterion such as "I need an escape route." Many possibilities can be generated, and one, just by chance, can be selected, whether in our brains or the brain of a tiger. The selected escape route is not utterly random because it had to be an escape route. But it is not determined because a different escape route might have been selected by chance. This is what Kane regards as freedom of action. Kane thinks we need more than this to have true free will because, unless we can intentionally reshape our characters, we are subject to the preceding regress.

To ground a type-2 libertarian free will, we can imagine, given the constraints imposed by our present character, the kind of future character that we want to achieve. This then sets criteria for the, in part, random fulfillment of those character-defining criteria. For example, in light of our present character, which, let us say, is disgusted by our present lack of integrity in late December, we might set a criterion that we need to resolve to increase our level of integrity somehow. Now, in part just by chance, we might resolve to be more honest and set that as our New Year's resolution.

But we could have just as easily decided, again by chance, to be more kind, or to keep our word better, or to be less impetuous, or less jealous, or less greedy, or more reliable, or to henceforth be a better friend, or be more organized. Let us say, just by chance, the possible resolution to be more punctual arose from unconscious information processing to the level of consciousness for consideration of adequacy. It might then be rejected as not adequately satisfying the criterion of increasing our integrity. Then a new possible resolution might come to the fore of consciousness, perhaps from unconscious processing—say, to keep our word better in the future. Let's say that this passes the threshold, after some conscious consideration, for meeting our integrity-enhancing criteria. So why did our New Year's resolution turn out to be "to be more honest" rather than "to be less selfish and greedy?" Well, just by chance, the first one passed our criteria for integrity enhancement first and became a self-forming resolution. This is not an utterly random resolution because it had to be an integrity-enhancing resolution. But it is not determined because it could easily have turned out otherwise, even though we could not change our character at the moment of threshold crossing. However, once a resolution is in place, in part just by chance, it will shape our actions come January, and, over several months, we may be able to accomplish the improved character we envisioned. Note that because we chose this resolution for self-improvement in part just by chance, we did not settle our resolution (i.e., determine which resolution we settle on); what we did settle were the criteria that possible self-forming resolutions would have to meet, and an adequate one passed threshold first, which we then went with. Type-2 libertarian free will is a slow and bootstrapping goal-directed sorites-like process because changing one's character takes effort and time.

Criterial causation can ground both type-1 and type-2 libertarian free will and is not rare, like torn decisions. We are constantly engaging in this kind of criterial decision-making at multiple levels and also constantly correcting deviations from our paths to our envisioned goals in a cybernetic process involving feedback. This is why both contingency and goal-directed agency are everywhere you look in human action and decision-making.

11. CRITERIAL CAUSATION OVERCOMES THE LUCK ARGUMENT AGAINST MORAL RESPONSIBILITY

In the previous section, I considered an alternative to the Kane-Balaguer strategy of trying to ground type-2 libertarian free will on a foundation of undetermined torn decisions. Balaguer thinks that it is an open empirical question whether torn decisions are made in the brain in a way that is undetermined. I think my account of neural criterial causation, if correct about how the brain works, would give his or Kane's related theories just the sort of empirical account they need to say that we are libertarian-free.

The Kane-Balaguer strategy of grounding libertarian free will in undetermined torn decisions is vulnerable to the criticism that it fails to overcome the argument from luck according to which, if a critical moral choice goes one way versus another due to randomness (say, amplification of quantum fluctuations to neural spike timing variability), then we cannot hold people responsible for the consequences that follow from that choice.

The argument from luck runs as follows: "If decisions or actions occur indeterministically, such that two or more alternative decisions might be made at t, each with a non-zero probability, and everything is exactly the same in world history until t, then there is nothing about the world or the decider prior to t that accounts for one decision being made over the other. Which gets chosen is just a matter of (perhaps weighted) chance, not a matter of agentic influence on specific outcomes. If one decision should turn out better or worse than another, well, that is just a matter of luck and not a matter of agentic choice. But if decisions and consequences are just a matter of luck, then the decider cannot be responsible for the decision made or for its consequences." Note that if the argument from luck works at all, it works not only against libertarians, but against everyone, including compatibilists; see Pereboom (2001, 2014), Levy (2011), and Caruso (2012, 2015).

Note that for Kane and Balaguer the important indeterminacy happens at the moment of choice, not before it. This means that people cannot bias the chance outcome toward the morally superior choice with their wills because, even if they could, their decision to bias one way versus the other would itself be subject to the argument from luck.

Neil Levy (2011) in *Hard Luck* argues that free will is ruled out or precluded by luck, regardless of the truth of determinism or indeterminism, making him a hard incompatibilist like Pereboom or Caruso, and a denier of free will. This entails a denial that anyone bears any moral responsibility whatsoever. According to Levy's account, even though Hitler chose to systematically annihilate all Jews he could find, he did so just by luck, whether because of inherited wicked character (constitutive luck, so not his fault) or because of present luck when deciding between options (so again, not his fault). As such, I find Levy's denial of moral responsibility a profoundly nihilistic view of human beings, their choices, and life in general.

Those who deny that information can be causal at all because they accept Kim's EA must accept that informational mental events such as willings cannot be causal either. For them, free will and moral responsibility vanish along with the disappearance of mental causation. So free will and moral responsibility vanish for Levy in at least two ways, via luck and via the idea that all causation seeps down to lowest level of physical causation, leaving no room for mental events like willings to make any difference to physical outcomes.

Levy's attack on libertarian free will is rooted in the traditional failure, he says, of libertarians to offer a contrastive account of choices (2011: 43, 90). Consider van Inwagen's (1983) potential thief pondering whether to steal from the church's poor box. He is in a classic torn state before the decision is made. He is torn between the motive to have money and the motive to honor a deathbed promise he made to his mother to live morally. Levy points out that libertarians fail to offer any explanation concerning why the thief decides to steal rather than not steal. It just happens. And had the decision, by chance, gone the other way, libertarians could not explain why the man refrained from stealing rather than steal. This allows Levy to say that (1) the character we start off with is a matter of (constitutive) luck, so we are not responsible for it or the choices, acts, and consequences that follow because of our character that was "foisted" on us by genetics, the environment, and their interaction; and (2) any so-called "self-forming act" or choice comes down to (present) luck, so we are also not responsible for it or its consequences either. He calls this his "luck pincer." It is really just a variant of the logical regress against the possibility of ultimate responsibility summarized in Section 10. This was of course the regress that drove Kane and Balaguer and other proponents of libertarian free will to rely on torn decisions in the first place. Because of luck, Levy says that torn decisions fail to afford moral responsibility because all choices come down to constitutive luck, present luck, or both.

Note that in Kane's and Balaguer's groundings of libertarian free will in undetermined torn decisions there is no higher governing basis for making a choice in a torn decision like van Inwagen's thief's; deciding to steal and then stealing the cash just happen. Had he decided not to steal, then not-stealing would have just happened. Levy's point is that if things just happen—and this is almost a Buddhist perspective—there can be no blame. Shit just happens.

The kind of self-forming New Year's resolution I considered in Section 10 would not be solely a matter of luck because it would have to meet the integrity-enhancing criteria set in place by the agent. Even though this resolution might have been chosen versus many possible others, it had to be one that met those criteria, so was not utterly random, so was not solely a matter of blind luck. In contrast with Kane and Balaguer's accounts of libertarian free will, in the case of criterial decision-making, there is a higher but non-determinative governing basis for making a choice. Yes, that the resolution ended up being "to be more honest this year" rather than "to be less greedy this year" was a matter of luck in the sense that the first proposal passed the threshold for adequate satisfaction of integrity-enhancing criteria first. But it is not an utterly random outcome, like choosing to steal the money or not, as in the thief's torn decision, or choosing to help someone in need or go to the meeting, as in Kane's businesswoman example of a torn decision. Under criterial causation, the choice is not utterly random because it had to be an integrity-enhancing resolution. It is also not determined, breaking the chain of sufficient causes that underlies the ultimate responsibility-destroying regress, because a different integrity-enhancing resolution might have won out. The regress is broken by adding indeterminism. But the luck argument is broken by forcing any choice or action to meet criteria set by the agent him- or herself. Kane's and Balaguer's accounts are vulnerable to the luck argument because there is no basis for choosing one self-forming path or another; one set of motives is randomly favored over the other. One half of our divided self wins over the other just because. On my account, the integrity-enhancing criteria specified by the agent imply that whatever resolution ends up being chosen was willed and not simply a matter of luck because the decision or choice had to meet the agent's self-forming criteria. Because agents play this criterial role in their self-forming decisions and continually adjust such criteria in a cybernetic or feedback-based process over years of self-formation (am I in fact getting closer to the envisioned future me?) agents are in part responsible for developing negative or positive characters over years of development. Yes, there is randomness in terms of which resolution will win, but there is not randomness at the level of the basis for choosing one option over another, as in Kane and Balaguer's undetermined torn decisions.

Concerning type-1 libertarian free will, we are in part responsible for our actions because we set these criteria versus others that we did not set. And we set these versus others because of the kind of agent who we are. We are not completely responsible because we are not responsible for the particular way those criteria were met (say we chose this escape route versus another) because this was a matter of chance or luck. So we are responsible for choosing an escape route although not fully responsible for the particularities of choosing this escape route versus others that we might have picked had it not proved adequate first.

Similarly, concerning type-2 libertarian free will or meta-free will, we are in part responsible for our characters because we set these criteria for self-forming resolutions versus others that we did not set. And we set these versus others because of the kind of agent who we are. But we are not completely responsible for our characters because the initial characters or capacities we inherited were a matter of constitutive luck, and the particular way in which the criteria that we set were met (say we chose to be more honest in the new year, rather than less greedy) was a matter of present luck. So we are responsible for choosing to make a New Year's resolution to improve our character, though not fully responsible for the particularities of choosing this character-forming resolution versus others that we might have picked had it not proved adequate first.

But even if we are only in part responsible for our actions and characters, criterial causation offers both a grounding for libertarian free will of both types 1 and 2 and a degree, therefore, of moral responsibility.

12. CONCLUSION

Assuming indeterminism, it is possible to be a physicalist who adheres to a libertarian conception of free will. On this view, mental and brain events really can turn out otherwise, yet are not utterly random. Prior neuronally realized information parameterizes what subsequent neuronally realized informational states will pass presently set physical/informational criteria for firing. This does not mean that we are utterly free to choose what we want to want. Some wants and criteria are innate, such as what smells good or bad. However, given a set of such innate parameters, the brain can generate and play out options, then select an option that adequately meets criteria or generate further options. This process is closely tied to voluntary attentional manipulation in working memory, more commonly thought of as deliberation or imagination.

Imagination is where the action is in free will. It allows animals not only to consider possible courses of present action (type-1 libertarian free will), but also, at least for the case of humans, it allows us to consider what kinds of choosers we want to strive to become (type-2 libertarian free will). It allows us to imagine learning this language or that, and then, once a choice has been made to become, with effort and practice, a new kind of nervous system that can eventually speak that language. It allows us to lay in bed and imagine flying machines, then go build one that we have imagined and thereby change the physical universe forever. It allows us to imagine a better self that we can then set about realizing through practice. And that future self will be able to make new kinds of choices that are not yet open to us now.

REFERENCES

Balaguer, M. 2009. *Free will as an open scientific problem*. Cambridge, MA: MIT Press.
Caruso, G. D. 2012. *Free will and consciousness: A determinist account of the illusion of free will*. Lanham, MD: Lexington Books.
Caruso, G. D. 2015. Kane is not able: A reply to Vicens' "Self-Forming Actions and Conflicts of Intention." *Southwest Philosophy Review* 31(2): 21–26.
Dowe, P. 1992. Wesley Salmon's process theory of causality and the conserved quantity theory. *Philosophy of Science* 59: 195–216.
Hume, D. 1739. *A treatise on human nature*. London: Clarendon Press.
Kane, R. 2007. Libertarianism. In J. M. Fischer, R. Kane, D. Pereboom, and M. Vargas (Eds.), *Four views on free will*, pp. 5–43. Blackwell Publishing.
Kim, J. 1993. The non-reductivist's troubles with mental causation. In J. Heil and A. Mele (Eds.), *Mental causation*, pp. 189–210. Oxford: Oxford University Press.
Kim, J. 1993. *Supervenience and mind: Selected philosophical essays*. Cambridge, MA University Press.
Kim, J. 1996. *Philosophy of mind*. Boulder, CO: Westview Press.

Kim, J. 2005. *Physicalism, or something near enough*. Princeton, NJ: Princeton University Press.

Levy, N. 2011. *Hard luck: How luck undermines free will and moral responsibility*. New York: Oxford University Press.

Papineau, D. 2009. The causal closure of the physical and naturalism. In A. Beckermann, B. P. McLaughlin, and S. Walter (Eds.), *The Oxford handbook of philosophy of mind*, pp. 53–65.

Pearl, J. 2000. *Causality*. New York: Cambridge University Press.

Pereboom, D. 2001. *Living without free will*. Cambridge, UK: Cambridge University Press.

Pereboom, D. 2014. *Free will, agency, and meaning in life*. Oxford: Oxford University Press.

Tse, P. U. 2013. *The neural basis of free will: Criterial causation*. Cambridge, MA: MIT Press.

Van Inwagen, P. 1983. *An essay on free will*. New York: Oxford University Press.

Woodward, J. 2003. *Making things happen: A theory of causal explanation*. Oxford: Oxford University Press.

PART III
Free Will, Moral Responsibility, and Meaning in Life

PART II

CHAPTER 11

Hard-Incompatibilist Existentialism

Neuroscience, Punishment, and Meaning in Life

DERK PEREBOOM AND GREGG D. CARUSO

As philosophical and scientific arguments for free will skepticism continue to gain traction, we are likely to see a fundamental shift in the way people think about free will and moral responsibility. Such shifts raise important practical and existential concerns: What if we came to disbelieve in free will? What would this mean for our interpersonal relationships, society, morality, meaning, and the law? What would it do to our standing as human beings? Would it cause nihilism and despair, as some maintain, or would it rather have a humanizing effect on our practices and policies, freeing us from the negative effects of belief in free will? In this chapter, we consider the practical implications of free will skepticism and argue that life without free will and basic desert moral responsibility would not be as destructive as many people believe. We argue that prospects of finding meaning in life or of sustaining good interpersonal relationships, for example, would not be threatened. On treatment of criminals, we argue that although retributivism and severe punishment, such as the death penalty, would be ruled out, preventive detention and rehabilitation programs would still be justified. While we will touch on all these issues herein, our focus will be primarily on this last issue.

We begin in Section 1 by considering two different routes to free will skepticism. The first denies the causal efficacy of the types of willing required for free will and receives its contemporary impetus from pioneering work in neuroscience by Benjamin Libet, Daniel Wegner, and John Dylan Haynes. The second, which is more common in the philosophical literature, does not deny the causal efficacy of the will but instead claims that whether this causal efficacy is deterministic or indeterministic, it does not achieve the level of control to count as free will by the standards of the historical debate. We argue that while there are compelling objections to the first route (e.g., Al Mele [2009], Eddy Nahmias [2002, 2011], and Neil Levy [2005]), the second route to free will skepticism remains intact. In Section 2, we argue that free will skepticism allows for a workable morality and, rather than negatively impacting our personal relationships and meaning in life, may well improve our well-being and our relationships to others since it would tend to eradicate an often destructive form of moral anger. In Section 3, we argue that free will skepticism allows for adequate ways of responding to criminal behavior—in particular, incapacitation, rehabilitation, and alternation of relevant social conditions—and that these methods are both morally justified and sufficient for good social policy. We present and defend our own preferred model for dealing with dangerous criminals, an incapacitation account built on the right to self-protection analogous to the justification for quarantine (see Pereboom 2001, 2013, 2014a; Caruso 2016a), and we respond to recent objections to it by Michael Corrado and John Lemos.

1. TWO DIFFERENT ROUTES TO FREE WILL SKEPTICISM

In the historical debate, the variety of free will that is of central philosophical and practical importance is the sort required for moral responsibility in a particular but pervasive sense. This sense of moral responsibility is set apart by the notion of *basic desert* (Caruso and Morris 2017; Feinberg 1970; Fischer 2007; Pereboom 2001, 2014a; G. Strawson 1994) and is purely backward-looking and nonconsequentialist. For an agent to be morally responsible for an action in this sense is for it to be hers in such a way that she would deserve to be blamed if she understood that it was morally wrong, and she would deserve to be praised if she understood that it was morally exemplary. The desert at issue here is *basic* in the sense that the agent would deserve to be blamed or praised just because she has performed the action, given an understanding of its moral status, and not, for example, merely by virtue of consequentialist or contractualist considerations (see Pereboom 2001, 2014a).

Free will skeptics reject this sort of moral responsibility. Rejecting basic desert moral responsibility, however, still leaves other senses intact. For instance, forward-looking accounts of moral responsibility would not be threatened (Pereboom 2014a), nor would the *answerability* sense of moral responsibility defended by Thomas Scanlon (1998) and Hilary Bok (1998). When we encounter apparently immoral behavior, for example, it is perfectly legitimate to ask the agent, "Why did you decide to do that?" or "Do you think it was the right thing to do?" If the reasons given in response to such questions are morally unsatisfactory, we regard it as justified to invite the agent to evaluate critically what his actions indicate about his intentions and character, to demand apology, or to request reform. Engaging in such interactions is reasonable in light of the right of those harmed or threatened to protect themselves from immoral behavior and its consequences. In addition, we might have a stake in reconciliation with the wrongdoer, and calling him to account in this way can function as a step toward realizing this objective. We also have an interest in his moral formation, and the address described naturally functions as a stage in this process (Pereboom 2012). The thesis of free will skepticism should therefore be understood as the claim that what we do, and the way we are, is ultimately the result of factors beyond our control, and, because of this, we are never morally responsible for our actions in the *basic desert* sense, not in these other senses.

In the literature, two prominent routes to free will skepticism are identifiable. The first, which is more prominent among scientific skeptics, maintains that recent findings in neuroscience reveal that unconscious brain activity causally initiates action prior to the conscious awareness of the intention to act and that this indicates conscious will is an illusion (e.g., Benjamin Libet, John-Dylan Haynes, Daniel Wegner). The pioneering work in this area was done by Benjamin Libet and his colleagues. In their groundbreaking study on the neuroscience of movement, Libet et al. (1983) investigated the timing of brain processes and compared them to the timing of consciousness will in relation to self-initiated voluntary acts and found that the conscious intention to move (which they labeled W) came 200 milliseconds before the motor act, but 350–400 milliseconds after the *readiness potential* (RP)—a ramplike buildup of electrical activity that occurs in the brain and precedes actual movement. Libet and others have interpreted this as showing that the conscious intention or decision to move cannot be the cause of action because it comes too late in the neuropsychological sequence (see Libet 1985, 1999). According to Libet, since we become aware of an intention to act only after the onset of preparatory brain activity, the conscious intention cannot be the true cause of the action (see also Haggard and Eimer 1999; Obhi and Haggard 2004; Pockett 2004; Roediger, Goode, and Zaromb 2008; Soon et al. 2008; Wegner 2002).

Libet's findings, in conjunction with additional findings by John Dylan Haynes (Soon et al. 2008) and others, have led some theorists to conclude that conscious will is an illusion and plays no important causal role in how we act. Haynes and his colleagues, for example, were able to build on Libet's work by using functional magnetic resonance imaging (fMRI) to predict with 60 percent accuracy whether subjects would press a button with either their right or left hand up to ten seconds before the subject became aware of having made that choice (Soon et al. 2008). For some, the findings of Libet and Haynes are enough to threaten our conception of ourselves as free and responsible agents since they appear to undermine the causal efficacy of the types of willing required for free will. We contend, however, that there are at least three reasons for thinking that these neuroscientific arguments for free will skepticism are unsuccessful.[1]

First, there is no direct way to tell which conscious phenomena, if any, correspond to which neural events. In particular, in the Libet studies, it is difficult to determine what the RP corresponds to—for example, is it an intention formation or decision, or is it merely an urge of some sort? Al Mele (2009) has argued that the RP that precedes action by a half-second or more need not be construed as the cause of the action. Instead, it may simply mark the beginning of forming an *intention* to act. According to Mele, "it is much more likely that what emerges around −500 ms is a *potential cause* of a proximal intention or decision than a proximal intention or decision itself" (2009: 51). On this interpretation, the RP is more accurately characterized as an "urge" to act or a preparation to act. That is, it is more accurately characterized as the advent of items in what Mele calls the *preproximal-intention group* (or PPG). We agree with Mele that this leaves open the possibility that conscious intentions can still be causes—that is, if the RP does not correspond to the formation of an intention or decision, but rather an urge, then it remains open that the intention formation or decision is a conscious event.

Second, almost everyone on the contemporary scene who believes we have free will, whether compatibilist or libertarian, also maintains that freely willed actions are caused by virtue of a chain of events that stretch backward in time indefinitely. At some point in time, these events will be such that the agent is not conscious of them. Thus, all free actions are caused, at some point in time, by unconscious events. However, as Eddy Nahmias (2011) correctly points out, the concern for free will raised by Libet's work is that *all* of the relevant causing of action is (typically) nonconscious, and consciousness is not causally efficacious in producing action. Given determinist compatibilism, however, it is not possible to establish this conclusion by showing that nonconscious events that precede conscious choice causally determine action since such compatibilists hold that every case of action will feature such events and that this is compatible with free will. And, given most incompatibilist libertarianisms, it is also impossible to establish this conclusion by showing that there are nonconscious events that render actions more probable than not by a factor of 10 percent chance (Soon et al. 2008) since almost all such libertarians hold that free will is compatible with such indeterminist causation by unconscious events at some point in the causal chain (De Caro 2011).

Furthermore, Neil Levy raises a related objection when he criticizes Libet's *impossible demand* (2005) that only consciously initiated actions could be free. Levy correctly argues that this presupposition places a condition upon

---

[1]. Some of the criticisms to follow were first made in Bjornsson and Pereboom (2014).

freedom of action which is in principle impossible to fill for reasons that are entirely conceptual and have nothing to do, per se, with Libet's empirical findings. As Levy notes, "Exercising this kind of control would require that we control our control system, which would simply cause the same problem to arise at a higher-level or initiate an infinite regress of controllings" (2005: 67). If the unconscious initiation of actions is incompatible with control over them, then free will is impossible on conceptual grounds. Thus, Libet's experiments do not constitute a separate, empirical challenge to our freedom (see Levy 2005).

Finally, several critics have correctly noted the unusual nature of the Libet-style experimental situation—that is, one in which a conscious intention to flex at some time in the near future is already in place, and what is tested for is the specific implementation of this general decision. Nahmias (2011), for example, convincingly points out that it is often the case—when, for instance, we drive or play sports or cook meals—that we form a conscious intention to perform an action of a general sort, and subsequent specific implementation are not preceded by more specific conscious intentions. But, in such cases, the general conscious intention is very plausibly playing a key causal role. In Libet-style situations, when the instructions are given, subjects form conscious intentions to flex at some time or other, and, if it turns out that the specific implementations of these general intentions are not in fact preceded by specific conscious intentions, this would be just like the kinds of driving and cooking cases Nahmias cites. It seems that these objections cast serious doubts on the potential for neuroscientific studies to undermine the claim that we have the sort of free will at issue.

Before moving on to the second route to free will skepticism, it is worth quickly noting that there are other scientific threats to free will in addition to those posed by neuroscience. Recent work in psychology and social psychology on *automaticity, situationism*, and the *adaptive unconscious*, for instance, has shown that the causes that move us are often less transparent to ourselves than we might assume—diverging in many cases from the conscious reasons we provide to explain and/or justify our actions (e.g., Bargh 1997, 2008; Bargh and Chartrand 1999; Bargh and Ferguson 2000; Doris 2002; Nisbett and Wilson 1977; Wilson 2002). These findings reveal just how wide open our internal psychological processes are to the influence of external stimuli and events in our immediate environment, without knowledge or awareness of such influence. They also reveal the extent to which our decisions and behaviors are driven by implicit biases (see Greenwald, McGhee, and Schwartz 1998; Kang et al. 2010; Nosek et al. 2007; Uhlmann and Cohen 2005). No longer is it believed that only "lower level" or "dumb" processes can be carried out nonconsciously. We now know that the higher mental processes that have traditionally served as quintessential examples of "free will"—such as evaluation and judgment, reasoning and problem-solving, and interpersonal behavior—can and often do occur in the absence of conscious choice and guidance (Bargh and Ferguson 2000: 926; see also Wilson 2002).

While these findings would not be enough, on their own, to establish global skepticism about free will and basic desert moral responsibility (see Levy 2014), they represent a potential threat to our everyday folk understanding of ourselves as conscious, rational, responsible agents since they indicate that the conscious mind exercises less control over our behavior than we have traditionally assumed. Even some compatibilists now admit that because of these findings "free will is at best an occasional phenomenon" (Baumeister 2008: 17; see also Nahmias). This is an important concession because it acknowledges that the *threat of shrinking agency*—as Thomas Nadelhoffer (2011) calls it—remains a serious one independent of the neuroscientific concerns just discussed. The deflationary view of consciousness which emerges from these empirical findings, including the fact that we often lack transparent awareness of our true motivational states, is potentially agency-undermining and could shrink the realm of morally responsible action (see Caruso 2012, 2015, 2016b; King and Carruthers 2012; Levy 2014; Nadelhoffer 2011; Sie and Wouters 2010). It is important therefore that accounts of moral responsibility that require, for instance, reasons-responsiveness or evaluation of personal-level attitudes (including beliefs, commitments, and goals) make explicit the role they see consciousness playing and the extent to which automaticity, situationism, and implicit bias may limit or restrict morally responsible behavior. For our purposes, however, we are going to table these concerns for the remainder of this chapter to focus on a second route to free will skepticism—one, which we maintain, is more successful at establishing a global skepticism about free will and basic desert moral responsibility.

In the past, the standard argument for free will skepticism was based on the notion of *determinism*—the thesis that every event or action, including human action, is the inevitable result of preceding events and actions and the laws of nature. *Hard determinists* argued that determinism is true and incompatible with free will and basic desert moral responsibility, either because it precludes the *ability to do otherwise* (leeway incompatibilism) or because it is inconsistent with one's being the "ultimate source" of action (source incompatibilism). While hard determinism had its classic statement in the time when Newtonian physics reigned, it has very few defenders today—largely because the standard interpretation of quantum mechanics has been taken by many to undermine, or at least throw into doubt, the thesis of universal determinism. We nonetheless maintain that even if you allow some indeterminacy to exist at the microlevel of our existence—the level studied by quantum mechanics—the sort of free will at issue in the historical debate would still be threatened. Our view differs from hard determinism, then, in that it maintains that, whatever the fundamental nature of reality, we would still lack free will. A more accurate name for our position would therefore be *harm incompatibilism* (see Pereboom 2001, 2014a), to differentiate it from hard determinism. Hard incompatibilism does not deny the causal efficacy of the will but instead claims that whether this causal efficacy is deterministic *or* indeterministic, it does not achieve the level of control required for basic desert moral responsibility.

Hard incompatibilism amounts to a rejection of both compatibilism and libertarianism. It maintains that the sort of free will required for basic desert moral responsibility is incompatible with causal determination by factors beyond the agent's control and *also* with the kind of indeterminacy in action required by the most plausible versions of libertarianism. Against the view that free will is compatible with the causal determination of our actions by natural factors beyond our control, we argue that there is no relevant difference between this prospect and our actions being causally determined by manipulators (see Pereboom 2001, 2014a). Against event causal libertarianism, we advance the disappearing agent objection, according to which agents are left unable to settle whether a decision occurs and hence cannot have the control required for moral responsibility (Caruso 2012; Pereboom 2001, 2014a). The same problem, we contend, arises for noncausal libertarian accounts, which also fail to provide agents with the control in action required for basic desert moral responsibility. While agent-causal libertarianism could, in theory, supply this sort of control, we argue that it cannot be reconciled with our best physical theories (Pereboom 2001, 2014a) and faces additional problems accounting for mental causation (Caruso 2012). Since this exhausts the options for views on which we have the sort of free will at issue, we conclude that free will skepticism is the only remaining position.

Since the arguments for hard incompatibilism have been spelled out and defended at great length elsewhere (see, e.g., Caruso 2012, 2014; Pereboom 2001, 2014a, 2014b), and no solid refutations of them have yet been offered (cf. Fischer 2014; Nelkin 2014; for a reply Pereboom 2014b), we will not elaborate on them further here. Instead, will now shift our attention to exploring the practical implications of free will skepticism. For many, it is not the philosophical arguments for free will skepticism that are the problem, it is the existential angst they create and the fear that relinquishing belief in free will and basic desert moral responsibility would undermine morality, negatively affect our interpersonal relationships, and leave us unable to adequately deal with criminal behavior. To these concerns we now turn.

2. MORAL AND PERSONAL IMPLICATIONS

If the argument for free will skepticism is convincing, one can conclude that we lack the sort of free will required for moral responsibility in the basic desert sense. The concern for the skeptical position is not that there is considerable empirical evidence that it is false or that there is a challenging argument for its incoherence. The main question it raises is instead practical: Can we live with the belief that it is true? A number of free will skeptics, including Honderich (1988), Pereboom (1995, 2001, 2014a), Levy (2011), and Caruso (2012, 2017a, forthcoming) argue that we, in fact, can. We will begin by briefly addressing two practical issues. The first concerns the extent to which the skeptic can retain our ordinary conception of morality and responsibility, the second the degree to which it coheres with the emotions required for the kinds of personal relationships we value. We will then discuss at length the implications of the view for treatment of criminals.

2.1. Free Will Skepticism, Morality, and Responsibility

Accepting free will skepticism requires rejecting our ordinary view of ourselves as blameworthy or praiseworthy in the basic desert sense. A critic might first object that if we gave up this belief, we could no longer count actions as morally bad or good. In response, even if we came to hold that a serial killer was not blameworthy due to a degenerative brain disease, we could still justifiably agree that his actions are morally bad. Still, secondly, the critic might ask, if determinism precluded basic desert blameworthiness, would it not also undercut judgments of moral obligation? If "ought" implies "can," and if because determinism is true an agent could not have avoided acting badly, it would be false that she ought to have acted otherwise. Furthermore, if an action is wrong for an agent just in case she is morally obligated not to perform it, determinism would also undermine judgments of moral wrongness (Haji 1998). In response, we contend that even if the skeptic were to accept all of this (and she might resist at various points; cf. Pereboom 2014a: ch. 6; Waller 2011), axiological judgments of moral goodness and badness would not be affected (Haji 1998; Pereboom 2001). So, in general, free will skepticism can accommodate judgments of moral goodness and badness, which are arguably sufficient for moral practice.

Third, the critic might object that if we stopped considering agents as blameworthy in the basic desert sense, we would be left with insufficient resources for addressing immoral behavior (e.g., Nichols 2007). However, the skeptic might turn instead to other senses of moral responsibility that have not been a focus of the free will debate (Pereboom 2013; 2014a: ch. 6). Our moral practice features a number of senses of moral responsibility, some of which do not invoke basic desert. For instance, when we encounter immoral action, we might ask the agent to consider what his actions indicate about his intentions and character, to demand apology, or to request reform, thereby having him consider reasons to behave differently in the future. Engaging in such interactions counts as reasonable in view of the right of those wronged or threatened by wrongdoing to protect themselves from bad behavior and its consequences. Our practice also features an interest in the wrongdoer's moral formation, and the address described naturally functions as a step in this process. Moreover, our practice also has a stake in our reconciliation with the wrongdoer, and calling him to account plausibly serves as a stage in securing this aim. Such interactions, because they address the agent's capacity to consider and respond to reasons, manifest respect for her as a rational agent. The main thread of the historical free will debate does not pose causal determination as a challenge to this sense of moral responsibility, and thus this is an aspect of our practice that the free will skeptic can endorse.

2.2. Personal Relationships and Meaning in Life

Is the assumption that we are morally responsible in the basic desert sense required for the sorts of personal relationships we value? The considerations raised by P. F. Strawson in his essay "Freedom and Resentment" (1962) suggest a positive answer. In his view, our justification for claims of blameworthiness and praiseworthiness is grounded in the system of human reactive attitudes, such as moral resentment, indignation, guilt, and gratitude. Strawson contends that because our moral responsibility practice is grounded in this way, the truth or falsity of causal determinism is not relevant to whether we justifiably hold each other and ourselves morally responsible. Moreover, if causal determinism were true and did threaten these attitudes, as the free will skeptic is apt to maintain, we would face instead the prospect of the cold and calculating objectivity of attitude, a stance that relinquishes the reactive attitudes. In Strawson's view, adopting this stance would rule out the possibility of the meaningful sorts of personal relationships we value.

Strawson may be right to contend that adopting the objective attitude would seriously hinder our personal relationships (for a contrary perspective, see Sommers 2007). However, a case can be made that it would be wrong to claim that this stance would be appropriate if determinism did pose a genuine threat to the reactive attitudes (Pereboom 1995, 2001, 2014a). While, for instance, kinds of moral anger such as resentment and indignation might be undercut if free will skepticism were true, these attitudes may be suboptimal relative to alternative attitudes available to us, such as moral concern, disappointment, sorrow, and moral resolve. The proposal is that the attitudes that we would want to retain either are not undermined by a skeptical conviction because they do not have presuppositions that conflict with this view, or else they have alternatives that are not under threat. And what remains does not amount to Strawson's objectivity of attitude and is sufficient to sustain the personal relationships we value.

Guilt is also imperiled by free will skepticism, and this consequence would seem to be more difficult to accommodate. The skeptic's view stands to undercut guilt because it would seem to involve the supposition that one is blameworthy in the basic desert sense for an immoral action one has performed. There is much at stake here, the critic might contend, because, absent guilt, we would not be motivated to moral improvement after acting badly, and we would be kept from reconciliation in impaired relationships. In addition, the critic continues, because guilt is undermined by the skeptical view, repentance is also ruled out because feeling guilty is a prerequisite for a repentant attitude. In response, suppose instead you acknowledge that you have acted immorally, and, as Bruce Waller advocates, you feel deep sorrow for what you have done (Waller 1990: 165–166; cf. Bok 1998); as a result, you are motivated to eradicate your disposition to behave in this bad way. This response can secure the good that guilt can also secure, and it is wholly compatible with the free will skeptic's view.

Gratitude arguably presupposes that the person to whom one is grateful is praiseworthy in the basic desert sense for a beneficial act (cf. Honderich 1988: 518–519). But even if this is so, certain aspects of gratitude would not be undercut, and these aspects would seem to provide what is required for the personal relationships we value. Gratitude involves being thankful toward the person who has acted beneficially. This aspect of gratitude is in the clear; one can be thankful to a young child for some kindness without supposing that she is praiseworthy in the basic desert sense. Gratitude typically also involves joy as a response to what someone has done, and free will skepticism does not yield a challenge to being joyful and expressing joy when others act beneficially.

Perhaps some of the recommended transformations in emotional attitudes may not be possible for us. For example, in certain situations, refraining from moral anger may be beyond our power, and thus even the committed skeptic might not be able to make the change the skeptical view suggests. At this point Shaun Nichols (2007) invokes the distinction between *narrow-profile* emotional responses—local or immediate emotional reactions to situations—and *wide-profile* responses, which are not immediate and involve rational reflection. We might expect to be unable to appreciably reduce narrow-profile moral anger as an immediate reaction upon being deeply hurt in an intimate personal relationship. In wide-profile cases, however, diminishing or even eliminating moral anger is open, or at least disavowing it in the sense of rejecting any force it may be assumed to have in justifying a harmful response to wrongdoing. This modification of moral anger might well be advantageous for our valuable personal relationships, and it stands to bring about the equanimity that Spinoza thought free will skepticism, more generally, would secure.

3. FREE WILL SKEPTICISM AND CRIMINAL BEHAVIOR

One of the most frequently voiced criticisms of free will skepticism is that it is unable to adequately deal with criminal behavior and that the responses it would permit as justified are insufficient for acceptable social policy. This concern is fueled by two factors. The first is that one of the most prominent justifications for punishing criminals, retributivism, is incompatible with free will skepticism. The second is that alternative justifications that are not ruled out by the skeptical view per se face significant independent moral objections. Yet, despite this concern, we maintain that free will skepticism leaves intact other ways to respond to criminal behavior—in particular incapacitation, rehabilitation, and alteration of relevant social conditions—and that these methods are both morally justifiable and sufficient for good social policy. In this section, we present and defend our preferred model for dealing with dangerous criminals, an incapacitation account built on the right to self-protection analogous to the justification for quarantine (see Caruso 2016a, 2017b; Pereboom 2001, 2013, 2014a), and respond to objections to it by John Lemos (2016) and Michael Corrado (2016).

To begin, we need to recognize that retributive punishment is incompatible with free will skepticism because it maintains that punishment of a wrongdoer is justified for the reason that he *deserves* something bad to happen to him just because he has knowingly done wrong—this could include pain, deprivation, or death. As Douglas Husak puts it, "Punishment is justified only when and to the extent it is deserved" (2000: 82). And Mitchell Berman writes, "A person who unjustifiably and inexcusably causes or risks harm to others or to significant social interests deserves to suffer for that choice, and he deserves to suffer in proportion to the extent to which his regard or concern for others falls short of what is properly demanded of him" (2008: 269). Furthermore, for the retributivist, it is the *basic* desert attached to the criminal's immoral action alone that provides the justification for punishment. The desert the retributivist invokes is basic in the sense that justifications for punishment that appeal to it are not reducible to consequentialist considerations nor to goods such as the safety of society or the moral improvement of the criminal.

Free will skepticism undermines this justification for punishment because it does away with the idea of basic desert. If agents do not deserve blame just because they have knowingly done wrong, neither do they deserve punishment just because they have knowingly done wrong. The challenge facing free will skepticism, then, is to explain how we can adequately deal with criminal behavior without the justification provided by retributivism and basic desert. While some critics contend this cannot be done, free will skeptics point out that there are several alternative ways of justifying criminal punishment (and dealing with criminal behavior more generally) that do not appeal to the notion of basic desert and are thus not threatened by free will skepticism.

These include moral education theories, deterrence theories, punishment justified by the right to harm in self-defense, and incapacitation theories. While we maintain the first two approaches face independent moral objections—objections that, though perhaps not devastating, make them less desirable than their alternative—we argue that an incapacitation account built on the right to harm in self-defense provides the best option for justifying a policy for treatment of criminals consistent with free will skepticism. Before turning to our positive account, let us briefly say something about the first two alternative approaches.

Moral education theories draw an analogy with justification of the punishment of children. Children are typically not punished to exact retribution, but rather to educate them morally. Since moral education is a generally acceptable goal, a justification for criminal punishment based on this analogy is one the free will skeptic can potentially accept. Despite its consistency with free will skepticism, though, a serious concern for this type of theory is that it is far from evident that punishing adult criminals is similarly likely to result in moral improvement. Children and adult criminals differ in significant respects. For example, adult criminals, unlike children, typically understand the moral code accepted in their society. Furthermore, children are generally more psychologically malleable than are adult criminals. For these and other reasons, we see this approach as less desirable than an alternative incapacitation account (see Pereboom 2014a: ch. 7).

Deterrence theories, especially utilitarian deterrence theories, have probably been the most discussed alternative to retributivism. According to deterrence theories, the prevention of criminal wrongdoing serves as the good on the basis of which punishment is justified. The classic deterrence theory is Jeremy Bentham's. In his conception, the state's policy on criminal behavior should aim at maximizing utility, and punishment is legitimately administered if and only if it does so. The pain or unhappiness produced by punishment results from the restriction on freedom that ensues from the threat of punishment, the anticipation of punishment by the person who has been sentenced, the pain of actual punishment, and the sympathetic pain felt by others such as the friends and family of the criminal (Bentham 1823/1948). The most significant pleasure or happiness that results from punishment derives from the security of those who benefit from its capacity to deter.

While deterrence theories are completely compatible with free will skepticism, there are three general moral objections against them. The first is that they will justify punishments that are intuitively too severe. For example, it would seem that, in certain cases, harsh punishment would be a more effective deterrent than milder forms, while the harsh punishments are intuitively too severe to be fair. The second concern is that such accounts would seem to justify punishing the innocent. If, for instance, after a series of horrible crimes the actual perpetrator is not caught, potential criminals might come to believe that they can get away with serious wrongdoing. Under such circumstances it might maximize utility to frame and punish an innocent person. Last, there is the "use" objection, which is a problem for utilitarianism more generally. Utilitarianism sometimes requires people to be harmed severely, without their consent, in order to benefit others, and this is often intuitively wrong. While some skeptics believe these objections can be met, we recommend that free will skeptics seek a different alternative to retributivism.

There is, however, a legitimate theory for prevention of especially dangerous crime that is neither undercut by free will skepticism nor by other moral considerations. This theory is based on an analogy with quarantine and draws on a comparison between treatment of dangerous criminals and treatment of carriers of dangerous diseases. The free will skeptic claims that criminals are not morally responsible for their actions in the basic desert sense. Plainly, many carriers of dangerous diseases are not responsible in this or in any other sense for having contracted these diseases. Yet we generally agree that it is sometimes permissible to quarantine them, and the justification for doing so is the right to self-protection and the prevention of harm to others. For similar justificatory reasons, we argue, even if a dangerous criminal is not morally responsible for his crimes in the basic desert sense (perhaps because no one is ever in this way morally responsible) it could be as legitimate to preventatively detain him as to quarantine the nonresponsible carrier of a serious communicable disease.

One might justify both quarantine in the case of disease and incapacitation of dangerous criminals on purely utilitarian or consequentialist grounds. But we want to resist this strategy. Instead, on our view, incapacitation of the dangerous is justified on the ground of the right to harm in self-defense and defense of others. That we have this right has broad appeal—much broader than utilitarianism or consequentialism has. In addition, this makes the view more resilient to objection, as will become clear in what follows.

It is important to see that this analogy places several constraints on the treatment of criminals. First, as less dangerous diseases justify only preventative measures less restrictive than quarantine, so less dangerous criminal tendencies justify only more moderate restraints. In fact, for certain minor crimes perhaps only some degree of monitoring could be defended. Second, the incapacitation account that results from this analogy demands a degree of concern for the rehabilitation and well-being of the criminal that would alter much of current practice. Just as fairness recommends that we seek to cure the diseased we quarantine, so fairness would counsel that we attempt to rehabilitate the criminals we detain (cf. D'Angelo 1968: 56–59). Third, if a criminal cannot be rehabilitated, and our safety requires his indefinite confinement, this account provides no justification for making his life more miserable than would be required to guard against the danger he poses. Finally, there are measures for preventing crime more generally, such as providing for adequate education and mental health care, which the free will skeptic can readily endorse.

We contend that this account provides a more resilient proposal for justifying criminal sanctions than either the moral education or deterrence theories. One advantage this approach has over the utilitarian deterrence theory is that it has more restrictions placed on it with regard to using people merely as a means. For instance, as it is illegitimate to treat carriers of a disease more harmfully than is necessary to neutralize the danger they pose, treating those with violent criminal tendencies more harshly than is required to protect society will be illegitimate as well (Pereboom 2001, 2013, 2014a). Our account therefore maintains the *principle of least infringement*, which holds that the least restrictive measures should be taken to protect public health and safety. This ensures that criminal sanctions will be proportionate to the danger posed by an individual, and any sanctions that exceed this upper bound will be unjustified. Furthermore, the less dangerous the disease, the less invasive the justified prevention methods would be, and similarly, the less dangerous the criminal, the less invasive the justified forms of incapacitation would be.

In addition to these restrictions on harsh and unnecessary treatment, our account also advocates for a broader approach to criminal behavior that moves beyond the narrow focus on sanctions. Consider, for example, the recent proposal by Caruso (2016a, 2017b) to place the quarantine analogy within the broad justificatory framework of *public health ethics*. Public health ethics not only justifies quarantining carriers of infectious diseases on the grounds that it is necessary to protect public health, it also requires that we take active steps to *prevent* such outbreaks from occurring in the first place. Quarantine is only needed when the public health system fails in its primary function. Since no system is perfect, quarantine will likely be needed for the foreseeable future, but it should *not* be the primary means of dealing with public health. The analogous claim holds for incapacitation. Taking a public health approach to criminal behavior would allow us to justify the incapacitation of dangerous criminals when needed, but it would also make prevention a *primary function* of the criminal justice system. If we care about public health and safety, the focus should always be on preventing crime from occurring in the first place by addressing the systemic causes of crime. Prevention is always preferable to incapacitation.

Furthermore, public health ethics sees *social justice* as a foundational cornerstone to public health and safety (Caruso 2017b). In public health ethics, a failure on the part of public health institutions to ensure the social conditions necessary to achieve a sufficient level of health is considered a grave injustice. An important task of public health ethics, then, is to identify which inequalities in health are the most egregious and thus which should be given the highest priority in public health policy and practice. The public health approach to criminal behavior likewise maintains that a core moral function of the criminal justice system is to identify and remedy social and economic inequalities responsible for crime. Just as public health is negatively affected by poverty, racism, and systematic inequality, so, too, is public safety. This broader approach to criminal justice therefore places issues of social justice at the forefront. It sees racism, sexism, poverty, and systemic disadvantage as serious threats to public safety, and it prioritizes the reduction of such inequalities (see Caruso 2017b).

Summarizing our account, then, the core idea is that the right to harm in self-defense and defense of others justifies incapacitating the criminally dangerous with the minimum harm required for adequate protection. The resulting account would not justify the sort of criminal punishment whose legitimacy is most dubious, such as death or confinement in the most common kinds of prisons in our society. Our account also specifies attention to the well-being of criminals, which would change much of current policy. Furthermore, free will skeptics would continue to endorse measures for reducing crime that aim at altering social conditions, such as improving education, increasing opportunities for fulfilling employment, and enhancing care for the mentally ill. This combined approach to dealing with criminal behavior, we argue, is sufficient for dealing with dangerous criminals, leads to a more humane and effective social policy, and is actually preferable to the harsh and often excessive forms of punishment that typically come with retributivism.

Michael Corrado raises three objections to this incapacitation account, which leads him to reject it in favor of a compromise view, which he calls Correction. This position, while denying basic desert moral responsibility, endorses hard treatment of reasons-responsive criminals on the ground of moral educational benefit to the criminal and deterrence of future crime. Corrado's first objection is that our view, unlike his, makes no distinction between people who are dangerous and yet have the sort of control captured by the reasons-responsiveness condition, and those who are dangerous but lack this sort of control, and instead treats all criminals on the model of illness. The second is that, given our view, too many people will be drawn into the criminal justice system since merely posing a danger is sufficient to make one a candidate for incapacitation. The third objection is that those who are incapacitated would need to be compensated, and this would be prohibitively costly.

On the first concern, in *Living Without Free Will,* Pereboom distinguished his position from views according to which criminal tendencies are exclusively psychological illnesses, modeled on physical illness (Pereboom 2001). It is true that on our view policies for making a detained criminal safe for release would address a condition in the offender that results in the criminal behavior. But such conditions are not restricted to psychological illnesses; they also include conditions that are not plausibly classified as illness, such as insufficient sympathy for others or a strong tendency to assign blame to others and not to oneself when something goes wrong. What unites policies for treatment of criminals on our view is not that they assume that they are psychologically ill and therefore in need of psychiatric treatment. Instead, they all aim to bring about moral change in an offender by nonpunitively addressing conditions that underlie criminal behavior.

What sets the illness model apart is that proposed treatment does not address the criminal's capacity to respond to reasons, but circumvents such capacities. For example, consider the Ludovico method, made famous by Anthony Burgess's book and Stanley Kubrick's film *A Clockwork Orange*. Alex, a violent criminal, is injected with a drug that makes him nauseous while at the same time he is made to watch films depicting the kind of violence to which he is disposed. The goal of the method is that the violent behavior be eliminated by generating an association between violence and nausea. Herbert Morris's objection to therapy of this sort is that the criminal is not changed by being presented with reasons for altering his behavior which he would autonomously and rationally accept. But Pereboom (2001) cites a number of programs for treating criminals that are not in accord with the illness model. The Oregon Learning Center, for instance, aims to train parents and families to formulate clear rules, monitor behavior, and to set out fair and consistent procedures for establishing positive and negative incentives. The method involves presentation of reasons for acting and strategies for realizing aims in accord with these reasons. This program is successful: in one study, youth in ten families showed reductions of 60 percent in aggressive behavior compared to a 15 percent drop in untreated control families.[2]

Pereboom also cites therapeutic programs designed to address problems for the offender's cognitive functioning. A number of cognitive therapy programs are inspired by S. Yochelson and S. Samenow's influential work *The Criminal Personality* (1976), which argues that certain kinds of cognitive distortions generate and sustain criminal behavior. Kris Henning and Christopher Frueh provide some examples of such cognitive distortions:

> Car thieves would be more likely to continue with their antisocial activities if they reasoned that *stealing cars isn't as bad as robbing people* (minimization of offense) or I deserve to make a couple of bucks after all the cops put me through last time (taking the role of the victim). Similarly, a rapist who convinces himself, she shouldn't have been wearing that dress if she didn't want me to touch her (denial of responsibility), would probably be at greater risk to reoffend than someone who accepts responsibility for his actions. (1996: 525)

---

2. Patterson, Chamberlain, and Reid (1982), cited in Walters (1992: 143). Cf. Patterson (1982), Alexander and Parsons (1982). For a review of studies on family therapy, see Gendreau and Ross (1979).

In 1988, the state of Vermont put in place a therapeutic program inspired by the Yochelson and Samenow's cognitive distortion model. The Cognitive Self-Change Program was initially designed as group treatment for imprisoned male offenders with a history of interpersonal aggression, and it later included imprisoned nonviolent offenders. Henning and Frueh provide a description of the procedure:

> Treatment groups met 3–5 times per week. During each session, a single offender was identified to present a "thinking report" to the group. Typically, these reports documented prior incidents of anti-social behavior, although more current incidents were reported on when appropriate. At the beginning of each session, the offender would provide the group with an objective description of the incident. He would then list all of the thoughts and feelings he had before, during, and after the event. After the report was delivered, the group worked with the offender to identify the cognitive distortions that may have precipitated the antisocial response to the situation. Role plays sometimes were used during these sessions to develop a better understanding of the cognitions and emotions that led up to the offender's behavior. Once an offender learned to identify his primary criminogenic thought patterns, intervention strategies were discussed in the group to help him prevent such distortions from occurring in the future. These might include cognitive strategies (e.g. challenging one's cognitions, cognitive redirection) and/or behavioral interventions (e.g. avoidance of high-risk situations; discussion of cognitions and feelings with therapist, friend, or partner). (1996: 525)

Henning and Frueh found that in a group of 28 who had participated in this program, 50 percent (14) were charged with a new crime following their release. In a control group of 96 who had not participated, 70.8 percent (68) were charged with a new offense. Twenty-five percent of offenders who had participated received a new criminal charge within one year, 38 percent within two years, and 46 percent within three. By contrast, in the comparison group, 46 percent had been charged with a new crime within one year, 67 percent within two, and 75 percent within three. These results were found to be statistically significant.

Models of *restorative justice* proved another alternative for rehabilitating criminals in a way that respects the reasons-responsiveness of agents. It also has the additional benefit of addressing the rights of victims by having the criminal admit the wrong done, acknowledge the harm caused, and agree to work toward reconciliation with the victim or the victim's families. Models of restorative justice are perfectly consistent with free will skepticism as long as they are employed in a way that does not appeal to backward-looking blame in the restorative process. Consider, for instance, the recent success of schools in using restorative methods as an alternative to school suspension.[3] In traditional school discipline programs, students face an escalating scale of punishment for infractions that can ultimately lead to expulsion. There is now strong research, however, that shows pulling students out of class as punishment can hurt their long-term academic prospects (Losen et al. 2015; Losen, Martinez, and Okelola 2014; Richmond 2015). Furthermore, data show that punishments are often distributed unequally. More black students, for example, are suspended nationally than white students (Richmond 2015; US Department of Education Office for Civil Rights 2014).

As an alternative, public schools from Maine to Oregon have begun to employ restorative justice programs designed to keep students in school while addressing infractions in a way that benefits both the offender and the offended. Here is one description of such a program:

> Lower-level offenses can be redirected to the justice committee, which is made up of student mediators, with school administrators and teachers serving as advisors. The goal is to provide a nonconfrontational forum for students to talk through their problems, addressing their underlying reasons for their own behaviors, and make amends both to individuals who have been affected as well as to the larger school community. (Richmond 2015)

Students are often given the option of participating in these alternative programs or accepting traditional discipline, including suspension. As reported on in *The Atlantic*, "Early adopters of the practice report dramatic declines in school-discipline problems, as well as improved climates on campuses and even gains in student achievement" (Richmond 2015). Programs like this reveal that the more punitive option—for example, expulsion rather than restorative processes—is often less effective from the perspective of future protection, future reconciliation, and future moral formation. They also reveal how rehabilitating individuals can be done in a way that appeals directly to their reasons-responsive capacities.

Contrary to Corrado's concerns, then, we maintain that methods of therapy that engage reasons-responsive abilities should be preferred. On the forward-looking account of moral responsibility we endorse (Pereboom 2013; 2014a: ch. 6), when we call an agent to account for immoral behavior, at the stage of moral address we request an explanation with the intent of having the agent acknowledge a disposition to act badly, and then, if she has in fact so acted without excuse or justification, we aim for her to come to see that the disposition issuing in the action is best eliminated. In normal cases, this

---

3. See "Alternative to School Suspension Explored Through Restorative Justice" (Associated Press), December 17, 2014; "When Restorative Justice in Schools Works" (*The Atlantic*), December 29, 2015 (Richmond 2015).

change is produced by way of the agent's recognition of moral reasons to eliminate the disposition. Accordingly, it is an agent's responsiveness to reasons—together with the fact that we have a moral interest in our protection, her moral formation, and our reconciliation with her—that explains why she is an appropriate recipient of blame in this forward-looking sense. While many compatibilists see some type of attunement to reasons as the key condition for basic desert moral responsibility, we instead view it as the most significant condition for a notion of responsibility that focuses on future protection, future reconciliation, and future moral formation.

Still, a concern for many forms of therapy proposed for altering criminal tendencies is that they circumvent, rather than address, the criminal's capacity to respond to reasons. On our view, forms of treatment that do address reasons-responsiveness are to be preferred. However, the fact that a mode of therapy circumvents rather than addresses the capacities that confer dignity on us should not all by itself make it illegitimate for agents who are in general responsive to reasons but not in particular respects. Imagine such an agent who is beset by bouts of violent anger that he cannot control in some pertinent sense. Certain studies suggest that this tendency is due to deficiencies in serotonin and that it can sometimes be alleviated by antidepressants.[4] It would seem mistaken to claim that such a mode of treatment is illegitimate because it circumvents capacities for rational and autonomous response. In fact, this sort of treatment often produces responsiveness to reasons where it was previously absent (Pereboom 2001). A person beset by violent anger will typically not be responsive to certain kinds of reasons to which he would be responsive if he were not suffering from this problem. Therapy of this sort can thus increase reasons-responsiveness. By analogy, one standard form of treatment for alcoholism—which many alcoholics voluntarily undergo—involves the use of a drug, Antabuse, which makes one violently ill after the ingestion of alcohol. By counteracting addictive alcoholism, this drug can result in enhanced reasons-responsiveness.

Furthermore, suppose that despite serious attempts at moral rehabilitation that do not circumvent the criminal's rational capacities, and despite procedures that mechanically increase the agent's capacities for reasons-responsiveness, the criminal still displays dangerously violent tendencies. Imagine that the choice is now between indefinite confinement without hope for release and behavioristic therapy that does not increase the agent's capacity for reasons-responsiveness. It is not obvious that here the behavioristic therapy should be ruled out as morally illegitimate. One must assess the appropriateness of therapy of this kind by comparing it with the other options. Suppose, for example, that the only legitimate alternative to confinement for life is application of some behavioristic therapy. It is not clear that,

---

4. Burlington Free Press (Associated Press), December 15, 1997, p. 1.

under such circumstances, the moral problems with such a therapy are not outweighed—especially if it is carried out in a way that respects autonomy by leaving the decision up to the criminal.

Behavioristic therapies, however, are almost always suboptimal when compared with their alternatives—that is, methods that directly appeal to a criminal's rational capacities or, when these fail, therapies that mechanically increase the agent's capacities for reasons-responsiveness. There are also additional alternatives to behavioral therapy that, at least in the future, may prove more successful in rehabilitating criminals. The use of neurofeedback, for instance, in correctional settings has been suggested as "an innovative approach that may ultimately lessen criminal behavior, prevent violence, and lower recidivism" (Gkotsi and Benaroyo 2012: 3; see also Evans 2006; Quirk 1995; Smith and Sams 2005). As Gkotsi and Benaroyo describe:

> Neurofeedback or neurotherapy is a relatively new, noninvasive method which is based on the possibility of training and adjusting the speed of brainwaves, which normally occur at various frequencies (Hammond, 2011). An overabundance, or deficiency in one of these frequencies, often correlates with conditions such as depression, and emotional disturbances and learning disabilities, such as Attention Deficit Hyperactivity Disorder (ADHD) (Greteman, 2009).... Therapists attach electrodes to the patients' head and a device records electrical impulses in the brain. These impulses are sorted into different types of brain waves. Using a program similar to a computer game, patients learn to control the video display by achieving the mental state that produces increases in the desired brain wave activity. Neurofeedback has gained recognition for its potential benefits for children with ADHD, alcoholics and drug addicts. It can also enhance athlete and musician performance as well as improve elderly people's cognitive function (Greteman 2009)

Douglas Quirk, a Canadian researcher, tested the effects of a neurofeedback treatment program on seventy-seven dangerous offenders in an Ontario correctional institute who suffered from deep-brain epileptic activity. The results demonstrated reduction in the subjects' criminal recidivism and suggested that "a subgroup of dangerous offenders can be identified, understood and successfully treated using this kind of biofeedback conditioning program" (Quirk 1995; as quoted by Gkotsi and Benaroyo 2012: 3). Additional studies by Smith and Sams (2005) on juvenile offenders with significant psychopathology and electroencephalographic abnormalities, and by Martin and Johnson (2005) on male adolescents diagnosed with ADHD also demonstrated reduced recidivism, improved cognitive performance, improved emotional and behavioral reactions, and inhibition of inappropriate responses.

More invasive than neurofeedback is another potential treatment: deep brain stimulation (DBS). DBS has been used as a last-resort treatment of neuropsychological disorders including schizophrenia, Parkinson's disease, dystonia, Tourette's syndrome, pain, depression, and obsessive compulsive disorder. It involves the surgical placement of a device in the brain that sends electrical impulses to target areas that have been linked to the particular condition. Some neurologists and neuroscientists have recently proposed that DBS can be used for the rehabilitation of criminal psychopaths (Center for Science and Law 2012; Hoeprich 2011).

Since there are very few options currently available for the effective rehabilitation of psychopaths, which often leaves continued incapacitation as society's sole means to protection, some have argued that DBS may provide a better and more effective alternative (see Hoeprich 2011). As the Center for Science and Law describe:

> Psychopaths have been shown to have neurophysiological deficiencies in various brain structures compared to healthy human subjects. These structures include the amygdala (an important center for processing of emotionally-charged and stimulus-reward situations) and the ventromedial prefrontal cortex (suppression of emotional reactions and decision-making). DBS can potentially be used, then, in these areas to see if psychopathic tendencies can be suppressed. (2012)

There are, however, important ethical concerns with regard to the use of DBS for rehabilitating psychopaths (see Gkotsi and Benaroyo 2012), especially since it is highly experimental, with many reported negative side effects, and is far more invasive than neurofeedback, which is generally believed to be a fairly safe procedure.

We propose, then, that rehabilitation methods that directly appeal to a criminal's rational capacities should always be preferred and attempted first. When these fail, we contend that it is sometimes acceptable to employ therapies that mechanically increase an agent's capacities for reasons-responsiveness but that these therapies should involve the participation of the subject to the greatest extent possible (e.g., talk therapies in conjunction with other forms of treatment), should involve the consent of the subject, and should be ordered such that noninvasive methods are prioritized. When all else fails and only more invasive methods are left—for example, DBS for psychopaths—important ethical questions need to be considered and answers weighed, but leaving the final choice up to the subject is an attractive option.

3.1. The Scope Issue

Corrado's second objection is that too many people will be drawn into the criminal justice system on our account. First, Corrado intimates that many more people would be detained than is the case currently. Second, there is the issue of incapacitating those who pose threats but have not yet committed a crime. Corrado is reasonably concerned about the prospects of such a policy.

On the first issue, in all of our writings on this topic we have in effect advocated the principle of least infringement, which specifies that the least restrictive measures should be taken to protect public health and safety. While we do believe that we should indefinitely detain mass murderers and serial rapists who cannot be rehabilitated and remain threats, we do not believe that nonviolent shoplifters who remain threats and cannot be rehabilitated should be preventatively detained at all, by contrast with being monitored, for example. Our view does not prescribe that all dangerous people be detained until they are no longer dangerous. Certain kinds of persisting threats can be dealt with by monitoring, in contrast with detention. Moreover, other behavior that is currently considered criminal might not require incapacitation at all. Our view is consistent, for example, with the decriminalization of nonviolent behavior such as recreational drug use and thus is consistent with many fewer people being detained than in the United States currently.

In addition to monitoring and decriminalization, monetary fines could also serve as suitable sanctions for low-level crimes. When someone fails to heed a stop sign, for example, they put at risk the potential safety of others. The right of self-protection and the prevention of harm to others justify liberty-limiting laws backed by the threat of sanctions, but the sanctions in this case would need to be significantly low since our account prohibits treating individuals more harshly than is required to protect society. Just as it is illegitimate to treat carriers of a disease more harmfully than is necessary to neutralize the danger they pose, treating criminals more harshly than is required to protect society will be illegitimate as well. A forwarding-looking conception of moral responsibility grounding in future protection and moral formation could justify a suitable fine here, but not more punitive measures. Such small infractions are analogous to common colds. While they do put at risk the health of others, the harm they represent is not significant enough to justify quarantine. Of course, with regard to running a stop sign, we might want to distinguish between first offense and habitual behavior since per incident risk is probably low but aggregates to a high probability of serious harms. Perhaps, then, we could justify increased sanctions over time for repeat offenders, including higher fines and eventually loss of one's drivers license.

On Corrado's second issue, the incapacitation of the dangerous who haven't committed a crime, on our view there are several moral reasons that count against such a policy. As Ferdinand Schoeman (1979) has argued, and Caruso has stressed (2016a), the right to liberty must carry weight in this context, as should the concern for using people merely as means. In addition, the risk posed by a state policy that allows for preventative detention of nonoffenders needs to be taken into serious consideration. In a broad range of societies, allowing the state this option stands to result in much more harm than good because misuse would be likely. Schoeman also points out that while the kinds of testing required to determine whether someone is a carrier of a communicable disease may often not be unacceptably invasive, the type of screening necessary for determining whether someone has violent criminal tendencies might well be invasive in respects that raise serious moral issues. Moreover, available psychiatric methods for discerning whether an agent is likely to be a violent criminal are not especially reliable, and, as Stephen Morse points out, detaining someone on the basis of a screening method that frequently yields false positives is seriously morally objectionable (Morse 1999; Nadelhoffer et al. 2012).

However, there is reason to think that impressive neural tests for violent tendencies are being developed (Nadelhoffer et al. 2012). We may in the near future be able to determine with reasonable accuracy on the basis of neural factors whether someone is likely to commit violent crimes. Would our account endorse someone's preventative detention even if he has not yet manifested such violence, on the supposition that the violence would be serious and highly likely in his normal environment and that less invasive measures such as effective monitoring or drug therapy are unavailable? Perhaps it would. But this should not count as a strong objection to our view because virtually everyone would agree that preventative detention of nonoffenders is legitimate under certain possible conditions. Imagine that someone has involuntarily been given a drug that makes it virtually certain that he will brutally murder at least one person during the one-week period he is under its influence. There is no known antidote, and because he is especially strong, mere monitoring would be ineffective. Almost everyone would affirm that it would be at least prima facie permissible to preventatively detain him for the week. Now suppose that reliable neural screening reveals that an agent, if left in his normal environment, is virtually certain to engage in rape and murder in the near future. There is no known viable drug therapy, and mere monitoring would be ineffective. Should he be preventatively detained? Here, it is important to understand that the incapacitation account will specify that the circumstances of such detention would not be harsh and that allowing the agent to be reasonably comfortable and to pursue fulfilling projects would be given high priority. But, even here, there are countervailing moral considerations that must be taken into account. In many societies, the danger of misuse posed by allowing the state to preventatively detain even highly dangerous nonoffenders is a grave concern that stands to outweigh the value of the safety provided by such a policy.

A further worry is raised by Lemos, who argues that if criminals were no more deserving of punishment than noncriminals, it would be unjust to expect criminals alone to bear the burden of violent crime prevention, and this tells in favor of lowering the evidentiary standard for criminality, thereby preventing more crime. The result would involve more people subject to criminal treatment. We offer two responses. The first is that we do not set out our position in a strict consequentialist theoretical context. Rather, we justify incapacitation on the ground of the right to self-defense and defense of others. That right does not extend to people who are non-threats. Thus the aim of protection is justified by a right with clear bounds, and not by a consequentialist theory on which the bounds are unclear. Second, we noted earlier that we endorse a theory of blame and moral responsibility that does not feature basic desert. Its aims are protection, moral formation, and reconciliation. Moral formation and reconciliation are not relevant in the case of the innocent, but they are for those who have done wrong. To the extent that treatment of criminals also aims at moral formation and reconciliation, these considerations also serve to set the innocent apart from the guilty. It is not clear how these various factors weigh out, but if, taking all of them into consideration, evidentiary standards for criminal liability are lowered, they would not be lowered by much. And this is offset by the fact that criminals would not be treated as harshly as they are in our current system.

On the third issue, cost to society, when a person with cholera is quarantined, she is typically made to experience deprivation she does not deserve. Society benefits by this deprivation. It is a matter of fairness that society do what it can, within reasonable bounds, to make the victim safe for release as quickly as possible, and this will have a cost. If we quarantined cholera victims but were unwilling to provide medical care for them because it would require a modest increase in taxation, then we would be acting unfairly. Similarly, when a dangerous agent, whether or not he has already committed a crime, is preventatively detained, then, supposing that the free will skeptic is right, he is made to experience a deprivation he does not fundamentally deserve and from which society benefits. By analogy with the cholera case, here also it is a matter of fairness for us to do what we can, within reasonable bounds, to rehabilitate him and make him safe for release, and this, too, will have a cost. For a society or state to oppose programs for rehabilitation because it is unwilling to fund them would involve serious unfairness. Corrado suggests that this cost would be prohibitively high. It is hard to see why it would be more costly than the current system, which is massively expensive.

Saul Smilansky (2011) sets out a version of this objection, and Corrado expresses approval. This may explain Corrado's concern about cost:

> Hard determinists cannot, however, permit incarceration in institutions of punishment such as those that currently prevail. Instead of punishment, they must opt for funishment. Funishment would resemble punishment in that criminals would be incarcerated apart from lawful society; and institutions of funishment

> would also need to be as secure as current prisons, to prevent criminals from escaping. But here the similarity ends. For institutions of funishment would also need to be as delightful as possible.... Since hard determinism holds that no one deserves the hardship of being separated from regular society, this hardship needs to be compensated for. (Smilansky 2011: 355)

Smilansky then argues that such a policy will be extremely expensive: "the cost of funishment will be incomparably higher than that of punishment," in fact so high that it will be intolerable.

First of all, because the free will skeptic rejects basic desert, no basic desert requirement to compensate those who are preventatively detained will be in effect (cf. Levy 2012; Pereboom 2013, 2014a). The specifics of the skeptic's reply to Smilansky's objection depend on which general moral theory she prefers to adopt. Suppose she endorsed an axiological moral theory which includes better consequences as valuable, where morally fundamental rights being honored and not violated count among the good consequences. Neil Levy (2012) points out that a consequentialist of this sort has a good response to Smilansky: "A consequentialist who is a moral responsibility skeptic will naturally hold that no one should be treated any worse than is needed to bring about the best consequences, with all agents' welfare—including the welfare of criminals—taken into account." So, first, the preventatively detained would not be treated worse than needed to protect against the danger they pose. In addition, the right to live a fulfilling life is in play and weighs heavily, and we would thus have a significant moral interest in providing those who are preventatively detained with the requisite opportunities and conditions. On the issue of cost, providing these sorts of opportunities may add expense to our system for dealing with criminal behavior, but not the expense required to provide all of those detained with "five-star hotel" accommodations (Smilansky 2011). Furthermore, as Levy (2012) contends, "rejecting the notion that some agents deserve punishment opens the way for us to adopt policies that respond to crime at much lower costs, economically, socially and morally." Mark Kleiman (2009) proposes and discusses many such less costly nonretributive policies, and he argues persuasively and in detail that adopting them instead of what we in the United States have in place would lead to highly beneficial consequences.

4. CONCLUSION

We have here considered two different routes to free will skepticism and argued that while the first route (the one that denies the causal efficacy of the types of willing required for free will) fails for a number of reasons, the second route based on hard incompatibilism is sound. We then considered the main concern critics have with this view (i.e., that we cannot live as if it is true) and argued that it can be answered. More specifically, we first argued that there are forward-looking aspects of our practice of holding responsible that don't presuppose basic desert that the skeptic can retain (i.e., those that aim at moral formation, protection, and reconciliation). We further maintained that emotional attitudes that don't implicate basic desert are sufficient for good human relationships. Last, and our primary focus, we argued that a nonretributivist set of practices for treating criminals, which highlight rehabilitation and incapacitation, are in fact workable.

REFERENCES

Alexander, J. F., and B. V. Parsons. 1982. *Functional family therapy*. Monterrey: Brooks/Cole.

Bargh, J. A. 1997. The automaticity of everyday life. In R. S. Wyer, Jr. (Ed.), *The automaticity of everyday life: Advances in social cognition* (vol. 10), pp. 1–61. Mahwah, NJ: Erlbaum.

Bargh, J. A. 2008. Free will is un-natural. In J. Baer, J. C. Kaufman, and R. F. Baumeister (Eds.), *Are we free? Psychology and free will*, pp. 128–154. New York: Oxford University Press.

Bargh, J. A., and T. L. Chartrand. 1999. The unbearable automaticity of being. *American Psychologist* 54 (7): 462–479.

Bargh, J. A., and M. J. Ferguson. 2000. Beyond behaviorism: On the automaticity of higher mental processes. *Psychological Bulletin* 126 (6): 925–945.

Bentham, J. 1823/1948. *An introduction to the principles of morals and legislation*. New York: Macmillan.

Berman, M. N. 2008. Punishment and justification. *Ethics* 18: 258–290.

Baumeister, R. F. 2008. Free will in scientific psychology. *Perspectives of Psychological Science* 3(1): 14–19.

Bok, H. 1998. *Descartes: His moral philosophy and psychology*. New York: New York University Press.

Bjornsson, G., and D. Pereboom. 2014. Free will skepticism and bypassing. In W. Sinnott-Armstrong (Ed.), *Moral psychology* (vol. 4). Cambridge, MA: MIT Press.

Caruso, G. D. 2012. *Free will and consciousness: A determinist account of the illusion of free will*. Lanham, MD: Lexington Books.

Caruso, G. D. 2014. Précis of Derk Pereboom's *Free will, agency, and meaning in life*. *Science, Religion and Culture* 1(3): 178–201.

Caruso, G. D. 2015. If consciousness is necessary for moral responsibility, then people are less responsible than we think. *Journal of Consciousness Studies* 22(7–8): 49–60.

Caruso, G. D. 2016a. Free will skepticism and criminal behavior: A public health-quarantine model. *Southwest Philosophy Review* 32(1): 25–48.

Caruso, G. D. 2016b. Consciousness, free will, and moral responsibility. (December 21.) Unedited version. https://ssrn.com/abstract=2888513 Edited version in *Routledge handbook of consciousness*, ed. Rocco J, Gennaro, forthcoming.

Caruso, G. D. 2017a. Free will skepticism and the question of creativity: Creativity, desert, and self-creation. *Ergo* 3(23): 23–39.

Caruso, G. D. 2017b. *Public health and safety: The social determinants of health and criminal behavior*. UK: ResearchLinks Books.

Caruso, G. D., and S. Morris. 2017. Compatibilism and retributivist desert moral responsibility: On what is of central philosophical and practical importance. *Erkenntnis*. doi: 10.1007/s10670-016-9846-2

Center for Science and Law. 2012. Deep brain stimulation in rehabilitating criminal psychopaths. http://www.neulaw.org/blog/1034-class-blog/3972-deep-brain-stimulation-in-rehabilitating-criminal-psychopaths

Corrado, M. L. 2016. Two models of criminal justice. http://ssrn.com/abstract=2757078

D'Angelo, E. 1968. *The problem of free will and determinism*. Columbia: University of Missouri Press.

De Caro, M. 2011. Is emergence refuted by the neurosciences? The case of free will. In A. Corradini and T. O'Connor (Eds.), *Emergence in science and philosophy*, pp. 190–221. London: Routledge.

Doris, J. M. 2002. *Lack of character: Personality and moral behavior*. Cambridge: Cambridge University Press.

Evans, J. R. (Ed.). 2006. *Forensic applications of QEEG and neurotherapy*. Informa Healthcare. https://www.amazon.com/Forensic-Applications-Neurotherapy-James-Evans/dp/0789030799

Feinberg, J. 1970. Justice and personal desert. In his *Doing and deserving*. Princeton, NJ: Princeton University Press.

Fischer, J. M. 2007. Compatibilism. In J. Fischer, R. Kane, D. Pereboom, and M. Vargas (Eds.), *Four views on free will*, pp. 44–84. Hoboken, NJ: Wiley-Blackwell Publishing.

Fischer, J. M. 2014. Review of *Free will, agency, and meaning in life*, by Derk Pereboom. *Science, Religion and Culture* 1(3): 202–208.

Gendreau, P., and R. Ross. 1979. Effective correction treatment: Bibliography for cynics. *Crime and Delinquency* 25: 463–489.

Gkotsi, G-M., and L. Benaroyo. 2012. Neuroscience and the treatment of mentally ill criminal offenders: Some ethical issues. *Journal of Ethics in Mental Health* 6: 1–7.

Greenwald, A. G., D. E. McGhee, and J. L. K. Schwartz. 1998. Measuring individual differences in implicit cognition: The implicit association test. *Journal of Personality and Social Psychology*, 74(6): 1464–1480.

Greteman, B. 2009. Improve mental health with neurofeedback. *Odewire Magazine*. http://odewire.com/61556/improve-mental-health-withneurofeedback.html

Haggard, P., and M. Eimer. 1999. On the relation between brain potentials and the awareness of voluntary movements. *Experimental Brain Research* 126(1): 128–133.

Haji, I. 1998. *Moral accountability*. New York: Oxford University Press.

Hammond, C. D. 2011. What is neurofeedback: An update. *Journal of Neurotherapy* 15: 305–336.

Hoeprich, M. R. 2011. An analysis of the proposal of deep brain stimulation for the rehabilitation of criminal psychopaths. Presented at Michigan Association of Neurological Surgeons. http://www.destinationmi.com/documents/2011MANSpresentation_MarkHoeprich.pdf

Honderich, T. 1988. *A theory of determinism: The mind, neuroscience, and life-hopes*. Oxford: Oxford University Press. Republished in two volumes: Mind and brain and The consequences of determinism.

Husak, D. 2000. Holistic retributivism. *California Law Review* 88: 991–1000.

Kang, J., and K. Lane. 2010. Seeing through colorblindness: Implicit bias and the law. *UCLA Law Review* 58(2): 465–520.

King, M., and P. Carruthers. 2012. Moral responsibility and consciousness. *Journal of Moral Philosophy* 9: 200–228.

Kleiman, M. 2009. *When brute force fails: How to have less crime and less punishment.* Princeton, NJ: Princeton University Press.

Lemos, J. 2016. Moral concerns about responsibility denial and the quarantine of violent criminals. *Law and Philosophy* 35(5): 461–483.

Levy, N. 2005. Libet's impossible demand. *Journal of Consciousness Studies* 12(12): 67–76.

Levy, N. 2011. *Hard luck: How luck undermines free will and moral responsibility.* New York: Oxford University Press.

Levy, N. 2012. Skepticism and sanctions: The benefits of rejecting moral responsibility. *Law and Philosophy* 31: 477–493.

Levy, N. 2014. *Consciousness and moral responsibility.* New York: Oxford University Press.

Libet, B. 1985. Unconscious cerebral initiative and the role of conscious will in voluntary action. *Behavioral and Brain Science* 8:529–566.

Libet, B. 1999. Do we have free will? *Journal of Consciousness Studies* 6(8–9): 47–57. Reprinted in R. Kane (Ed.), *The Oxford handbook of free will*, pp. 551–564. New York: Oxford University Press, 2002.

Libet, B., C. A. Gleason, E. W. Wright, and D. K. Pearl. 1983. Time of conscious intention to act in relation to onset of cerebral activity (readiness-potential): The unconscious initiation of a freely voluntary act. *Brain* 106: 623–642.

Losen, D., C. Hodson, M. A. Keith II, K. Morrison, and S. Belway. 2015. Are we closing the school discipline gap? *The Center for Civil Rights Remedies*, February 23.

Losen, D., T. Martinez, and V. Okelola. 2014. Keeping California's kids in school: Fewer students of color missing school for minor misbehavior. *The Center for Civil Rights Rememedies*, June 10.

Martin, G., and C. L. Johnson. 2005. The boys totem town neurofeedback project: A pilot study of EEG biofeedback with incarcerated juvenile felons. *Journal of Neurotherapy* 9(3): 71–86.

Mele, A. 2009. *Effective intentions.* New York: Oxford University Press.

Morse, S. 1999. Neither desert nor disease. *Legal Theory* 5: 265–309.

Nadelhoffer, T. 2011. The threat of shrinking agency and free will disillusionism. In L. Nadel and W. Sinnott-Armstrong (Eds.), *Conscious will and responsibility: A tribute to Benjamin Libet*, pp. 173–188. New York: Oxford University Press.

Nadelhoffer, T., S. Bibas, S. Grafton, K. A. Kiehl, A. Mansfield, W. Sinnott-Armstrong, and M. Gazzaniga. 2012. Neuroprediction, violence, and the law: Setting the stage. *Neuroethics* 5: 67–99.

Nahmias, E. 2002. When consciousness matters: A critical review of Daniel Wegner's *The illusion of conscious will*. *Philosophical Psychology* 15(4): 527–541.

Nahmias, E. 2011. Intuitions about free will, determinism, and bypassing. In R. Kane (Ed.), *The Oxford handbook of free will* 2nd ed., pp. 555–576. New York: Oxford University Press.

Nelkin, D. K. 2014. Free will skepticism and obligation skepticism: Comments on Derk Pereboom's *Free will, agency, and meaning in life*. *Science, Religion and Culture* 1(3): 209–217.

Nichols, S. 2007. After compatibilism: A naturalistic defense of the reactive attitudes. *Philosophical Perspectives* 21: 405–528.

Nisbett, R. E., and T. D. Wilson. 1977. Telling more than we can know: Verbal reports on mental processes. *Psychological Review* 84: 231–259.

Nosek, B. A., F. L. Smyth, J. J. Hansen, T. Devos, N. M Linder, K. A. Ranganath, et al. 2007. Pervasiveness and correlates of implicit attitudes and stereotypes. *European Review of Social Psychology* 18: 36–88.

Obhi, S. S., and P. Haggard. 2004. Free will and free won't: Motor activity in the brain precedes our awareness of the intention to move, so how is it that we perceive control? *American Scientist* 92(July–August): 358–365.

Patterson, G. R., P. Chamberlain, and J. Reid. 1982. A comparative evaluation of a parent training program. *Behavior Therapy* 13: 638–650.

Pereboom, D. 1995. Determinism *al dente*. *Nous* 29: 21–45.

Pereboom, D. 2001. *Living without free will*. New York: Cambridge University Press.

Pereboom, D. 2012. On John Fischer's *Our stories*. *Philosophical Studies* 158: 523–528.

Pereboom, D. 2013. Free will skepticism and criminal punishment. In Thomas Nadelhoffer (Ed.), *The future of punishment*, pp. 49–78. New York: Oxford University Press.

Pereboom, D. 2014a. *Free will, agency, and meaning in life*. Oxford: Oxford University Press.

Pereboom, D. 2014b. Replies to John Fischer and Dana Nelkin. *Science, Religion, and Culture* 1: 218–225.

Pockett, S. 2004. Does consciousness cause behavior? *Journal of Consciousness Studies* 11: 23–40.

Quirk, D. A. 1995. Composite biofeedback conditioning and dangerous offenders: III. *Journal of Neurotherapy* 1(2): 44–54.

Richmond, E. 2015 December 29. When restorative justice in schools works. *The Atlantic*.

Roediger III, H. L., M. K. Goode, and F. M. Zaromb. 2008. Free will and the control of action. In J. Baer, J. Kaufman, and R. F. Baumeister (Eds.), *Are we free? Psychology and free will*, pp. 205–225. New York: Oxford University Press.

Schoeman, F. 1979. On incapacitating the dangerous. *American Philosophical Quarterly* 16: 27–35.

Scanlon, T. 1998. *What we owe to each other*. Cambridge, MA: Harvard University Press.

Sie, M., and A. Wouters. 2010. The BCN challenge to compatibilist free will and personal responsibility. *Neuroethics* 3(2): 121–133.

Smilansky, S. 2011. Hard determinism and punishment: A practical reductio. *Law and Philosophy* 39: 353–367.

Smith, P. N., and M. W. Sams. 2005. Neurofeedback with juvenile offenders: A pilot study in the use of QEEG-based and analog-based remedial neurofeedback training. *Journal of Neurotherapy* 9(3): 87–99.

Sommers, T. 2007. The objective attitudes. *Philosophical Quarterly* 57: 321–341.

Soon, C. S., M. Brass, H.-J. Heinze, and J.-D. Haynes. 2008. Unconscious determinants of free decisions in the human brain. *Nature Neuroscience* 11(5): 543–545.

Strawson, G. 1994. The impossibility of moral responsibility. *Philosophical Studies* 75(1): 5–24.

Strawson, P. F. 1962. Freedom and resentment. *Proceedings of the British Academy* 48:1–25. Reprinted in D. Pereboom (Ed.), *Free will*, pp. 119–142. Indianapolis, IN: Hackett.

Uhlmann, E. L., and G. L. Cohen. 2005. Constructed criteria: Redefining merit to justify discrimination. *Psychological Science* 16: 474–480.

US Department of Education Office for Civil Rights. 2014 March 1. Civil rights data collection: Data snapshot: School Discipline. Issue Brief No. 1. http://ocrdata.ed.gov/Downloads/CRDC-School-Discipline-Snapshot.pdf

Waller, B. 1990. *Freedom without responsibility*. Philadelphia: Temple University Press.

Waller, B. 2011. *Against moral responsibility*. Cambridge, MA: MIT Press.

Walters, G. D. 1992. *Foundations of criminal science*. New York: Praeger.

Wegner, D. M. 2002. *The illusion of conscious will*. Cambridge, MA: Bradford Books, MIT Press.

Wilson, T. 2002. *Strangers to ourselves: Discovering the adaptive unconscious*. Cambridge, MA: The Belknap Press of Harvard University Press.

Yochelson, S., and S. Samenow. 1976. *The criminal personality: A profile for change*. New York: Aronson.

CHAPTER 12

On Determinism and Human Responsibility

MICHAEL S. GAZZANIGA

Let's face it. We are big animals with brains that carry out every single action automatically and outside our ability to describe how it works. We are a soup of dispositions controlled by genetic mechanisms, some weakly and some strongly expressed in each of us. Now, here comes the good news. We humans have something called the *interpreter*, located in our left brain, that weaves a story about why we feel and act the way we do. That becomes our narrative, and each story is unique and full of sparkle. So my question is: What's wrong with being that—just that? Why is everybody so uptight about our cool machine? Does an intricate pocket watch want to do more than keep time? Being self-aware narrators is what human brains do. And, for those who invent stories that we are more spiritual, that is fine, too, because in the final analysis we all remain personally responsible for our actions. We remain responsible because responsibility arises out of each person's interaction with the social layer he or she is embedded in. Responsibility is not to be found in the brain.

1. INTRODUCTION

It was more than fifty years ago when my mentor Roger Sperry spoke on the occasion of the Pontifical Academy of Sciences symposium *Brain and Conscious Experience*. I remember the event well because Sperry was speaking about our research on the original "split-brain" patients. In rereading that paper, it is interesting to note that the participants only commented on those studies, not on his rather extensive arguments dealing with the problem of mind and free will! This is a shame because his thoughts on free will were quite clear and indeed set the stage for many discussions since that time.

Sperry segued from discussing the patients to the issue of free will by suggesting that, with the slice of a surgeon's knife, one brain might become two, each with its own set of controls. This suggestion was immediately challenged by two fellow neuroscientists, Sir John Eccles and Donald MacKay. Both at that conference and in the years that followed, Eccles argued that the right hemisphere had a limited kind of self-consciousness, but not enough to bestow personhood, which resided in the left hemisphere. Donald Mackay was not satisfied with the idea either and commented in his Gifford lecture some ten years later, "But I would say that the idea that you can create two individuals merely by splitting the organizing system at the level of the corpus callosum which links the cerebral hemispheres is unwarranted by any of the evidence so far. . . . It is also in a very important sense implausible" (1991).

These concerns were all about the meaning of split-brain research, not about the issue of determinism and free will. As I have reviewed elsewhere (Gazzaniga 2011), many early interpretations of the meaning of split-brain work have been modified, leaving this aspect of the debate moot. On the larger question of determinism and free will, Sperry's original thoughts remain clear. It is worthwhile to remind ourselves of what he said.

> Unlike *mind, consciousness,* and *instinct, free will* has not made any notable comeback in behavioral science in recent years. Most behavioral scientists will refuse to recognize the presence of free will in brain function. Every advance in the science of behavior, whether it has come from the psychiatrist's couch, from microelectrode recording, from brain splitting, from the use of psychomimetic drugs, or from the running of cannibalistic flatworms, seems only to reinforce that old suspicion that free will is just an illusion, like the rise and setting of the sun. The more we study and learn about the brain and behavior, the more deterministic, lawful, and causal it appears.
>
> In other words, behavioral science tells us that there is no reason to think that any of us here today had any real choice to be anywhere else, nor even to believe in principle that our presence here was not already in the cards, so to speak, five, ten, or fifteen years ago. I do not like or feel comfortable about this kind of thinking any more than you do, but so far I have not found any satisfactory way around it. Alternatives to the rule of causal determinism in behavior that I have seen proposed so far, as for example, the inferred unlawfulness in the dance of subatomic particles, seem decidedly more to be deplored as a solution than desired.
>
> This is not to say that in the practice of behavioral science we have to regard the brain as just a pawn of the physical and chemical forces that play in and around it. Far from it. Recall that a molecule in many respects is the master of its inner

atoms and electrons. The latter are hauled and forced about in chemical interactions by the overall configurational properties of the whole molecule. At the same time, if our given molecule is itself part of a single-celled organism like paramecium, it in turn is obliged, with all its parts and its partners, to follow along a trail of events in time and space determined largely by the extrinsic overall dynamics of *Paramecium caudatum*. And similarly, when it comes to brains, remember always that the simpler electric, atomic, molecular, and cellular forces and laws, though still present and operating, have all been superseded in brain dynamics by the configurational forces of higher level mechanisms. At the top, in the human brain, these include the powers of perception, cognition, memory, reason, judgment, and the like, the operational, causal effects of forces which are equally or more potent in brain dynamics than are the outclassed inner chemical forces.

You sense the underlying rationalization we are leading to here: "If you can't lick 'em, join 'em." If we cannot avoid determinism, accept and work with it. There may be worse "fates" than causal determinism. Maybe, after all, it is better to be properly imbedded in the causal flow of cosmic forces, as an integral part thereof, than to be on the loose and out of contact, free-floating, as it were, with behavioral possibilities that have no antecedent cause and hence no reason or any reliability for future plans or predictions. (Sperry 1966)

Sperry captures many ideas in his summary. Overall, he articulates well that elements of all kinds become something else when configurational issues are accounted for. Some kind of other complexity arises out of interacting parts, and that new layer can constrain the very elements that produce it.

Still, this formulation was not widely accepted at the time and other participants at the conference contested Sperry's view and for markedly different reasons. Again, Eccles and Mackay challenged his ideas. Eccles was a dualist and believed the mental inserted itself into the brain in the left supplementary motor area! Mackay, on the other hand, agreed that the brain "was as mechanical as clock work." However, he believed there was what he called a "logical indeterminacy" that kept free will alive. This was the concept that, in order for something to be true, it had to be true for everybody at all times. Thus, if a super brain scientist made a prediction about my future actions, all I would have to do to negate the prediction is not carry out the act at a prescribed time. If the super brain scientist wrote down the prediction and sealed it in an envelope and, sure enough, I did do what he predicted, it doesn't count since for something to be true and valid, it has to be known to all at all times.

While this debate raged on for years, it was somewhat local to neuroscience. The philosophers, by and large, were and still are coming at the problem from different angles, far too many to review here. What is relevant to the current effort is the strong belief among philosophers that it is difficult to separate the issue of free will from the issue of responsibility. This traditional position, which is well represented by Daniel Dennett, finds people viewing the determined brain as in fact an exemplar of a "free" system. As Dennett recently stated:

> When, on the other hand, we have our wits about us, and are not massively misinformed or otherwise manipulated, then *there is no important sense* in which the outcome of all the interactions in the many levels or layers of "machinery" is not a free choice. That's what a free choice *is!* It's the undistorted, unhindered outcome of a cognitive/conative/emotive process of exquisite subtlety, capable of canvassing the options with good judgment and then acting with fairly full knowledge of what is at stake and what is likely to transpire. (Dennett 2013)

With this kind of definition of what it means to be "free," the yoking of free will and responsibility remain intact. While other philosophers don't see this need and indeed claim the two concepts are dissociable (Fischer and Ravissa 1999), it is a key issue. In short, does a deterministic view of brain function make nonsense out of the idea of responsibility?

I join with those who believe one can hold a deterministic view and still maintain that humans are personally responsible for their actions. In what follows, I will make this argument by suggesting a layered view of human decision-making that incorporates the social network within which we live, which makes the idea not only plausible but inevitable and necessary. It is this perspective that advances the ideas of Sperry's contribution of almost fifty years ago. While incorporating the mental realm in the casual flow is important, it doesn't liberate one from the reality of determinism. By recognizing the existence of the social layer, however, one does introduce another level of explanation which imposes the constraint of personal responsibility. In short, responsibility is established by participating in the social network, not in the brain per se.

2. TOWARD LAYERED AND DYNAMICAL VIEWS OF BRAIN–MIND FUNCTION

Much of what follows I have presented elsewhere (Bassett and Gazzaniga 2011; Gazzaniga 2011, 2013). In those efforts, I reviewed neuroscientific data which support the modular view of brain organization, now widely established, along with a possible understanding of why our subjective life seems largely unified.

From today's vantage point: it's all about the brain—what it does and does not do. First, how is that thing built and connected, and how does it work? Is it a bowl of mush shaped by its environment, like a wheelbarrow full of wet concrete being poured into a form? Or does the brain arrive on the scene preformed, to some extent, and then await experience to place the final touches on its mature shape? More importantly, for the purposes of this discussion, does it matter how it is built?

It does. We are born with an intricate brain slowly developing under genetic control, with refinements being made under the influence of epigenetic factors and activity-dependent learning. It displays structured, not random, complexity, with automatic processing, with particular skill sets, with constraints, and with a capacity to generalize. All these evolved through natural selection and provide the foundation for myriad cognitive abilities that are separated and represented in different parts of the brain. These parts feature distinct but interrelated neural networks and systems. In short, the brain has distributed systems running simultaneously and in parallel. It has multiple control systems, not just one. Our personal narrative comes from this brain and how it interprets the outside world within which it lives.

This overall neural architecture has been unearthed at many levels of examination. While developmental neurobiologists have revealed how the brain gets built, cognitive neuroscientists have studied the brain in healthy maturity and often when it is damaged. My colleagues and I used these insights to confirm that there are modularized, and frequently localized, processes in the functioning, fully developed brain. Classic studies on neurologic patients by Broca and others supported the idea that brain injury can lead to the loss of specific cognitive abilities. This notion has been the backbone of behavioral neurology. Split-brain research—studies of patients who had undergone epilepsy surgery separating the two halves of the brain—complemented this work. It showed what happened when one processing system was disconnected from others, even though it was still present and functioning. And what did happen? It just went on functioning, outside the realm of awareness of the other systems. The right brain was able to go about its business normally while the left brain didn't have the slightest idea what the right brain was doing—and vice versa.

Still, this emerging knowledge of how our brain is organized was hard to square with ordinary experience. People feel integrated, whole and purposeful, not modularized and multiple. How can our sense of being singular and responsible come from a neural architecture like ours?

3. THE INTERPRETER OF EXPERIENCE

Years ago, we unearthed a special capacity, a module in the left hemisphere that we called the "interpreter." Studies of split-brain patients demonstrated that each side of the brain could respond separately to queries about what it perceives by having the hand it controls point to answers in a multiple-choice task. So flash a picture of a chicken claw to the left brain, and the right hand could choose a picture of a chicken out of a group of pictures (each side of the brain controls the opposite side of the body). If the right brain was at the same time shown a picture of a snow scene, it could guide the left hand to select a picture of a snow shovel from a different set of pictures. It took us years to figure out the key question to ask after a split-brain patient performed this task: "Why did you do that?"

We arranged for one patient's left hemisphere (which controls speech) to watch the left and right hand pointing to two different pictures while not allowing the left brain to see the snow scene. Of course, the left hemisphere knew why the hand it controlled had pointed to the chicken, but it had no access to information about why the patient's left hand, controlled by the right hemisphere, had pointed to the shovel. Nonetheless, immediately upon being asked our key question, the left hemisphere made up a story—an interpretation—of why the left hand, controlled by a separated brain module, did what it did. The patient answered, "Oh, the chicken claw goes with the chicken and you need a shovel to clean out the chicken shed."

Years of research have confirmed that there is a system that builds a narrative in each of us about why we do things we do, even though our behaviors are the product of a highly modularized and automatic brain working at several different layers of function (Gazzaniga 2000). Our dispositions, quick emotional reactions, and past learned behavior are all fodder for the interpreter to observe. The interpreter finds causes and builds our story, our sense of self. It asks, for example, "Who is in charge?" and, in the end, concludes, "Well, it looks like I am."

Additionally, neuroscientists have continued to examine when the brain carries out its work that is associated with behavior or even conscious activity itself. Ever since the classic work of Benjamin Libet, it has been believed that the neural events associated with a phenotypic response occur long before one is consciously aware of even wanting to act. Libet stimulated the brain of an awake patient during the course of a neurosurgical procedure and found that there was a time lapse between the stimulation of the cortical surface that represents the hand and when the patient was conscious of the sensation in the hand (Libet et al. 1979). In later experiments, brain activity involved in the initiation of an action (pushing a button) occurred about 500 milliseconds *before* the action. What was surprising was that there was increasing brain activity related to the action as many as 300 milliseconds *before* the *conscious intention* to act, according to subject reports. The buildup of electrical charge within the brain that preceded what were considered conscious decisions was called a *Bereitschafts* potential or more simply, the readiness potential (Libet et al. 1983). Using more sophisticated functional magnetic resonance imaging (fMRI) techniques, John-Dylan Haynes (Soon et al. 2008) recently showed that the outcomes of an inclination can be encoded in brain activity up to ten seconds before it enters awareness! Furthermore, the brain scan can be used to make a prediction about what the person is going to do. The implications of this result appear definitive.

These sorts of findings, however, can be interpreted differently when the brain is viewed as a multilayered system, as is commonly seen in information systems (see Bachman et al. 2000; Hillis 1999; also see Doyle and Csete 2011). Simply put, layered systems use layers to separate different units of functionality. Each layer preferentially communicates with the layer above and the layer below. Each layer *uses* the layer below to perform its function. "A *Layer* is a design construct. It is implemented by any number of classes or modules that behave like they are all in the same layer. That means that they only communicate with classes in layers immediately above or below their layer and with themselves" (Van Bergen).

The framing of how the brain manages its tasks will undoubtedly be modified and extended in the years to come. Still, this suggested informational/functional assessment of how the brain does its work is liberating because it frees us from the linear assumptions of bottom-up causality. The traditional reductionist/constructionist approach with its claim of linearity on how the brain produces mental states leaves little apparent room for the role of mental life in human destiny. On the surface that seems absurd. This was Sperry's point fifty years ago, and it is as valid today as it was then.

Clearly, we humans enjoy mental states that arise from our underlying neuronal, cell-to-cell interactions. Mental states do not exist without those interactions. However, as argued in the foregoing, mental states cannot be defined or understood by knowing only the cellular interactions. Mental states that emerge from our neural actions do constrain the very brain activity that gave rise to them, just as "a molecule in many respects is the master of its inner atoms and electrons." Mental states, such as beliefs, thoughts, and desires, represent a layer, and that layer arises from brain activity and, in turn, can and does influence our decisions to act one way or another. Ultimately, these interactions will only be understood with a new vocabulary that captures the fact that two different layers of stuff are interacting in such a way that existing alone animates neither.

Yet this interpretation of the problem, where both upward and downward causation are discussed, comes with warning signs. As John Doyle puts the issue (see Gazzaniga 2011):

> [T]he standard problem is illustrated with hardware and software; software depends on hardware to work, but is also in some sense more "fundamental" in that it is what delivers function. So what causes what? Nothing is mysterious here, but using the language of "cause" seems to muddle it. We should probably come up with new and appropriate language rather than try to get into some Aristotelian categories. (Personal communication)

Understanding this nexus and finding the right language to describe it represents, as Doyle says, "the hardest and most unique problem in science" (Personal communication). The freedom that is represented in a choice not to eat the jelly donut comes from a mental-layer belief about health and weight, and it can trump the pull to eat the donut because of its taste. The bottom-up pull sometimes loses out to a top-down belief in the battle to initiate an action. And yet the top layer does not function alone or without the participation of the bottom layer.

The desire for a unique language, which has yet to be developed, is needed to capture the thing that happens when mental processes constrain the brain and vice versa. The action is at the interface between those layers. In one kind of vocabulary, it is where downward causation meets upward causation. In still another perspective, it is not only there but also in the space between brains that are interacting with each other. Overall, it is what happens at the interface of our layered hierarchical existence that holds the answer to our quest for understanding mind–brain relationships. How are we to describe that? That mind–brain layers' interacting has its own time course, and that time course is current with the actions taking place. In short, it is the abstract interactions between the mind–brain layers that make us current in time, real and accountable to our past mental experiences. The whole business about the brain doing it before we are conscious of it becomes moot and inconsequential from the vantage point of a layered interacting system. Again, and as I have discussed elsewhere: Once a mental state exists, is there downward causation? Can a thought constrain the very brain that produced it? Does the whole constrain its parts? This is the $64,000 question in this business. The classic puzzle is usually put this way: There is a physical state, P1, at time 1, which produces a mental state, M1. Then, after a bit of time—now time 2—there is another physical state, P2, which produces another mental state, M2. How do we get from M1 to M2? This is the conundrum. We know that mental states are produced from processes in the brain so that M1 does not directly generate M2 without involving the brain. If we just go from P1 to P2 then to M2, then our mental life is doing no work and we are truly just along for the ride. No one really likes that notion. The tough question is: Does M1, in some downward constraining process, guide P2, thus affecting M2?

We may get a little help with this question from the geneticists. They used to think gene replication was a simple upwardly causal system: genes were like beads on a string that make up a chromosome that replicates and produces identical copies of itself. Now, they know that genes are not that simple; there is a multiplicity of events going on. Our systems control guy, Howard Pattee, finds that a good example of upward and downward causation is the genotype-phenotype mapping of description to construction. It "requires the gene to describe the sequence of parts forming enzymes, and that description, in turn, requires the enzymes to read the description.... In its simplest logical form, the parts represented by symbols (codons) are, in part, controlling the construction of the whole (enzymes), but the whole is, in part, controlling the identification of the parts (translation) and the construction itself (protein synthesis)." And, once again, Pattee wags his finger at extreme positions that champion which is more important, upward or downward. As a teenager would sum it up, "Duh, they are, like, complementary."

It is this sort of analysis that finds me realizing the reasoning trap we can all too easily fall into when we look to the Libet kind of fact, that the brain does something before we are consciously aware of it. With the arrow of time always moving in one direction, with the notion that everything is caused by something before it, we lose a grip on the concept of complementarity. What difference does it make if brain activity goes on before we are consciously aware of something? Consciousness is its own abstraction on its own time scale, and that time scale is current with respect to it. Thus, the Libet thinking is wrong-headed. That is not where the action is, anymore than a transistor is where the software action is.

Setting a course of action is automatic, deterministic, modularized, and driven not by one physical system at any one time but by hundreds, thousand, and perhaps millions of systems. The course of action taken appears to us as a matter of "choice," but the fact is that it is the result of a particular emergent mental state being selected by the complex interacting surrounding milieu. Action is made up of complementary components arising from within and without. That is how the machine (brain) works. Thus, the idea of downward causation might be confusing our understanding. As John Doyle says, "where is the cause?" What is going on is the match between ever-present multiple mental states and the impinging contextual forces within which the brain functions. Our interpreter then claims we freely made a choice (Gazzaniga 2011).

It is also true that viewing the mind–brain interface from this perspective reveals a certain truth: the brain is a dynamical system. Instead of working in a simple linear way, where one thing produces another, it works in a dynamic way in which two layers interact to produce a function. Hardware and software interact to produce the PowerPoint image. Mental states interact with neuronal states to produce conscious states. Starting the clock on what happens when, in trying to analyze the flow of events during conscious activity, doesn't start with neurons firing because those events might reflect little more than the brain warming up for its participation in the dynamic events. The time line starts at the moment of the interaction between layers. At the level of human experience, that would mean we are all online when we are thinking about whatever we are thinking about. Thought is not a delay after action. It also leads to the question of whether or not mental beliefs can be in the flow of events determining ultimate action (Posner and Rothbart 2012). I think so.

4. MOVING FORWARD: EMERGENCE, HUMAN RESPONSIBILITY, AND FREEDOM

In one sense, the concept of multiple levels has a long-standing history in the study of the brain and mind. For literally thousands of years, philosophers have argued about whether the mind and body are one entity or two. The compelling idea that people are more than just a body, that there is an essence, a spirit, or mind, has been around a long time. What has not been fully appreciated, however, is that viewing the mind–brain system as a layered system sets the stage for understanding how the system actually works. As reviewed in the preceding pages, it also allows for understanding the role of how beliefs and mental states stay part of our determined system. With that understanding comes the insight that layers exist both below the mind–brain layers and above them as well. Indeed, there is a *social layer*, and it is in the context of interactions with that layer that we can begin to understand concepts such as personal responsibility.

I believe that we neuroscientists are looking at the concept of responsibility at the wrong organizational level. Put simply, we are examining it at the level of the individual brain when perhaps responsibility is a property of social groups of many brains interacting. Mario Bunge makes the point that we neuroscientists should heed: "we must place the thing of interest in its context instead of treating it as a solitary individual." By placing such concepts as personal responsibility in the social layer, it removes us from the quagmire of understanding how determined brain states negatively influence responsibility for our actions. Being personally responsible is a social rule of a group, not a mechanism of a single brain.

Sperry did not introduce the idea of the social layer. He fully accepted and indeed implored us all to "join 'em if you can't lick 'em." He did feel better about determinism by conceptualizing it as the mind layer intervening and becoming a part of the causal chain that constrains the neural elements that built the mind. With the present view, adding the social layer to the human condition completely restores the idea of personal and therefore moral responsibility no matter how stringent a deterministic stance one adopts. Responsibility comes out of the agreement that humans have with each other to live in the social world. The human social network is like any other kind of network. The participants have to be held accountable for their actions—their participation. Without that rule, nothing works.

Brains are automatic machines following hierarchical decision pathways, and analyzing single brains in isolation cannot illuminate the capacity to be responsible. Again, responsibility is a dimension of life that comes from social exchange, and social exchange requires more than one brain. When more than one brain interacts, a new set of rules comes into play, and new properties—such as personal responsibility—begin to emerge. The properties of responsibility are found in the space between brains, in the interactions between people.

Finally, neuroscience is happy to accept that human behavior is the product of a determined system that is guided by experience. But how is that experience doing the guiding? If the brain is a decision-making device and gathers information to inform those decisions, then can a mental state that is the result of some experience or the result of some social interaction affect or constrain the brain and, with it, future mental states?

We humans are about becoming less dumb, about making better decisions to cope with and adapt to the world we live in. That is what our brain is for and what it does. It makes decisions based on experience, innate biases, and much more. Our "freedom" is to be found in developing more options for our computing brains to choose among. As we move though time and space, we are constantly generating new thoughts, ideas, and beliefs. All of these mental states provide a rich array of possible actions for us. The couch potato simply does not have the same array as the explorer. Just as Daniel Dennett suggests, even though we live in a determined world, new experience provides the window into more choices, and that is what freedom truly means.

Personal responsibility is another matter. My argument is that it is real, the product of the social rules that people establish when living together and that are the fabric of social life. Personal responsibility is not to be found in the brain, any more than traffic can be understood by knowing about everything inside a car. All networks, whether they are neural or artifactual like the Internet, can operate only if accountability—cause and effect, action and consequence—is built into their functioning. Human society is the same.

Responsibility is a needed consequence of more than one individual interacting with another. It is established by people. Researchers might study the mechanistic ways of the mind–brain interface forever, with each year yielding more insights. Yet none of their research will threaten the central value of human life. It is because we have a contract with our social milieu that we are responsible for our actions.

ACKNOWLEDGMENTS

The majority of this chapter appears elsewhere in a publication of the Vatican (see Gazzaniga 2013). It is reprinted here with the permission of the Pontificia Accademia delle Scienze.

REFERENCES

Gazzaniga, M. S. (ed.) 2000. *The new cognitive neuroscience* (2nd. Edition). Cambridge, MA: MIT Press.
Hillis, W. D. 1999. *The pattern on the stone: The simple ideas that make computers work.* New York: Basic Books.

Libet, B., C. A. Gleason, E. W. Wright, and D. K. Pearl. 1983. Time of conscious intention to act in relation to onset of cerebral activity (readiness potential): The unconscious initiation of a freely voluntary act. *Brain* 106(3): 623–642.

Libet, B., E. W. Wright, B. Feinstein, and D. K. Pearl. 1979. Subjective referral of the timing for a conscious sensory experience: A functional role for the somatosensory specific projection system in man. *Brain* 102(1): 193–224.

MacKay, D. M. 1991. *Behind the eye.* Oxford: Basil Blackwell.

Posner, M., and M. Rothbart. 2012. Willpower and brain networks. *Bulletin of the International Society for the Study of Behavioural Development* (ISSBD) (1): 7–10.

Soon, C. S., M. Brass, H-J. Heinze, and J-D. Haynes. 2008. Unconscious determinants of free decision in the human brain. *Nature Neuroscience* 11(5): 543–545.

Sperry, R. W. 1966. Brain bisection and mechanisms of consciousness. In J. C. Eccles (Ed.), *Brain and conscious experience*, pp. 298–313. Heidelberg: Springer-Verlag.

Van Bergen, P. Garfixia software architectures. http://www.dossier-andreas.net/software_architecture/index.html.

CHAPTER 13

Free Will Skepticism, Freedom, and Criminal Behavior

FARAH FOCQUAERT, ANDREA L. GLENN, AND ADRIAN RAINE

1. INTRODUCTION

According to existentialist views, human interaction cannot be properly understood by resorting to the natural sciences such as physics, biology, psychology, and other sciences. We need to go beyond those reductionist categories. Philosophical approaches to human behavior should not rely on the objective sciences if we wish to provide a proper understanding of human behavior and should instead focus on the ways in which we experience human interaction. When it comes to free will and freedom, individuals experience their choices and behavior as *free choices* and *freely willed behaviors*. We experience ourselves as *agents* that have control over our own decisions and behaviors.

The philosophical debate on free will focuses to a large extent on the legitimation of moral responsibility and the conviction that we need a concept of "voluntary control/behavior" (i.e., free will) in order to justify its use: "it is a presupposition that some sort of control is a necessary condition of morally responsible agency" (Fischer and Ravizza 1998: 20; Waller 2015). The crucial question, then, seems to be which kind of control is required for us to hold others morally responsible. McKenna (2009) states that "what most everyone is hunting for, both in *Four Views* [including Kane, Pereboom, Fischer, and Vargas], and in the wider philosophical arena, is the sort of moral responsibility that is desert entailing, the kind that makes blaming and punishing as well as praising and rewarding justified" (McKenna 2009: 12). Caruso similarly argues that the sort of *free will* that is at stake in the free will debate "refers to the control in action required for a core sense of moral responsibility" (Caruso 2016: 26). According to Caruso and many others, this core sense of moral responsibility incorporates a notion of *basic desert* (see Clarke 2005; Feinberg 1970; Fischer et al. 2007; Pereboom 2014; Scanlon 2013; Strawson 1994).

In our society, the notion of *guilt* is so pervasive that it is hard to imagine a world without it. Free will skeptics fully acknowledge this reality but nevertheless argue that we have strong moral and scientific reasons to abandon *desert*-based moral responsibility (see Caruso 2013). The idea of *justice without retribution* evokes several pressing questions. What is the contribution of guilt to normal moral development and moral behavior? Is it a necessary element of normal moral functioning, or can we function well or perhaps better without it (see Nadelhoffer 2011; Philipse 2012; Smilansky 2000)? Does disbelief in free will increase immoral behavior, as some studies seem to suggest, or does it increase antisocial behavior because it promotes retaliation and revenge (e.g., Rigoni et al. 2012; Vohs and Schooler 2008)? Most, if not all victims, understandably experience strong feelings of revenge, and many members of society, without being a direct victim of a particular crime, will have the same feelings of disgust, hatred, and revenge toward offenders. Our intuitions urge us to retributively punish wrongdoers and to attribute guilt.

At first glance, existentialism and free will skepticism approaches to human behavior, including criminal behavior, seem to be diametrically opposed in the same way that retributive punishment and free will skepticism are. Free will skepticism and retributive punishment are incompatible because retribution entails that punishment of the wrongdoer is justified for the reason that he *deserves* to be harmed just because he has knowingly done wrong (Pereboom 2014: 157). "For the retributivist, it is the basic desert attached to the criminal's immoral action alone that provides the justification for punishment. This means that the retributivist position is not reducible to consequentialist considerations nor does it appeal to a good such as the safety of society or the moral improvement of the criminal in justifying punishment" (Caruso 2016).

Existentialist arguments will not convince free will skeptics of the need to hold on to desert-based moral responsibility. However, an existentialist perspective may offer free will skeptics valuable insights on how justice without retribution can be achieved in an ethically acceptable way. Existentialist approaches to human behavior urge us not to reduce an individual's behavior to its social and biological determinants, but to focus on human agency and on the needs of victims to find restoration and emotional closure. It urges us to carefully address feelings of vengeance and retribution. All of us live our lives as if we make *real choices*. Although free will skepticism denies the existence of *free* choices, it cannot deny the reality of our experiences: *we feel in charge* and understand ourselves as *agents* with the capacity to make autonomous, free choices. We feel sadness, hatred, and vengeance when confronted with injustice and especially so when we ourselves or our loved ones are the target of horrific crimes.

In this chapter, we explore whether free will skeptic views on criminal justice need to take seriously our experiential awareness of "freedom."

2. UNIVERSAL DENIAL OF MORAL RESPONSIBILITY: "EXCUSE-EXTENSIONISM" IS NOT THE ANSWER

How should we, as a society, deal with criminal behavior? If we wish to punish offenders by purposely inflicting harm, we need to justify such harm. All harm, including retributive punishment, requires a normative justification. Most philosophers and ethicists will agree that purposely inflicting harm on other individuals, including offenders, needs to meet the highest moral standards available to us (Double 2002; Levy 2013). This requires that our justification of the harm that is imposed needs to be consistent with our best scientific theories. According to free will skeptics, purposely inflicting harm cannot be fully justified on the mere possibility of free will, or on our experiential feeling of freedom, or solely on the basis of public safety or pragmatic grounds. It cannot draw upon non-naturalistic, miraculous phenomena. Compatibilist freedom (and the capacities that are necessary to attribute such freedom) may provide adequate grounds to establish whether or not individuals have *take-charge* responsibility (i.e., human agency and the capacity to change one's future behavior provided the necessary means are put in place), but such freedom does not qualify to answer the question of desert-based moral responsibility (Waller 2015). Moreover, consequentialist or existentialist approaches drawing on compatibilist *freedom* may not provide strong enough safeguards to prevent overly harsh punishments or other grave injustices within the criminal justice system (Caruso 2016).

Within compatibilist approaches, the denial of moral responsibility derives from a special excuse that basically strips human beings of having normal capacities for human agency (Waller 2015). Within criminal justice, such excuses can lead to mitigation and reduced sentencing or to civil commitment under the insanity defense and a nonguilty verdict. Various countries understand the insanity defense in various ways, but it typically draws on the offender having impaired cognitive and/or emotional capacities due to one or more serious mental disorders (e.g., intellectual disability, schizophrenia, addiction, borderline personality disorder, psychopathy). Typically, some kind of link is required between suffering from these impairments or lacking certain capacities for normal (moral) decision-making and the crime that was committed in the sense that these impairments likely contributed to the behavior in question. Waller (2015) argues that one of the reasons why compatibilist thinkers like Daniel Dennett argue that we should hold on to our current practices of attributing praise and blame is the worry that "the steady march of scientific understanding will erode away all the space required for free will and moral responsibility" (1978: 176). We will have nothing left. As we understand more and more about individuals' formative histories, we will have more and more excuses available to argue away all attributions of competency and human agency. We will be left with a system that denies moral responsibility based on the recognition of "universal flaws and a universal denial of individual competency." We will have argued away personal agency, personal strength, and all human achievements, to be left with a world of noncompetent automatons. Waller (2015) convincingly argues that there is no reason to hold that view. Waller (2015) strongly rejects what he coined the "*excuse-extensionist*" model": the idea that the denial of moral responsibility implies or necessitates a universal extension of competence-destroying excuses (Waller 2006: 81). Or, from an existentialist perspective, human agency–destroying excuses. On the contrary, the denial of moral responsibility entails that *no individual* is morally responsible in a desert-based manner no matter how much or little competency that individual possesses. No matter how much or little capacity for human agency that individual possesses. It doesn't exclude that an individual's capacities may or may not foster take-charge responsibility. We can use our scientific understanding of human behavior to identify whether or not an individual has the capacity for take-charge responsibility but not (desert-based) moral responsibility.

On the first day of his trial, Anders Breivik admitted that he killed sixty-nine innocent people at a summer camp on Utoya Island and that he planted a bomb in Oslo that killed another eight innocent individuals. He admitted to committing these crimes and is therefore considered guilty. He is obviously responsible for these horrific crimes in a causal sense because his actions are the causal factors that brought these crimes about. Is he morally guilty and/or is he legally responsible for committing these crimes? According to the court system in Norway, and in many other countries, if he was sane at the time of the crime, he should be held legally culpable because he possessed normal cognitive, motivational, and affective capacities, including the capacity for moral decision-making and human agency (i.e., implying that "he was able to do the right thing and refrain from committing such horrific acts"). If he is found to be insane by the court, based on an expert team of psychiatrists and psychologists, and likely so at the time of the crime, then he should not be held legally responsible and should be excused for his crimes. He would then be responsible in the "causal" sense but not in the "legal" sense.

The first psychiatric team concluded that Breivik was suffering from paranoid schizophrenia during and after the attacks, which means that he fulfilled the criteria for legal insanity. However, a second analysis by a different psychiatric team concluded that Breivik suffered from a narcissistic personality disorder, which means that he did not satisfy the criteria for legal insanity and could be considered legally sane. Based on the first diagnosis, Breivik is not eligible for punishment. Based on the second diagnosis, Breivik should be considered legally guilty and should be punished. A gradual difference in mental impairment, based on the behavioral diagnosis of different psychiatric teams, carries the weight of an important legal decision: to be considered mentally ill and thus legally insane versus mentally healthy, but still impaired, and thus legally sane.

A lot is at stake when individuals are deemed to be legally sane versus legally insane (e.g., when diagnosed with a serious psychiatric disorder such as schizophrenia or a "mere" personality disorder such as narcissistic personality disorder). In this chapter, we wish to highlight this difference and the extent to which it is (or is not) philosophically sound. If someone "foreseeably" or intentionally harms others, then our moral intuitions may tell us that this individual deserves to suffer for what he or she did and that the victim(s) deserve to be compensated by harshly punishing wrongdoers. Many individuals, even when they learn that the culprit has a serious psychiatric disorder, continue to have a strong intuition that the offender deserves and should be punished. Most victims, understandably, have strong feelings of revenge, and many members of society, without being a direct victim of a particular crime, will have the same feelings of disgust, hatred, and revenge toward offenders. Our intuitions urge us to punish wrongdoers and to attribute guilt. Jens Breivik, Breivik's father, wrote a book about his son's horrific crimes with the title *My Fault?*

Some individuals, when learning that an offender had a brain tumor that likely caused or contributed to the individual's deviant behavior, might conclude that the offender was not (fully) responsibly for his behavior; not in the legal sense, but equally not in the moral sense (Gilbert and Focquaert 2015). Similarly for individuals with mental retardation or severe mental disorders such as schizophrenia (involving psychosis). A small percentage of the population, including some prominent philosophers and neuroscientists, will argue that neither the individual with the brain tumor, nor the psychopath, nor the individual who can be considered mentally sane are morally responsible in the basic desert sense or perhaps in any sense. As Miller frames the issue at hand: "are agents ever really responsible for the outcomes they produce" (2007: 86) and, if so, under what conditions?

Philosophically speaking, there's a difference between free will, as for example discussed by Sam Harris (2012), and *freedom*, as for example defended by compatibilists such as Daniel Dennett (2003) and Fischer and Ravizza (1998). Freedom and free will are different concepts. The reality of our experience of "freedom" and human agency provides us with an existentialist, experiential understanding of human behavior, whereas the existence or denial of "free will" directly impacts on the justified or unjustified attribution of desert-based responsibility. Our experience of "freedom" is important, and we have various reasons to underscore its importance, but we equally have strong reasons to not identify this "freedom" and our capacity for human agency, or the lack thereof, as the rationale to justify claims of moral/legal responsibility versus moral/legal excuse. The attribution of human agency is central when thinking about the normal development of (moral) decision-making skills and moral education in general as well as rehabilitation, restoration, and emotional closure in particular. It need not and cannot justify retributive punishment and mass incarceration practices (Waller 2015).

Free will skeptics argue that the relevant question is whether an offender could have acted differently in precisely those circumstances with precisely the powers and limitations he or she actually had. Could he or she *actually* have acted differently? Free will skeptics hold that this is the kind of free will that is needed to attribute "moral guilt" and to justify retributive punishment, and it is entirely different from compatibilist *freedom*. Compatibilist freedom distinguishes between individuals with and without normal capacities for moral decision-making and human agency within a deterministic worldview. Such a difference is normatively relevant and may be scientifically identifiable. However, if agents with and without these capacities are considered inhabitants of a world in which behavior is fully determined by antecedent factors, can it ever be justified to purposively inflict harm on offenders on the ground that they acted *freely*? There is a huge difference between feeling *in charge* of one's actions and having the (contra-causal) capacity to make a different choice at a given moment in time. Whereas compatibilists are willing to use such freedom as the basis of moral responsibility attributions, free will skeptics argue that this capacity cannot justify desert-based moral responsibility or a concept of legal responsibility and punishment drawing upon desert.

According to free will skeptics, we lack the kind of free will that is needed to justify retribution. The kind of free will that would justify moral responsibility in the basic desert sense is scientifically questionable and justifying desert-based moral responsibility on the mere possibility of such free will is normatively questionable. Science informs us that all behavior is the result of deterministic and/or indeterministic causal processes over which we ultimately do not have control (i.e., the kind of control that would justify desert). All behavior, not just deviant behavior, is caused by a complex interplay among our genes, the environment, and (likely/potentially) some added randomness. All free will skeptics adhere to the belief that "what we do, and the way we are, is ultimately the result of factors beyond our control and because of this we are never morally responsible for our actions in the *basic desert* sense—the sense that would make us *truly deserving* of blame or praise" (Caruso 2016). Can retribution be morally justified if we live in a world in which all human behavior is the result of the complex interplay among genes, the environment, and (potentially) some added randomness?

Freedom exists in gradations and refers to psychological capacities that can be described by our existing scientific knowledge. Compatibilist thinkers typically understand freedom as a backward-looking notion. Individuals with a normal capacity for moral decision-making and behavior are considered free and therefore morally responsible for their behavior. Free will skeptics may also attribute some notion of freedom to individuals, but their understanding of freedom will be forward-looking. Forward-looking freedom, human agency, and take-charge responsibility are interrelated concepts within a free will skeptic account. Having or lacking human agency and a capacity for take-charge responsibility implies having or lacking the freedom to change one's future behavior if given the means to do so. Whether individuals possess or lack normal capacities for human agency and take-charge responsibility is therefore important with respect to rehabilitation and leading a crime-free life. These capacities are based on cognitive, motivational, and emotional processes that can be enhanced to a greater or lesser extent if found to be lacking or impaired in a given individual. The latter is the goal of moral enhancement, which can be achieved by traditional means such as education and moral upbringing or, potentially, by biomedical means. As individuals, we do not have the free will to act differently at a given moment in time, but we do possess the freedom or *take-charge* responsibility to change our future behavior, provided adequate means to accomplish such changes are provided. Such means can focus on changing the environment by addressing structural impediments to leading a crime-free life (e.g., addressing poverty, addiction, unemployment, incarceration) or on changing the individual in question (i.e., restoring or enhancing an individuals' decision-making capacities and behavior through behavioral and/or neurobiological interventions).

3. ALL HUMAN BEHAVIOR IS DETERMINED: "DEVIANT" AND NONDEVIANT ALIKE

Several decades of research on criminal behavior has demonstrated that a multitude of environmental factors contribute to criminal behavior. In addition, a growing body of research shows that, on average, individuals who engage in crime demonstrate differences in brain structure and functioning (Glenn, Raine, and Schug 2009; Yang and Raine 2009) and hormone and neurotransmitter levels (Glenn et al. 2011; O'Leary, Taylor, and Eckel 2010). Behavioral genetics studies also indicate that about 40–60 percent of the variance in criminal behavior is due to genetic influences (Raine 2013). These environmental and biological factors that increase risk for criminal behavior are highly connected. Biological predispositions influence the ways in which individuals react to the environment. In turn, environmental factors can affect gene expression, hormone and neurotransmitter levels, and, ultimately, brain structure and functioning.

This research highlights the fact that not all individuals have the same capacity for behaving morally. However, attempting to draw a line between moral/legal responsibility and nonresponsibility may not be practical. Trying to assess and weigh the biological and environmental risk factors at play in a particular individual, as well as evaluate his or her cognitive and psychological functioning, and to consider all of this within the circumstances of a particular crime would be nearly impossible. Furthermore, we know that biological and environmental factors influence *everyone*'s behavior—not just those who are persistently criminal. Thus, we argue that it is more important to use the information gained from research on the biological and environmental factors influencing criminal behavior to try to offer better solutions to the problem. In light of the fact that our behavior is influenced by factors beyond our control, a focus on (moral) responsibility is less important than a more forward-thinking system that focuses on addressing the risk factors for criminal behavior and on rehabilitating offenders (Eagleman 2011). If an offender could be humanely and effectively treated and released, this would provide the most benefit for society.

Although the idea that biological predispositions increase risk for criminal behavior may imply that some individuals are destined to a life of crime, the knowledge we have gained about how the environment can influence biology suggests otherwise. A child who may have a genetic predisposition to be less emotional or less sensitive to punishment may be labeled from an early age as "bad" and may subsequently develop maladaptive cognitions and narratives. Being able to intervene early and in positive ways may go a long way in terms of preventing these narratives from developing. For example, we know that in addition to biologically focused forms of treatment such as medication, psychosocial and behavioral interventions can also produce changes in brain functioning (Felmingham et al. 2007; Paquette et al. 2003) and hormone levels (Brotman et al. 2007). Different forms of brain training, such as mindfulness training or attention training, also have the potential to alter the functioning of brain regions. For example, eight weeks of cognitively based compassion training, a program based on Tibetan Buddhist compassion meditation practices, resulted in a trend increase in amygdala functioning (Desbordes et al. 2012), an area of the brain that has consistently been found to function less in psychopathic individuals (Birbaumer et al. 2005; Glenn et al. 2009; Kiehl et al. 2001). Finally, the fact that environmental factors such as early child maltreatment and trauma influence the development of the brain and other biological systems makes improving environments an even more important goal.

Furthermore, even if there are deficits present that we cannot address directly, there may still be ways in which behavioral outcomes can be improved. For example, it may prove to be very difficult to increase the emotional responsiveness of a psychopathic individual, particularly in adulthood. However, research suggests that these individuals are still sensitive to reward. Thus, it may be possible to provide programs that essentially work around the deficits (e.g., providing incentives for moral behavior). This could be thought of as a form of cognitive prosthesis. Indeed, one program designed to rehabilitate highly aggressive delinquent males has been found to be successful by providing readily available, rapidly increasing incentives for compliance and positive participation in treatment and has been found to result in significant reductions in psychopathic traits (Caldwell et al. 2012).

4. SUCCESSFUL (RE-)INTEGRATION IN SOCIETY REQUIRES A NOTION OF FREEDOM AND HUMAN AGENCY

A free will skeptic understanding of free will and human behavior shows us the importance of rejecting desert-based moral responsibility and retributive punishment that is justified by reference to an individual's "moral guilt." It does not necessarily involve the rejection of forward-looking notions of freedom. Some free will skeptics argue that "freedom" is essential for human interaction and flourishing and the normal development of moral decision-making, both with respect to offenders and law-abiding citizens. Waller (2015) warns us that learned helplessness and depression are waiting around the corner if we do not take seriously an individual's capacity for take-charge responsibility. Moreover, respecting victims' rights means respecting their need for emotional closure and restorative justice. For many individuals, this may require the identification of the crime as a *freely willed* behavior for which the offender is held (morally) responsible and for which the offender may or may not experience (strong) feelings of remorse. If holding offenders morally responsible, in the society in which we live, unavoidably leads to overly harsh punishments and mass incarceration practices which likely exacerbate rather than reduce recidivism, then, for reasons of public safety and with regard to the rights of offenders, we may be better off holding offenders accountable rather than morally responsible. Whether or not a forward-looking concept of freedom (i.e., having the capacity for human agency and take-charge responsibility) rather than a backward-looking notion of freedom (i.e., having the capacity for moral agency and moral responsibility) should be identified in legal contexts and during which, if any, phases of a criminal trial is a difficult question that we do not aim to answer here. Free will skeptics obviously reject backward-looking notions of freedom that foster attributions of guilt but may have reasons to adopt forward-looking notions of freedom. We argue that there may be moral benefits in addressing and enhancing offenders' capacities for *take-charge* responsibility.

Within treatment settings, Pickard (2011, 2014) has argued that holding individuals responsible may be necessary for the success of therapy as it provides a sense of agency. Individuals may not be able to learn to change their behavior if they and those who work with them do not believe it is in their power to do so. She distinguishes between the concepts of responsibility and blame, suggesting that responsibility focuses on the perpetrator and whether he or she is accountable, whereas blame is about our emotions, judgments, and actions toward them (e.g., condemning them as bad or worthless individuals). In a treatment setting, it may be most effective to hold individuals responsible for their actions but not to blame them (Pickard 2011). Thus, even if we determine that an individual is not responsible in a legal context, it may be beneficial to treat them as such in a rehabilitation setting. Holding the individual responsible for his immoral actions, without condemnation of his character (e.g., suggesting he is bad or worthless) may be appropriate for motivating behavioral change. It may be a necessary prerequisite to tune in to an offender's capacity for take-charge responsibility and effectively achieve a crime-free life. Other studies have provided support for the idea that enabling individuals to see themselves as capable of change may have a positive effect. In a study of juvenile offenders with psychopathic features, Salekin, Lester, and Sellers (2010) compared the effects of providing two different messages to youth regarding intelligence and the brain. Youth who were given a message that the brain is dynamic, forms new connections, and grows every time new knowledge is acquired performed better on a task measuring fluency and flexibility of thought than did youth who were informed that intelligence is static and were given no information about brain development based on learning. Thus, just by enabling the youth to see themselves as capable of improving, their behavior changed.

An existentialist understanding of human freedom similarly shows us the importance of addressing human behavior in terms of our capacities for human agency and take-charge responsibility, from the perspective of both the offender and victims. However, free will skeptics agree that the presence or absence of normal capacities for human agency and/or take-charge responsibility should not be invoked when attributing moral guilt versus excuse. Within the criminal justice system, we argue that this difference may nevertheless be important with respect to (1) rehabilitation and recidivism and (2) victims' need for restoration and emotional closure. Whether or not an individual possesses normal capacities for human agency and/or take-charge responsibility is important when considering rehabilitation and the most successful ways in which offenders can be integrated or reintegrated into society without recidivating. It is also important with respect to victims' rights and finding adequate ways to achieve restorative justice. Although attributing moral/legal guilt to offenders who possess normal capacities for moral decision-making and human agency might be necessary for many individuals, free will skepticism accounts of human behavior suggest that society at large could gain from a nonretributive, non–desert-based approach to punishment.

Recent free will skepticism approaches to criminal justice aim to provide nonretributive alternatives that can effectively address recidivism while respecting the rights of victims and their need for restoration, as well as adequately securing the rights of offenders (e.g., Caruso 2016; Pereboom 2013, 2014). Although retributive punishment involving mass incarceration of offenders is defended by many as society's most appealing answer to crime, such an approach is vulnerable to human rights violations and has the potential to increase rather than decrease recidivism rates (see Focquaert, Glenn, and Raine 2013). About 95 percent of prisoners are eventually released back into society. A major problem that prisoners face upon reentry is that their time spent in prison has eroded their life both professionally and privately. Stigmatization acts as an indirect, enduring form of punishment that aggravates this situation and prevents successful reentry. Most prisons worldwide lack the necessary resources to provide adequate somatic and mental health care and adequate opportunities for rehabilitation. Research shows that imprisonment is counterproductive in terms of rehabilitation and reintegration into society for low-level offenders and for those in need of special care (e.g., individuals with mental health and substance abuse problems). The lack of resources to provide adequate physical and mental health care, overcrowding, and a shortage of effective alternative sanctions gives rise to growing numbers of human rights violations within prisons (Focquaert and Raine 2012; Vandevelde et al. 2011).

Prisons are often described as the nation's hidden mental health asylums. According to a large international systematic review, 65 percent of male prisoners and 42 percent of female prisoners suffer from serious mental disorders (psychotic illnesses, major depression, and personality disorders) (Fazel and Danesh 2002) and 70–100 percent of incarcerated youth suffer from at least one mental health disorder and 20 percent suffer from a serious mental health disorder (Odgers et al. 2005). Moreover, a large retrospective cohort study reveals that offenders with serious psychiatric disorders (major depressive disorder, bipolar disorders, schizophrenia, and nonschizophrenic psychotic disorders) have a substantially increased risk of multiple incarcerations over a six-year follow-up (Baillargeon et al. 2009). A systematic international review shows that reoffending rates are higher for prisoners than for forensic psychiatric patients (Fazel et al. 2016). Although many factors likely contribute to the difference, the lack of *adequate* mental health care in prisons may account for part of this difference. According to forensic psychiatrists, adequate mental health care and successful rehabilitation requires the attribution of agency and the ability to foster self-change by focusing on the strengths and the capacities that individuals possess and may develop over time (Ward, Mann, and Gannon 2007). It is important to understand rehabilitation as a holistic process in which all aspects of successful human functioning are given the necessary attention (e.g., education, skill training, sports, recreation, housing opportunities, family relations, etc.). Such an approach is in line with existentialist approaches to human behavior that understand our identity as a project of self-definition through freedom, choice, and commitment (Crowell 2015).

There is a strong need for caution when considering the idea that prevention and/or rehabilitation measures are "better" than (retributive) punishment, especially with respect to forensic mental health treatment. While a lot of hope is vested in the (early) detection of neurobiological risk and protective factors and the implementation of prevention measures for antisocial and criminal behavior, we need to bear in mind that practices of (early) detection and prevention may unintentionally result in undesirable ethical, social, and legal consequences or be deliberately misused for social control purposes (Horskötter, Berghmans, and de Wert 2014). As mentioned earlier, such undesirable consequences may result from labeling individuals who have never committed a crime but may be more likely to commit crimes once identified as at risk. Moreover, individuals receiving a psychiatric label run the risk of being stigmatized and may experience fear of rejection and mistreatment, may be subject to discrimination and prejudice, or may develop low self-esteem and internalize self-blame, which may prevent successful treatment outcomes in the long run. The risk of false positives inherent to medical diagnoses in general and psychiatric diagnoses in particular urges us to be very careful when focusing on (early) detection, prevention, and rehabilitation (Glenn, Focquaert, and Raine 2015). We should be mindful that certain behaviors are easily misclassified as behavioral indications of an underlying disorder while in reality being expressions of normal variation in personality traits and behaviors. The risk of false positives is especially worrisome when faced with children and adolescents living in low socioeconomic-level, deprived neighborhoods and attending schools with high delinquency rates. Behavior that reflects normal survival and coping strategies in such an environment may be misunderstood as exemplifying underlying disorders.

5. CONCLUSION

Our experience of freedom suggests that an offender's capacity for human agency and take-charge responsibility is important with respect to offender rehabilitation and restorative justice. Offenders are obviously causally responsible for their crimes and should be held accountable for these crimes. If and how this causal responsibility relates to normal capacities for human agency and moral decision-making matters in a forward-looking sense and should be incorporated within restorative justice practices as well as rehabilitation efforts. Nonretributive sanctions (or some form of punitive measures) are needed to deter crime, protect the public, reduce recidivism, allow for restorative justice, and foster rehabilitation. From a free will skeptic perspective, it is less clear whether we should use backward-looking notions of freedom as the rationale to justify desert-based moral responsibility claims and retributive punishment practices.

As all human behavior results from the same cumulative and interactive social, psychological, and neurobiological causal factors leading up to the behavior in question, drawing a guilty versus nonguilty distinction between offenders with and without normal capacities for human agency and moral decision-making does not make sense. Whether or not we have those capacities at a given moment in time is not under our control. Whether or not an individual possesses normal capacities for human agency and moral decision-making should therefore not be understood within a backward-looking account of freedom that involves moral guilt attributions. This knowledge can inform forward-looking notions of freedom in which offenders can be held accountable and society can rightfully impose nonretributive sanctions that address the extent to which an offender has or lacks such capacities. In our view, unlike retributive punishment, a free will skeptic approach to crime that sanctions (or punishes) in order to prevent future crime and restore past crimes can meet our highest moral standards and be equally effective in protecting society without causing unnecessary harms, such as overly lengthy sentences and harsh prison environments.

ACKNOWLEDGMENTS

We would like to thank Bruce N. Waller for his comments on an earlier draft.

REFERENCES

Baillargeon, J., I. A. Binswanger, J. V. Penn, B. A. Williams, and O. J. Murray. 2009. Psychiatric disorders and repeat incarcerations: The revolving prison door. *American Journal of Psychiatry* 166(1): 103–109.

Birbaumer, N., R. Viet, M. Lotze, M. Erb, C. Hermann, W. Grodd, and H. Flor. 2005. Deficient fear conditioning in psychopathy: A functional magnetic resonance imaging study. *Archives of General Psychiatry* 62(7): 799–805.

Brotman, L. M., K. K. Gouley, K. Y. Huang, D. Kamboukos, C. Fratto, and D. S. Pine. 2007. Effects of a psychosocial family-based preventative intervention on cortisol response to a social challenge in preschoolers at high risk for antisocial behavior. *Archives of General Psychiatry* 64: 1172–1179.

Caldwell, M. F., D. McCormick, J. Wolfe, and D. Umstead. 2012. Treatment-related changes in psychopathy features and behavior in adolescent offenders. *Criminal Justice and Behavior* 39(2): 144–155. doi:10.1177/0093854811429542

Caruso, G. D. (Ed.). 2013. *Exploring the illusion of free will and moral responsibility*. Plymouth, UK: Lexington Books.

Caruso, G. D. 2016. Free will skepticism and criminal behavior: A public health-quarantine model. *Southwest Philosophy Review* 32(1): 25–48.

Clarke, R. 2005. On an argument for the impossibility of moral responsibility. *Midwest Studies in Philosophy* 29: 13–24.

Crowell, S. 2015. Existentialism. In E. N. Zalta (Ed.), *The Stanford Encyclopedia of Philosophy* (Spring 2016 edition). http://plato.stanford.edu/archives/spr2016/entries/existentialism/

Dennett, D. 1978. *Elbow room*. Cambridge, MA: MIT Press.

Dennett, D. 2003. *Freedom evolves*. New York: Viking.

Desbordes, G., L. T. Negi, T. W. W. Pace, B. A. Wallace, C. L. Raison, and E. L. Schwartz. 2012. Effects of mindful-attention and compassion meditation training on amygdala response to emotional stimuli in an ordinary, non-meditative state. *Frontiers of Human Neuroscience* 6: 292.

Double, R. 2002. The moral hardness of libertarianism. *Philo* 5(2): 226–234.

Eagleman, D. 2011. The brain on trial. *Atlantic Monthly* 308: 112–123.

Fazel, S., and J. Danesh. 2002. Serious mental disorder in 23,000 prisoners: A systematic review of 62 surveys. *The Lancet* 359(9306): 545–550.

Fazel, S., Z. Fiminska, C. Cocks, and J. Coid. 2016. Patient outcomes following discharge from secure psychiatric hospitals: A systematic review and meta-analysis. *The British Journal of Psychiatry* 208: 17–25.

Feinberg, J. 1970. Justice and personal desert. In J. Feinberg (Ed.), *Doing and deserving: Essays in the theory of responsibility*. Princeton, NJ: Princeton University Press.

Felmingham, K., A. Kemp, L. Williams, P. Das, G. Hughes, A. Peduto, and R. Bryant. 2007. Changes in anterior cingulate and amygdala after cognitive behavior therapy of posttraumatic stress disorder. *Psychological Science* 18: 127–129.

Fischer, J. M., R. Kane, D. Pereboom, and M. Vargas. 2007. *Four views on free will*. Malden, MA: Blackwell Publishing.

Fischer, J. M., and M. S. J. Ravizza. 1998. *Responsibility and control. A Theory of moral responsibility*. Cambridge, MA: Cambridge University Press.

Focquaert, F., A. L. Glenn, and A. Raine. 2013. Free will, responsibility, and the punishment of criminals. In T. Nadelhoffer (Ed.), *The future of punishment and retribution*, pp. 247–274. New York: Oxford University Press.

Focquaert, F., and A. Raine. 2012. Ethics of community-based sanctions. In S. Barton-Bellessa (Ed.), *Encyclopedia of community corrections*, pp. 144–148. Chicaco, IL: SAGE.

Gilbert, F., and F. Focquaert. 2015. Does diagnosis of acquired pedophilia relieve offenders of responsibility and call for mandatory treatment? *The International Journal of Law and Psychiatry* 38: 51–60.

Glenn, A. L., F. Focquaert, and A. Raine. 2015. Prediction of antisocial behavior: Review and ethical issues. In J. Clausen and N. Levy (Eds.), *Handbook of neuroethics*, pp. 1689–1701. New York: Springer.

Glenn, A. L., A. Raine, and R. A. Schug. 2009. The neural correlates of moral decision-making in psychopathy. *Molecular Psychiatry* 14: 5–6.

Glenn, A. L., R. A. Schug, Y. Gao, and D. A. Granger. 2011. Increased testosterone-to-cortisol ratio in psychopathy. *Journal of Abnormal Psychology* 120: 389–399.

Harris, S. 2012. *Free will*. New York: Free Press.

Horskötter, D., R. Berghmans, and G. de Wert. 2014. Early prevention of antisocial behavior (ASB): A comparative ethical analysis of psychosocial and biomedical approaches. *BioSocieties* 9(1): 60–83.

Kiehl, K. A., A. M. Smith, R. D. Hare, A. Mendrek, B. B. Forster, and J. Brink. 2001. Limbic abnormalities in affective processing by criminal psychopaths as revealed by functional magnetic resonance imaging. *Biological Psychiatry* 50: 677–684.

Levy, N. 2013. Be a skeptic, not a metaskeptic. In G. D. Caruso (Ed.), *Exploring the illusion of free will and moral responsibility*, pp. 87–102. Plymouth, UK: Lexington Books.

McKenna, M. 2009. Compatibilism and desert: Critical comments on *Four Views on Free Will*. *Philosophical Studies* 144: 3–13.

Miller, D. 2007. *National responsibility and global justice*. New York: Oxford University Press.

Nadelhoffer, T. A. 2011. The threat of shrinking agency and free will disillusionism. In L. Nadel and W. Sinnott-Armstrong (Eds.), *Conscious will and responsibility: A tribute to Benjamin Libet*, pp.173–188. New York: Oxford University Press.

Odgers, C. L., M. L. Burnette, P. Chauhan, M. L. Moretti, and D. Reppucci. 2005. Misdiagnosing the problem: Mental health profiles of incarcerated juveniles. *The Canadian Child and Adolescent Psychiatry Review* 14(1): 26–29.

O'Leary, M. M., J. Taylor, and L. A. Eckel. 2010. Psychopathic personality traits and cortisol response to stress: The role of sex, type of stressor, and menstrual phase. *Hormones and Behavior* 58: 250–256.

Paquette, V., J. Levesque, B. Mensour, J. M. Leroux, G. Beaudoin, P. Bourgouin, and M. Beauregard. 2003. Change the mind and you change the brain: Effects of cognitive-behavioral therapy on the neural correlates of spider phobia. *Neuroimage*, 18: 401–409.

Pereboom, D. 2013. Free will skepticism and criminal punishment. In T. Nadelhoffer (Ed.), *The future of punishment*, pp. 49–78. New York: Oxford University Press.

Pereboom, D. 2014. *Free will, agency, and meaning in life*. New York: Oxford University Press.

Philipse, H. 2012. *Neurofilosofie van de geest* (Neurophilosophy of the mind). Audio Course, Home Academy.

Pickard, H. 2011. Responsibility without blame: Empathy and the effective treatment of personality disorder. *Philosophy, Psychiatry, & Psychology* 18(3): 209–223.

Pickard, H. 2014. Responsibility without blame: Therapy, philosophy, law. *Prison Service Journal* 213: 10–16.

Raine, A. 2013. *The anatomy of violence: The biological roots of crime*. New York: Pantheon.

Rigoni D., S. Kühn, G. Gaudino, G. Sartori, and M. Brass. 2012. Reducing self-control by weakening belief in free will. *Consciousness and Cognition* 21: 1482–1490.

Salekin, R. T., C. Worley, and R. D. Grimes. 2010. Treatment of psychopathy: A review and brief introduction to the mental model approach for psychopathy. *Behavioral Sciences & the Law* 28(2): 235–266.

Scanlon, T. 2013. Giving desert its due. *Philosophical Explorations* 16(2): 101–116.

Smilansky, S. 2000. *Free will and illusion*. New York: Oxford University Press.

Strawson, G. 1994. The impossibility of moral responsibility. *Philosophical Studies* 75(1): 5–24.

Vandevelde, S., T. Vander Beken, V. Soyez, and E. Broekaert. 2011. Mentally ill offenders in prison: The Belgian case. *International Journal of Law and Psychiatry* 34(1): 71–78.

Vohs, K. D., and J. W. Schooler. 2008. The value of believing in free will: Encouraging a belief in determinism increases cheating. *Psychological Science* 19: 49–54.

Waller, B. N. 2006. Denying responsibility without making excuses. *American Philosophical Quarterly* 43(1): 81–90.

Waller, B. N. 2015. *The stubborn system of moral responsibility*. Cambridge, MA: MIT Press.

Ward, T., R. E. Mann, and T. A. Gannon. 2007. The good lives model of offender rehabilitation: Clinical implications. *Aggression and Violent Behavior* 12: 87–107.

Yang, Y., and A. Raine. 2009. Prefrontal structural and functional brain imaging findings in antisocial, violent, and psychopathic individuals: A meta-analysis. *Psychiatry Research* 174: 81–88.

CHAPTER 14

Your Brain as the Source of Free Will Worth Wanting

Understanding Free Will in the Age of Neuroscience

EDDY NAHMIAS

1. THREE REACTIONS TO NEURONATURALISM

Imagine you have an important decision to make about which of two job offers to accept. You must decide by 5:00 PM. The offers, A and B, have various competing attractions and drawbacks. Currently, there is no answer to the question of what you will decide. But there needs to be by 5:00 (option C of picking neither and being unemployed is not on the table). You have several more hours to consider your reasons for each option, to discuss them with friends and family, to imagine how your life will go if you choose A and how it will go if you choose B. This feels like an existential choice since your future depends upon it: your life will be significantly different depending on what you decide.[1] Some of these differences are evident to you—they are the ones you imagine and weigh against each other—others are unknowable, but you cannot do anything about those. You also realize that some of your reasons, and how important they seem to you, are influenced by factors you don't know about, some of which you wouldn't want to influence you. But you do the best you can to consider what you think *is* most relevant and to decide based on

---

1. I recently advised a student who had to decide whether or not to major in philosophy before he registered for classes. While this decision is not as significant as the one Sartre describes of the young man deciding between joining the Resistance and staying to care for his mother, the student I advised described his decision as "existential."

what really matters to you. Of course, this means you are also making some decisions along the way about what matters to you and how much. As the day wears on, you find yourself leaning toward option B. It's only 4:30, so you continue to deliberate, testing your reasons and your feelings about B, letting yourself plump for A to see how it feels. But just before 5:00, you make your decision final by sending an email to A to decline and by calling B to accept.

Not all decisions are like this, of course. Most are not so existential (e.g., choosing between soup and salad for lunch), many are made without such extensive conscious deliberation or with less rational consideration, and, alas, many are made without feeling as confident about what to do by the time the decision must be made. I hope you will fill in the example sketched here with an *actual* decision you've made that has these features: an important choice for which you imagined various alternative outcomes, evaluated your reasons and feelings, and eventually came to a relatively confident decision about what to do. (I'll wait here while you think about it.)

These decisions seem to represent one paradigm of free and responsible agency. But what if I told you that all of the mental processes involved in making your choice—imagining the options, evaluating them, making the decision—*all* of these processes happened . . . *in your brain*? Indeed, each of those mental processes just *is* (or is *realized in*) a complex set of neural processes which causally interact in accord with the laws of nature. Call this thesis about the relation of mental processes and neural processes "neuronaturalism."[2]

If you are like me (and forgive the pun), your mind is not blown by this assertion of neuronaturalism. It may instead seem banal, though at the same time a bit mysterious, since we do not yet understand how neural processes could achieve all of these remarkable conscious and rational decision-making tasks. I think my reaction is a common one, at least among contemporary educated people, and later I'll provide some evidence for this. I will call it the "natural reaction" to neuronaturalism.

Some people, however, find neuronaturalism patently absurd or impossible; for instance, they are committed to a dualistic conception of the mind and free will. They do not accept that *mere* neurobiological activity could explain consciousness, imagination, or decision-making, and hence they resist the possibility of neuronaturalism and take its assertion to be a threat to free will. Call this the "dualist reaction."

---

2. Neuronaturalism, as I'll use it, is meant to be compatible with various forms of physicalism in philosophy of mind, including both nonreductive and reductive varieties (Stoljar 2009). However, neuronaturalism does not commit one to a reductionistic ontological thesis that says the only things that really exist are whatever entities physics determines compose everything, nor to a reductionistic epistemological thesis that says the best explanations are always those offered by lower level sciences (e.g., physics or neuroscience).

Finally, others (call them "pessimists") take the neuronaturalist understanding of our mind and agency to be angst-inducing. On the one hand, they take most people to be wedded to the dualist *conception* of mind and agency, but, on the other hand, they accept the truth of neuronaturalism, typically a reductionistic brand of it. So, the pessimists think that if people could be induced to get their heads out of the sand, the truth would blind them. Most people, having a dualist understanding of self and free will, would fight to put their heads back in the sand or only painfully come to accept the truth. To be fair, most pessimists think that, even if the truth initially causes some angst, it will also rid us of some harmful illusions and, in the end, be beneficial.[3]

My goal in this chapter is to provide some diagnoses of these different reactions to neuronaturalism and provide some reasons to think that the natural reaction is both common and correct. Focusing on free will, I will offer reasons to think that a neuronaturalistic understanding of human nature does not take away the ground (or grounding) that supports most of our cherished beliefs about ourselves, any more than Copernicus's shifting the earth from the center of the universe took away the ground that supports us. It did require Galileo's theory of inertia to make sense of how the earth can be flying through space while we feel unmoving, supplemented by his helpful analogy with our feeling unmoving in the hull of a smooth-sailing, fast-moving ship. But with that explanation of our experience in place, most people could grow up learning the Copernican theory without existential angst. We *experience* the earth as unmoving, and we need Galileo's *theory* to make sense of that experience. But once that experience is accounted for, it was those most committed (for scientific, philosophical, or religious reasons) to the competing Ptolemaic or Aristotelian *theory* who felt the most angst about the Copernican revolution.[4]

---

3. The "pessimist" label is drawn from P. F. Strawson (1962), whose views I hope to reflect, if only dimly, here (as I do with some of Daniel Dennett's views in his 1984 book, whose subtitle I use in my title). Strawson's use of "pessimist" to refer to incompatibilists about free will and determinism is inapt in some of the same ways my use is, since many free will skeptics are *optimistic* about the benefits of our giving up outmoded views of free will (e.g., Pereboom [2014], and chapters in this volume by Pereboom and Caruso [Chapter 11] and Focquaert, Glenn, and Raine [Chapter 13]).

4. God has the power to move heaven and earth, and religions eventually moved their conception of earth to its actual place in the cosmos. Similarly, an all-powerful God would have the power to create persons in fully physical form. Religions can, and have, imagined their God or gods creating humans without nonphysical souls but with all the good stuff typically ascribed to souls, such as consciousness, identity, free will, a moral sense, even eternal life. The point is that religious belief is not wedded to dualistic belief, and dualistic religious tenets are not strong evidence of deep dualistic *intuitions* (or a dualistic folk psychology). Furthermore, the concept of a nonphysical soul does not *explain* how we have consciousness, free will, etc. Rather, it serves as a placeholder essence that is stipulated to have these properties but without explaining them. People's use of terms like "soul" and "mind" often occurs in a causal-explanatory framework and only rarely refer to any alleged nonphysical attributes of either.

Similarly, dualists and reductionists, committed to their competing theories, tend to think that neuronaturalism conflicts with people's self-conception. But, I will suggest, most people are "theory-lite" and amenable to whatever metaphysics makes sense of what matters to them. We do not yet have a theory like Galileo's to explain how neural activity can explain our conscious experiences. However, I predict that such a theory will have to make sense of how those neural processes involved in our consciously considering options and deliberating about them—for instance, about what job offer to take—are crucial causes of our decisions about which option to actualize. I will suggest later that interventionist theories of causation offer the best way to understand this essential causal role. The neuronaturalistic picture has already begun to seep into the public consciousness, and many people have the natural reaction, because they seem to assume that a future theory will be able to make sense of our experiences *within* the neuronaturalistic picture. Hence, their lack of angst. If and when such a theory actually emerges, then, even though it will establish that there is "nothing" more to us than our complex brains and bodies existing in a physical world, governed by the laws of nature, it likely will also make sense of how we can have a type of free will that can ground our being unique, creative, unpredictable, imaginative, autonomous agents who are the sources of our actions.

2. A TRANSPARENT BOTTLENECK ... OR NEXUS

In a much-discussed piece, Joshua Greene and Jonathan Cohen (2004) argue that neuroscience has vindicated a reductionistic form of neuronaturalism that will provide a window for people to see the threats it poses. They assume that people have deeply held implicit or explicit beliefs about the mind as a nonphysical entity and about free will as a libertarian power to make decisions ungoverned by natural laws. But Greene and Cohen think that the opaque metaphysical theses of naturalism and determinism are not vivid enough to pull people's heads out of the sand so they can see how their dualist and libertarian beliefs are challenged by these theses, and this explains why these theses have yet to shake up our moral and legal systems. Neuroscience, on the other hand, will illuminate the mechanisms of decision-making in a way that will challenge people's beliefs and hence challenge our moral and legal practices (notably, our retributive punitive system):

> [N]euroscience holds the promise of turning the black box of the mind into a *transparent bottleneck* ... your brain serves as a bottleneck for all the forces spread throughout the universe of your past that affect who you are and what

you do. Moreover, this bottleneck contains the events that are, intuitively, most critical for moral and legal responsibility. (2004: 1781 [emphasis original])

That's one way to look at it. But we can also flip this image on its head to recognize that neuroscience will open up the black box of the mind to illuminate how the very processes that we take to be critical to decision-making actually work. We can recognize the "transparent bottleneck" of the brain as the complex *nexus* that brings together a remarkable amount of information from both the past (including genes, upbringing, and learning) and the present (including external stimuli and internal beliefs, desires, goals, etc.). This nexus then serves as the source of the causal activity that integrates (some of) this information as we make decisions. These integrative processes, according to the natural reaction to the neuronaturalist picture, will indeed be the ones most critical for responsibility, such as our consciously imagining and weighing options and reasoning about what to do. Because each nexus of neural activity—that is, each of our brains—is the site of a unique causal history, this picture also helps to explain why each of us, along with our subjective experiences, is unique. Neuronaturalism can explain, in a way dualism can only vaguely assert, why you are *you* and why no one else is like you.

On the neuronaturalist view, neural activity also has to explain the existence of our conscious experiences as we make these decisions. Again, while we lack a theory that explains all the features of conscious experiences, especially its subjective or qualitative features, assuming (as I am here) that such a theory is forthcoming, it will presumably illuminate where and how the brain represents our conscious imagining of future options (like job offers A and B), of likely outcomes of choosing those options, and of evaluations of those outcomes, including emotional reactions to them. While the role of consciousness in agency is contentious (e.g., Caruso 2012; Levy 2014), it is plausible that conscious processes allow integration of a wide range of information, which seems crucial for the sort of imagination and evaluation described here (see Nahmias [forthcoming] and Sripada [2016] for discussion of the role of imagination and prospection in free will and of brain regions likely responsible for such processes).

The modern mind sciences have, of course, discovered that our decision-making is subject to external stimuli and nonconscious internal states that can lead us to make suboptimal choices, which we may then rationalize after the fact (see, e.g., Nahmias 2007). We are also learning that some genes and/or early experiences can have significant influences on how our brains are "wired" and hence on our decisions; and, in some cases, the result is "faulty wiring" and bad decisions. If neuronaturalism entailed that we are somehow unable to recognize genuine reasons for action or unable to control action in light of such reasons, then pessimism would be warranted. But for now, let us assume that these specific empirical challenges to our capacities for rational decision-making and self-control do not universalize—that is, in many cases, we are "wired" in a way that *explains* the proper functioning of these capacities rather than explaining them *away*. In that case, we can examine the less radical neuronaturalistic thesis that simply says that our conscious deliberation and rational decision-making, to whatever extent we actually possess them, are carried out by neurobiological processes (see Mele [2009] and Nahmias [2014] for responses to empirical evidence presented as challenging any causal role for conscious or rational processes).

To help us understand the neuronaturalistic possibility, Greene and Cohen ask us to imagine a time in the future when "we may have extremely high resolution scanners that can simultaneously track the neural activity and connectivity of every neuron in a human brain, along with computers and software that can analyse and organize these data" (2004: 1781). They ask you "to imagine watching a film of your brain choosing between soup and salad" for lunch. I will ask us instead to consider what we would see if we watched what occurs in our brains as we make the more consequential and more complex decision I described earlier (italics indicate where I have altered their text accordingly):

> The analysis software highlights the neurons pushing for *offer A* in red and the neurons pushing for *offer B* in blue *(i.e., the neuronal processes that realize your imagining the pros and cons of each job offer)*. You zoom in and slow down the film, allowing yourself to trace the cause-and-effect relationships between individual neurons—the mind's clockwork revealed in arbitrary detail. *After examining the neuronal processes involved in the extended and complicated mappings of the conscious deliberations as you imagined various consequences of, and reasons for, the two options*, you find the tipping-point moment, at which the blue neurons in your prefrontal cortex "out-fire" the red neurons, seizing control of your pre-motor cortex and causing you to send the email to reject offer A. (2004: 1781; compare Harris 2012: 10–11)

Notice that it is awkward and radically incomplete to try to describe in these terms the astronomical complexity of the neural activity that would actually have to be captured and analyzed to illuminate what occurs as we spend extended periods of time considering an important decision. If we try to consider the complexity of what we would see occurring in the "bottleneck" of our brains as we make such decisions, we would see a process unfolding in the nexus of our brain over time and space, and we would not "shrink under this scrutiny to an extensionless point" (to repurpose Nagel's memorable phrase [1979: 35]). Trying to imagine a more complete account of such complex decisions is important if we aim to diagnose the different reactions to the neuronaturalism that this futuristic brain-scanning is supposed to illuminate. We need to avoid selling our brains short. If neuronaturalism is true, then the "film" of your deliberations and decision about which job to take will not really be reducible to mere images of blue and red neurons, and it will not look like "clockwork."

3. AVOIDING THE BYPASSING INTUITION

Are pessimists like Greene and Cohen correct to predict that most people have deep dualist intuitions such that they will take neuronaturalism to conflict with free will and challenge our moral and legal practices? I will suggest they are not. But first, let's consider why these pessimists assume there is a conflict. I suspect some neuroscientists are especially prone to see a conflict between their reductionistic methodology, with its focus on the mechanisms causing human and animal behavior, and folk psychological explanations of behavior that do not refer to such mechanisms. They might also think that ordinary people, like scientists, have a substantive theory about the way human agency and minds work—namely, a dualist theory. And these scientists are especially likely to recognize the current absence of a scientific explanation of consciousness, like the pre-Galileo theorist who sees the lack of explanation for our experiences within the Copernican theory. As such, when these neuroscientists assume that neuroscientific explanations can explain and predict all human behavior, they may conclude that the unexplained conscious features of our mental life have no causal role to play—they are *bypassed*.[5]

But most people—at least those who do not delve into science, philosophy, or theology—do not have such substantive theories about how the mind or agency works. And the less substantive or specific their commitments to the

---

5. For examples of other pessimist neuroscientists and psychologists, see Nahmias (2014). However, not all neuroscientists take the pessimistic view toward discoveries of what the brain does during decision-making. Consider a recent study that seems to bring to life the fictional one Greene and Cohen describe by finding the neural activity associated with the "tipping point" in the brain when people made decisions about where to focus their attention (Gmeindl et al. 2016). The researchers found the activity (in specific areas of prefrontal cortex) that built up starting about three seconds before people shifted their attention, likely with their awareness of a decision to shift occurring during that buildup of activity. Like other functional magnetic resonance imaging (fMRI) studies using multivoxel pattern analysis (MVPA), the mapping had to be individualized to each participant's unique brain. The researchers do not take their approach as challenging free will, but instead as helping to discover how it works. In a media report titled "What Free Will Looks Like in the Brain" (seemingly an oxymoron for dualists and pessimists), one researcher says, the aim of the study is to "peek into people's brain and find out how we make choices . . . and what parts of the brain are involved in free will." Another says, "that by devising a way to detect brain events that are otherwise invisible—that is, a kind of high-tech 'mind reading'—we uncovered important information about what may be the neural underpinnings of free will" (http://www.eurekalert.org/pub_releases/2016-07/jhu-wfw071316.php).

underlying structure of the mind or the underlying causal processes that connect mental states to each other and to behavior, the less substance there is to be falsified by metaphysical or scientific theories. If people are "theory-lite" in this way, then although they may have a relatively nonnegotiable understanding of humans' basic capacities to make choices and control their actions, they may have relatively negotiable or revisable beliefs about what actually underlies or explains these capacities—that is, the metaphysical or scientific nature of the substance(s), processes, or sources of them. If so, then we should predict that people are not committed to a dualist understanding of free will that conflicts with neuronaturalism, and hence they may not have the reaction predicted by Greene and Cohen to the possibility of seeing all the neural processes responsible for decision-making, at least so long as those processes are described as the neural instantiation of the mental processes "that are, intuitively, most critical for moral and legal responsibility." That is, as long as they are not led to believe that a neuronaturalist picture entails bypassing of these critical processes.

In Nahmias, Shepard, and Reuter (2014), we presented people with a neuroprediction scenario inspired by the ones that pessimists like Greene and Cohen and Sam Harris predict will lead people to see the threat to free will posed by neuronaturalism.[6] Our scenario first describes future technology: "Neuroscientists can use brain scanners to detect all the activity in a person's brain and use that information to detect the activity that causes decisions and predict with 100% accuracy every single decision a person will make before the person is consciously aware of their decision." It states that in the future a woman named Jill agrees to wear the scanner for a month, during which time the neuroscientists are able to predict all of her decisions, even when she is trying to trick them, including decisions about whom to vote for in an election. And it ends with a statement meant to suggest neuronaturalism; in one version stating, "these experiments confirm that all human mental activity is entirely based on brain activity such that everything that any human thinks or does could be predicted ahead of time based on their earlier brain activity," and, in another, using the phrase "all human mental activity just *is* brain activity" (see Nahmias et al. [2014] for details of studies and results).

When asked whether it is possible for such technology to exist in the future, 80 percent said it was. Pessimists, it would seem, should predict that many more people would reject the possibility of this technology, since a dualist or libertarian should reject the possibility of obtaining from physical

---

6. Harris writes: "Imagine a perfect neuroimaging device that would allow us to detect and interpret the subtlest changes in brain function . . . . the experimenters knew what you would think and do just before you did it. You would, of course, continue to feel free in every present moment, but the fact that someone else could report what you were about to think and do would expose this feeling for what it is: an illusion" (2012: 10–11).

information complete information about a person's (nonphysical) mental processes during decision-making or the possibility of decisions being fully caused by prior neural activity, or both. In fact, almost none of our participants explained their responses to this possibility question with any mention of free will, souls, or the impossibility of understanding or predicting decisions based on brain processes.[7]

Furthermore, across a range of questions, 75–90 percent of participants responded that the technology would *not* conflict with free will or moral responsibility and that Jill was free and responsible even while having all her decisions perfectly predicted by the neuroscientists. They do not see the threat predicted by pessimists. The minority who said the technology would threaten free will also expressed bypassing intuitions—for example, agreeing that it would mean that people's reasons have no effect on what they do. The majority, however, did not express such bypassing intuitions. Instead, these participants seemed to be assuming that what the brain scanners are detecting (e.g., in Jill) is precisely the neural activity that instantiates the reasoning processes (e.g., as Jill considers what to do). That is, most people seem to interpret this scenario to mean that Jill's reasons and reasoning are both caused by her brain states and cause her decisions. To the extent that people's theory-lite view is being "filled in" by the scenario, then it is likely that they assume that the neuroscientists are predicting Jill's behavior based on those brain states that constitute her freely deciding what to do. They might also be implicitly assuming that a post-Galileo theory of mind will have been discovered, a future neuroscience that has figured out how "mere" neural activity can explain, rather than explain away, our imagining and evaluating future options and how those processes have the right sort of causal influence on our decisions.[8]

---

7. Instead, most who said it was possible referenced the remarkable advances of science and technology or the fact that our mental activity all occurs in our brains, whereas the 20 percent who said it was impossible referenced the likely technological glitches or political and ethical resistance to developing such technology. Granted, our participants were college students with at least some scientific background (and while quite diverse at my institution, still less religious than the general population). However, this actually supports my overall view, since I assume that dualist *beliefs* (and avowals) are prevalent, at least in Judeo-Christian cultures, yet also revisable without too much resistance, as long as our folk psychological explanations are not being undercut. It is helpful to remember that even though Gilbert Ryle (1949) calls Descartes's view the "Official Doctrine," he then argues at length that our ordinary talk and beliefs about mental phenomena suggest a behaviorist (or perhaps, better, functionalist) folk theory of mind, whereas substance dualism is driven by philosophical mistakes. While people may *not* think or talk of mental phenomena in physical terms, that does not mean that they think of mental phenomena as *non*-physical.

8. Another interpretation of these results is that people are so committed to a dualist or libertarian view of decision-making that they simply reject the stipulations of the scenario once they start thinking about a human who is making decisions (see Rose, Buckwalter, and Nichols 2015). Further research is required to test this alternative

Supposing most people are theory-lite in this way, such that a common response to neuronaturalism is what I am calling the natural reaction, we might then wonder whether this is the *correct* reaction—whether there is a way to make sense of decision-making, even free will, in a neuronaturalistic framework. I will now sketch a positive answer to that question.

4. CAUSAL SOURCEHOOD IN A NEURONATURALISTIC FRAMEWORK

Suppose that ordinary people, like scientists, think about causation in roughly the way suggested by interventionist theories of causation (see Lagnado, Gerstenberg, and Zultan 2013; Sloman 2005). On this view, to know whether one event, X, causally influences another event, Y, we consider what would happen to Y if X were different in various ways. More precisely, we consider interventions on the value of X while controlling for the other causal influences on Y, and we see what happens to the value of Y (for details, see Woodward 2003). Furthermore, we can compare the relative strength of causal influences on an outcome by seeing which among those influences has a more causally invariant relation with that outcome. For instance, consider variable W, which represents the presence of a particular gene in an organism, and variable X, which represents the presence of another gene. Assume that there is a causal invariance relationship between W and a further variable, Y, which represents how a particular phenotypic trait (such as bone density) is expressed in that organism, and there is also a causal invariance relationship between genes X and Y. We can analyze strength of causal invariance to conclude that W has a *stronger* causal invariance relation with Y than X does just in case:

1. Holding fixed relevant background conditions C (i.e., the other implicit causal influences on Y), interventions on the value of W result in a wider range of changes to the value of Y than do interventions on X—for instance, holding fixed the rest of the organism's genome and environment, changes to the value of W (mutations in that gene) influence Y's value (bone density) more than changes to the value of X (mutations in that gene) influence Y's value.
2. The causal relationship between W and Y remains stable across a wider range of relevant changes to the background conditions, C, than does the

theory. For further experimental work suggesting that most people do not have dualist intuitions about free will, see Monroe and Malle (2010) and Mele (2012). For further experimental work suggesting that people take determinism and mechanism to be threats to free will primarily when they (mis)interpret them to entail bypassing of relevant mental states, see Nahmias, Coates, and Kvaran (2007) and Murray and Nahmias (2014).

relationship between X and Y—for instance, the influence that W has on Y remains stable across a wider range of changes in other genes in the organism, or changes in the organism's environment, than does the influence of X on Y (see Deery and Nahmias 2016).

If we are looking for the *causal source* of a particular outcome, we would look for the causal influences of that outcome that have the strongest causal invariance relation with the outcome.[9] Goal-directed causes will typically have a strong causal invariance relation with their effects since they will lead to adjustments in response to varying conditions in order to bring about the desired outcome. For instance, whether Romeo ends up kissing Juliet (effect variable K) will have a strong causal invariance relation with his desiring to kiss her (causal variable J), since (1) interventions on J (e.g., such that Romeo desires to kiss Rosaline instead) would lead to very different outcomes than K, and (2) Romeo's desiring to kiss Juliet will lead to his actually kissing Juliet across many alterations of background conditions—Romeo's desire will lead him to reach Juliet despite a range of obstacles, such as walls to scale (see Lombrozo 2010, who uses the Romeo example from William James). Similarly, whether I have the goal of having a challenging job (represented by causal variable G) might have a stronger causal invariance relation with my decision to accept offer B than other factors because: (1) holding fixed relevant conditions C, varying G's value (e.g., cases where I care less about a challenging job) would alter my decision more than varying the value of any other factor; and (2) G influences my decision more than other factors across the widest range of relevant changes to C (e.g., no other causal factors lead to the *same* decision while altering conditions such as the relative salaries of the two jobs).

Now, imagine we're looking at Greene and Cohen's film of the neural activity as I'm making my decision about jobs, and allow me to oversimplify (though less so than they do). We're assuming a version of neuronaturalism that does not eliminate psychological variables, so there must be various complex neural processes that realize the factors influencing my final decision, including ones I knew about and recognize ("Oh, there's my goal of having a challenging job") and ones I did not know about ("Ah, now I can

---

9. We can also consider causal sourcehood as coming in degrees, such that we can say that W is the source of trait Y more than X is. On this view, much depends on how we are understanding the relevant outcome. For instance, my parents' conceiving me at the particular time they did (event X) may be the causal source of my existing (rather than not existing or rather than some other person existing). But, holding fixed my existence, event X is not the causal source of my deciding on job offer B (rather than A), at least not if we are considering X in relation to a variable such as my considering B to be a more challenging job than A would be, which has a stronger invariance relation with that particular decision (see later discussion).

see the influence of the guilt I felt when I thought about moving away from my mother"), along with many other neural variables that influence my decision, some realizing psychological variables, many not. We could not do the actual interventions on this actual decision to test the relative strength of causal relationships (since it only occurs once), but the interventionist theory does not require that such interventions actually are, or can be, done (see Woodward 2003). Presumably, the neuroscientists had to do many experimental interventions on me and many others to discover which neural pathways are relevant to which behavioral outcomes (including people's verbal reports about experiences while carrying out tasks). With futuristic optogenetic technology, perhaps they could intervene on specific neural processes to test various effects. Here, let us assume that, for some decisions, like the one about which job to take, some of the variables with the strongest invariance relations to my decision are the very ones that I considered important as I deliberated (like my goal to have a challenging job), the ones I am now observing in their "neural form." If so, two important consequences follow regarding the causal source of my decisions.

First, in many cases, psychological variables, such as my conscious imagining of options, will have a stronger invariance relation to my decision than neural variables, *even the ones that realize those psychological variables*, and hence they are plausibly understood to be the causal sources of my decision. As argued by Campbell (2010), List and Menzies (2009), and Woodward (2015) against the causal exclusion argument, interventionism suggests that psychological variables (e.g., beliefs or intentions) can be picked out as the causes of effects (such as decisions or actions) over the neural variables that realize them (or on which the psychological variables supervene). This is because (at least plausibly) the psychological variables could be realized by different neural variables, so interventions on the neural variables might not alter the effects, whereas interventions on the psychological variables would. For instance, holding fixed relevant background conditions, if my actually having the goal of having a challenging job (i.e., G's taking a particular value) is realized by neural variable N's taking a particular value, an intervention on N (such that it takes another one of the "neural" values that can also realize G) would *not* alter my decision for job B. Conversely, an intervention on G would. This argument does not require a commitment to mental states being multiply realized by computers, alien minds, or anything else besides brains (though it is consistent with that possibility). Rather, it only requires that a psychological variable's taking a particular value could be realized by a neural variable's taking multiple different values, which is certainly consistent with current neuroscientific practices. Indeed, most cognitive neuroscientific studies pick out the target neural processes by manipulating psychological and behavioral variables and then allow that the neural processes will vary slightly across participants (we're all unique) and even among the same participants over time (see Laumann et al. 2015).[10]

My application of this reasoning simply requires the plausible follow-up suggestion that, for cases of freely willed choices, some of these psychological variables will both have the strongest causal invariance relations with decisions and also be ones that, from our own first-person perspective, we would pick out as the sources of our decisions (and that we fulfill other plausible compatibilist conditions for free will). If the film of my decision to pick job B shows that my having a particular goal, in the context of my deliberations, played a crucial causal role, I will not react with surprise or angst. Conversely, if the film showed that some variable V, one that I did not know about and would not want to play a crucial role in my decision, in fact played a crucial causal role in my decision, my reaction would be quite different. If it showed, for instance, that the neural realizer of a nonconscious priming influence (such as an anchoring effect in the salary offer or the tone of voice of the person offering me the job) had a stronger causal invariance relation with my decision than my goals, I would see my decision as unfree (or less free) since I would not accept the influence of this variable V were I to know about it, yet I could not control for it since I did not know about it (see Nahmias 2007). Indeed, Greene and Cohen's thought experiment offers a useful way for us to imagine seeing the differences at the neural level that would explain the differences between free and unfree choices, between more or less free actions.

A second application of this interventionist understanding of causal sourcehood uses it to respond to the threat allegedly posed by causal determinism. In many cases, it is plausible that psychological variables, such as your conscious evaluation of a reason for one option over another, will have a stronger invariance relation with your decision than any of the many causal variables in the past that influences you. If determinism is true, then there is a

---

10. Nonetheless, neuroscientists will often be reductionistic in their study of neural mechanisms. So it is not surprising that some of them think that the neural variables they study have the strongest causal invariance relations with human behavior (for instance, they might assume that it is the Readiness Potential, RP, that is the cause of the wrist flex, not my decision to flex or whatever its neural realizers are; Libet 1999). But they are likely wrong to be reductionist in this way, even if neuronaturalism is true—that is, even if all mental states supervene on neural states. Furthermore, without a Galileo-style theory of consciousness and mental causation, some people may think a dualist metaphysics is the only way to understand the psychological causal interactions (how could a mere meat machine account for Romeo's experience of love?). But, of course, nonphysical minds or souls do nothing to make sense of these folk psychological explanations. To the extent that those psychological explanations are preserved with a future neuronaturalistic theory of mind, then we will not become strangers to ourselves. I hasten to add, however, that the details of future neuroscientific discoveries will certainly refine and correct the rough-hewn edges of our self-understanding. And they are likely to indicate that some of us, and some of our decisions, are typically less free than we think (see later discussion).

set of these past causal variables that, all together and in accord with the laws of nature, are sufficient for your decision (this set gets larger and "wider" the further back in time we look). Nonetheless, determinism does *not* entail that any of those variables within this ever-expanding set of conditions in the past is the causal source of your decision. Rather, in many cases, events occurring in the "bottleneck" or nexus of your brain, such as your evaluating one option as better than the other, can be picked out as the causal source of your decision. That deliberative event has a stronger causal invariance relation with your decision than any variables in your distant past. Or, being less precise, we might say that the integrated causal activity of your conscious deliberations, as instantiated in the complex nexus of relevant neural activity, is the causal source of your decision. Hence, even if there are causally sufficient conditions in the distant past for your decision, the *causal source* of your decision lies within *you*, not in any of the causal influences in your distant past.

This use of interventionist theories of causation can also be used to respond to the most powerful current argument for incompatibilism, the *manipulation argument*. This argument says that an agent who is manipulated, even in the distant past, so that he will decide to do B (while satisfying compatibilist conditions) is not free or morally responsible for that decision, but there is no relevant difference between such an agent and one who is causally determined to decide to do B, so the causally determined agent is not free either (Mele 2013; Pereboom 2014). Compatibilists typically respond to this argument by biting the counterintuitive bullet that the manipulated agent is free and responsible. However, the interventionist view of causal sourcehood allows us to uncover a principled distinction between determinism and manipulation. It allows us to see that the manipulated agent's decision has a causal source outside of him— namely, in the intention of the manipulator since, given her knowledge and power, her intention has the strongest invariance relation with his decision. If she intends for her dupe to do B, he will do B (across the widest set of changes to background conditions), and if she intends for her dupe to do A instead of B, he will do A (holding fixed background conditions). But for the agent in a deterministic universe, there is no causal variable outside of him (e.g., in his distant past) that is the causal source of his decision. There is no variable that can be intervened on to ensure that he does B while altering background conditions and that will cause him to A instead of B, holding fixed background conditions. Hence, the compatibilist can use a plausible and generally applicable analysis of causal sourcehood to respond to an incompatibilist argument that aims to conclude that determinism rules out the possibility of our being the causal source of our actions (see Deery and Nahmias 2016, for details).[11]

11. The use of causal interventionism might also be used to analyze what it means to say that we have the ability to decide otherwise, even in a deterministic universe, and

This way of understanding causal sourcehood in terms of interventionist causal invariance relations and of applying it to defuse potential worries posed by neuronaturalism or by determinism is technical (though only sketched briefly here). I am not suggesting people are explicitly thinking in these terms when they consider these issues or when confronted with a neuroprediction scenario as in Nahmias et al. (2014). After all, I think most people are "theory-lite." Nonetheless, it provides a technical way of unpacking our implicit causal cognition such that we can conclude not only that the natural reaction (a non–angst-ridden reaction) to neuronaturalism is, as I've argued, a common reaction, but it is also plausibly the correct reaction.

5. RESPONSIBILITY AND DESERT

I have not focused here on some issues that are likely the focus of some of the other chapters in this section: moral responsibility, desert, and punishment. I cannot defend here why the view of free will I've outlined secures the types of desert and punishment that some argue are ruled out by determinism and perhaps neuronaturalism (e.g., Pereboom 2014; Pereboom and Caruso Chapter 11 this volume). Obviously, a lot turns on how one defines free will and understands the relationship between free will and the relevant notions of desert and punishment. And just as the Copernican theory cannot secure *everything* we once believed, such as our central place in the universe, this view of free will cannot secure some beliefs, such as the misguided ideas that we can be ultimately responsible in some way that might make us deserving of eternal suffering (Strawson 1986) or that we can be uncaused causes in some way that is likely unintelligible (even on a dualist view).

Nonetheless, my own view is that the neuronaturalist understanding of free will can support a viable notion of desert that does not depart substantially from most ordinary beliefs and practices regarding moral and legal responsibility (which, unsurprisingly, I suspect are theory-lite as well). Namely, it can support the type of desert that justifies the reactive attitudes—our feelings of gratitude and resentment, pride and guilt—and the related communicative functions of punishment—e.g., holding responsible criminals, who freely do wrong because they *deserve* to be forced to understand the nature of their crime, to reform to avoid future crimes, to restore the harms they've done as much as possible, and also to express to victims and society the seriousness of those harms (see Nahmias in preparation). While this view does not advocate wrongdoers' suffering *for the sake of* suffering, as some define retributivist punishment, it does advocate that criminals deserve to suffer to the extent

to explain why we experience our future options as open and as causally dependent on what we decide (see Deery 2015).

that such suffering is a constitute feature of these communicative goals of punishing them—for instance, suffering may be a necessary feature (not just a side effect) of the process of coming to understand the harm one has done and feeling and demonstrating appropriate remorse for it.

At the same time, on a naturalistic view of free will, empirical discoveries can inform us about limitations in the relevant capacities and opportunities of humans in general and also of particular humans. It can thus explain why all of us may be less free and responsible than many tend to assume. And it can explain when and why some people with particular neural deficits (perhaps due to their genes or upbringing) thereby lack the cognitive and emotional capacities to evaluate relevant reasons or control their actions in such a way that it is appropriate to mitigate blame and punishment.

This limited-free-will view may have advantages over pessimism about free will. If many people, in a theory-lite way, associate free will with our capacities for choice and self-control, then when pessimists tell us that we have no free will at all, it risks undermining people's belief in those capacities necessary to advocate working hard to improve one's position, to take responsibility for one's failures, to exert willpower in the face of weariness, and to deliberate carefully among alternatives to make good choices—that is, to make personal and moral progress.[12] The limited-free-will view, on the other hand, provides room for such virtues while it also suggests increased tolerance and compassion for people unfortunate enough to lack sufficient capacities or opportunities for achieving them. This view can counter the unlimited-free-will view that some people, especially in America, seem to hold—one that suggests people completely deserve everything that happens to them, good or bad, as if they are untethered from the rest of the universe. Realism about the limits of free will, along with a realistic and empirically informed understanding of our capacities, is both more forgiving than an unrealistic theory of unlimited free will and more hopeful and explanatorily fruitful than a pessimism about free will that risks erasing useful distinctions between free and unfree (more or less free) actions.

ACKNOWLEDGMENTS

Portions of this chapter draw on ideas developed in Nahmias (forthcoming), Nahmias (2014), Nahmias and Thompson (2014), Nahmias, Shepard, and Reuter (2014), and Deery and Nahmias (2016). For helpful comments, I thank Gregg Caruso and Oisin Deery.

---

12. I think this possibility is plausible without depending on the existing empirical work that has suggested it but that also has various problems in both design and replication (see Schooler et al. 2014; Nadelhoffer et al. Chapter 15 this volume).

REFERENCES

Campbell, J. 2010. Control variables and mental causation. *Proceedings of the Aristotelian Society* 110: 15–30.

Caruso, G. 2012. *Free will and consciousness: A determinist account of the illusion of free will.* Plymouth, MA: Lexington Books.

Deery, O. 2015. Why people believe in indeterminist free will. *Philosophical Studies* 172(8): 2033–2054.

Deery, O., and E. Nahmias. 2016. Defeating manipulation arguments: Interventionist causation and compatibilist sourcehood. *Philosophical Studies.* doi:10.1007/s11098-016-0754-8

Greene, J., and J. Cohen. 2004. For the law, neuroscience changes nothing and everything. *Philosophical Transactions of the Royal Society of London B* 359: 1775–1778.

Gmeindl, L., Y-C. Chiu, M. S. Esterman, A. S. Greenberg, S. M. Courtney, and S. Yantis. 2016. Tracking the will to attend: Cortical activity indexes self-generated, voluntary shifts of attention. *Attention, Perception, & Psychophysics.* doi: 10.3758/s13414-016-1159-7

Harris, S. 2012. *Free will.* New York: Free Press.

Lagnado, D., T. Gerstenberg, and R. Zultan. 2013. Causal responsibility and counterfactuals. *Cognitive Science* 37: 1036–1073.

Laumann T. O., E. M. Gordon, B. Adeyemo, A. Z. Snyder, et al. 2015. Functional system and areal organization of a highly sampled individual human brain. *Neuron* 87(3): 657–670.

Levy, N. 2014. *Consciousness and moral responsibility.* Oxford: Oxford University Press.

Libet, B. 1999. Do we have free will? In B. Libet, A. Freeman, and K. Sutherland (Eds.), *The volitional brain*, pp. 47–57. Exeter, UK: Imprint Academic.

List, C., and P. Menzies. 2009. Nonreductive physicalism and the limits of the exclusion principle. *Journal of Philosophy* 106: 475–502.

Lombrozo, T. 2010. Causal-explanatory pluralism: How intentions, functions, and mechanisms influence causal ascriptions. *Cognitive Psychology* 61(4): 303–332.

Mele, A. 2009. *Effective intentions: The power of conscious will.* New York: Oxford University Press.

Mele, A. 2013. Manipulation, moral responsibility, and bullet biting. *Journal of Ethics* 17(3): 167–184.

Monroe, A., and B. Malle. 2010. From uncaused will to conscious choice: The need to study, not speculate about people's folk concept of free will. *Review of Philosophy and Psychology* 1: 211–224.

Murray, D., and E. Nahmias. 2014. Explaining away incompatibilist intuitions. *Philosophy and Phenomenological Research* 88(2): 434–467.

Nagel, T. 1979. *Mortal questions.* New York: Cambridge University Press.

Nahmias, E. 2007. Autonomous agency and social psychology. In M. Marraffa, M. Caro and F. Ferretti (Eds.), *Cartographies of the mind: Philosophy and psychology in intersection*, pp. 169–85. Dordrecht: Springer.

Nahmias, E. 2014. Is free will an illusion? Confronting challenges from the modern mind Sciences. In W. Sinnott-Armstrong (Ed.), *Moral Psychology* (vol. 4): *Freedom and responsibility*, pp. 1–26. Cambridge: MIT Press.

Nahmias, E. Forthcoming. Free will as a psychological accomplishment. In D. Schmitdz and C. Pavel (Eds.), *The Oxford handbook of freedom.* New York: Oxford University Press.

Nahmias, E., J. Coates, and T. Kvaran. 2007. Free will, moral responsibility, and mechanism: Experiments on folk intuitions. *Midwest Studies in Philosophy* 31: 214–232.

Nahmias, E., J. Shepard, and S. Reuter. 2014. It's OK if "my brain made me do it": People's intuitions about free will and neuroscientific prediction. *Cognition* 133(2): 502–513.

Nahmias, E., and M. Thompson. 2014. A naturalistic vision of free will. In E. Machery and E. O'Neill (Eds.), *Current controversies in experimental philosophy*, pp. 86–103. New York: Routledge.

Pereboom, D. 2014. *Free will, agency, and meaning in life*. Oxford: Oxford University Press.

Rose, D., W. Buckwalter, and S. Nichols. 2015. Neuroscientific prediction and the intrusion of intuitive metaphysics. *Cognitive Science* 39 (7): 482–502.

Ryle, G. 1949. *The concept of mind*. London: Hutchinson's.

Schooler, J., T. Nadelhoffer, E. Nahmias, and K. Vohs. 2014. Measuring and manipulating beliefs about free will and related concepts: The good, the bad, and the ugly. In A. Mele (Ed.), *Surrounding free will: Philosophy, psychology, neuroscience*, pp. 72–94. New York: Oxford University Press.

Sloman, S. A. 2005. *Causal models: How people think about the world and its alternatives*. New York: Oxford University Press.

Sripada, C. 2016. Free will and construction of options. *Philosophical Studies*. doi:10.1007/s11098-016-0643-1

Stoljar, D. 2009. Physicalism. *Stanford Encyclopedia of Philosophy*.

Strawson, P. 1962. Freedom and resentment. *Proceedings of the British Academy* 48: 1–25.

Strawson, G. 1986. *Freedom and belief*. Oxford: Clarendon Press.

Woodward, J. 2003. *Making things happen: A theory of causal explanation*. New York: Oxford University Press.

Woodward, J. 2015. Interventionism and causal exclusion. *Philosophy and Phenomenological Research* 91(2): 303–347.

CHAPTER 15

Humility, Free Will Beliefs, and Existential Angst

How We Got from a Preliminary Investigation to a Cautionary Tale

THOMAS NADELHOFFER AND JENNIFER COLE WRIGHT

More than twenty years ago, President George Bush Sr. presciently designated that the 1990s would be the "decade of the brain." Given the huge strides that were made in the wake of those remarks—for example, the first articles on the use of functional magnetic resonance imaging (fMRI) for "brain mapping" were published in 1992—President Bush was not far off the mark. Since the early 1990s, researchers in the brain sciences have made major advances ranging from improved imaging technologies to deepening our understanding of the neurochemical (and genetic) substrates of human thought. More recently, in 2013, President Barack Obama launched the BRAIN (Brain Research through Advancing Innovative Neurotechnologies) Initiative—a program designed to shed important new light on brain functioning. The promised funding for just fiscal year 2014 was $110 million—the type of financial commitment that will only further fuel the progress that has already been made when it comes to understanding the relationship between brain and behavior.

The rapid development in the brain sciences during the past few decades has been met by some with what might be called "neuroexuberance" and others with "neuroleeriness."[1] While some researchers think that recent (and future)

---

1. One common (and ironic) complaint in the neuroethics literature is that writers

advances in neuroscience will revolutionize how we understand the nature of agency and responsibility,[2] others take a much more deflationary or conservative stance.[3] Regardless of which side of this metaphysical debate draws one's theoretical allegiance, it is beyond question that the debate has captured the *public's attention*—an interest that is fueled by articles in the popular press with headlines such as "Neuroscience, Free Will, and Determinism: I'm Just a Machine,"[4] "The Neuroscience of Free Will and the Illusion of 'You,'"[5] and "Is Neuroscience the Death of Free Will?"[6] While some of these pieces advocate for the compatibility of free will with the findings from the brain sciences, others suggest that the gathering data from neuroscience challenge (or even completely undermine) traditional notions of moral agency.

To get a feel for the urgency that is often evoked in the popular press when it comes to the troubled relationship between free will and the brain sciences, consider the following remarks by Dennis Overbye in *The New York Times*: "The death of free will, or its exposure as a convenient illusion, some worry, could wreak havoc on our sense of moral and legal responsibility . . . it would mean that people are no more responsible for their actions than asteroids or planets. Anything would go" (2007). On this anxiety-inducing view, advances in the brain sciences could eventually show that we are mere biological automaton—reductive mechanisms with neither a soul nor free will (that merit neither praise nor blame). Purportedly, in a world like this, anything would go. It's easy to see why some people might be *worried* about the scientific status of free will. Trying to figure out whether we should have existential angst about free will and the brain sciences is a complicated interdisciplinary task that would take us too far afield.[7] So, we are going to focus instead on the potential *ramifications* of free will skepticism from the standpoint of positive psychology. For, regardless of whether the skeptics are correct about the purported metaphysical and meta-ethical tension between the brain sciences and free will, their public proclamations could still have a negative impact on

---

are too inclined to label things with the *neuro-* prefix—e.g., neurophilosophy, neuroprediction, neuroaesthetics, neuromarketing, neuroexistentialism, etc. So, we use "neuroexuberance" and "neuroleeriness" partly to provoke these critics.

2. See, e.g., Bargh (2008); Cashmore (2010); Greene and Cohen (2004); Harris (2012); Haynes (2011); Montague (2008); Wegner (2003, 2008).

3. See, e.g., Mele (2011, 2012); Morse (2011); Nahmias (2011, 2014).

4. http://www.telegraph.co.uk/news/science/8058541/Neuroscience-free-will-and-determinism-Im-just-a-machine.html

5. https://www.psychologytoday.com/blog/psych-unseen/201411/the-neuroscience-free-will-and-the-illusion-you

6. http://opinionator.blogs.nytimes.com/2011/11/13/is-neuroscience-the-death-of-free-will/

7. For discussion of some of the background issues, see, e.g., Mele (2009, 2010a, 2010b, 2010c, 2011, 2012); Nadelhoffer (2011, 2014); Nahmias (2011, 2014).

people's moral attitudes, beliefs, and behaviors. Consider, for instance, the following findings:

1. Vohs and Schooler (2008) show that participants who are exposed to anti–free will primes are more likely to cheat than participants exposed to pro–free will or neutral primes.
2. Baumeister, Masicampo, and DeWall (2009) show that participants who are exposed to anti–free will primes behave more aggressively than participants exposed to pro–free will or neutral primes. Moreover, these researchers show participants who naturally believe more strongly in free will are more likely to behave helpfully toward strangers.
3. Stillman, Baumeister, Vohs, Lambert, Finchman, and Brewer (2010) show that participants who naturally believe more strongly in free will have more positive careerist attitudes, and their bosses and co-workers give them better job performance ratings.
4. Stillman and Baumeister (2010) show that believing in free will facilitates learning from self-conscious emotions and that inducing disbelief in free will hinders learning.
5. Zhao, Liu, Zhang, Shi, and Huang (2014) show that people who are exposed to anti–free will primes are more prejudiced than people who are exposed to pro–free will primes.
6. Rigoni, Kuhn, Sartori, and Brass (2011) show that participants who are exposed to anti–free will primes display reduced activation in the motor cortex compared to participants who received a pro–free will or neutral prime.[8]

These (and similar) empirical results suggest that even if the skeptics are right that we don't have free will, believing in free will could nevertheless be a so-called positive illusion (Taylor and Brown 1988: 193).[9] If this is right, then perhaps the illusion of free will is one we should strive to keep in place—a controversial view called "illusionism" that has already been defended in the philosophical literature (Smilansky 2000, 2001, 2002).[10] In this respect, the

---

8. While there have been a number of conceptual replications of the early findings from Vohs and Schooler (2008), it is worth pointing out that researchers have been unable to replicate the original findings—which are part of the so-called *repligate scandal* in social psychology (see Open Science Collaboration 2015). For a discussion of some of the underlying problems with the empirical literature on free will beliefs, see Schooler, Nadelhoffer, Nahmias, and Vohs (2014). We will say more about these issues in the concluding section of this chapter.

9. For a detailed discussion of the possibility that believing in free will is a positive illusion, see Nadelhoffer and Matveeva (2009).

10. For a preliminary defense of disillusionism about free will, see Nadelhoffer (2011).

possible specter of widespread skepticism about free will is as much a public policy issue as it is a metaphysical debate.

However, we will not be taking a stand on these (and related) issues in this chapter.[11] Instead, the task we set before ourselves is more modest—namely, to shed light on the psychology of free will beliefs by combining our past work on free will beliefs with some of our present research on the psychology of humility. The specific issues we want to address are the following: First, is there a relationship between free will beliefs (or the lack thereof) and existential anxiety? If so, can trait humility serve as a "buffer" between the two? For instance, are people who are high in dispositional humility less likely to experience existential anxiety in the face of skepticism about free will? Given the perspectival and attitudinal nature of humility (see later discussion for details), we predict that humble people will be less anxious in the face of stories about the purported death of free will (or the reduction of the mind to the brain).

In the following pages, we will proceed as follows: first, we set the stage with a brief discussion of our recent empirical work on the psychology of humility. Second, we present the results from four recent studies that we designed to explore the relationship between humility, free will beliefs, and existential anxiety. Third, we highlight some limitations of our present work and explore some future avenues of research. Finally, we offer some concluding remarks and situate our current research within the broader debate about the potential costs of skepticism about free will. As we will see, while our efforts to test the buffering hypothesis about humility and existential anxiety were less successful than we'd hoped, we nevertheless shed some light along the way on the difficulties involved in trying to manipulate and measure people's beliefs in free will and related concepts. In this respect, our work should serve as a cautionary tale for philosophers, psychologists, and pundits who want to discuss the potential ramifications of the supposed death of free will. For while it's certainly possible for people to change their minds about free will, it's not clear that researchers have figured out effective, reliable, and stable methods for bringing these epistemic changes about (even temporarily).

1. THE PSYCHOLOGY OF HUMILITY

Humility is perhaps the oddest of the traditional moral virtues. If dictionaries are to be trusted, humility involves "having a modest or low view of one's

---

11. Some of the members of our present team have addressed these issues in our previous work—see, e.g., Nadelhoffer and Matveeva (2009); Nadelhoffer (2011, 2014); Nadelhoffer and Goya-Tocchetto (2013).

importance" (Oxford Dictionary) in addition to being "not proud or haughty," "not arrogant or assertive," "reflecting, expressing, or offered in a spirit of deference or submission," and even "ranking low in a hierarchy scale" (Merriam-Webster). As such, humility seems to be an important antidote to pride and hubris, yet also troubling from the standpoints of morality and positive psychology. Do we really want to count as a virtue a trait that recommends low self-esteem and submissive deference?

Given the somewhat paradoxical nature of humility, it is perhaps unsurprising that philosophers disagree when it comes to its normative status. For some philosophers, humility is the proper attitudinal stance to adopt when it comes to one's subsidiary relationship to God (or nature or even to other people)—a virtue that is purportedly "central to human life" (Wirzba 2008: 226). However, philosophers ranging from Spinoza and Hume to Nietzsche and Sidgwick (1874/2011) have adopted a markedly more critical stance toward humility. On this more negative view, to the extent that St. Aquinas and other theologians are right that being humble involves "self-abasement to the lowest place" (*ST*, II–II, Q. 161, Art. 1, ad. 2), humility is a trait that ought to be criticized rather than exalted.

As a result, the parties to the debate about the nature and value of humility find themselves at a dialectical stalemate. As is the case with many disagreements, where one ends up in this debate will depend in part on where one starts. Here, as elsewhere, the initial definitions we adopt tend to drive our normative conclusions. If we simply define humility innocuously in terms of having "an inclination to keep one's accomplishments, traits, abilities, etc., in perspective" (Richards 1988: 256), being humble seems virtuous. If we define humility instead in terms of self-abasement and self-degradation, then being humble looks more like a vice. But which definition should we adopt?

Based on our empirical research, we think humility is an important (if not foundational) moral virtue. On our view, humility is a *particular psychological positioning of oneself* within the larger context of the universe—one that is both *epistemically* and *ethically aligned*. By "epistemically aligned," we mean that humility is the understanding and experiencing of oneself *as one, in fact, is*—namely, as a finite and fallible being that is but a very small part of something much larger than oneself. This is often experienced spiritually as a connection to God or some higher power, though it can also be experienced through an awareness of one's place in, and connection to, the natural or cosmic order (a state of "existential awareness"). Operationalized, this is the dimension of *low self-focus*. By "ethically aligned," we mean that humility is the understanding and experiencing of oneself as only one among a host of other morally relevant beings whose interests are as legitimate, and as worthy of attention and concern, as one's own (a state of "extended compassion"). In this way, humility is a corrective to our natural tendency to strongly prioritize or privilege ourselves (our needs, interests, benefits, etc.)—that is, to seek "premium treatment" for ourselves, even at significant cost to others. Operationalized, this is the dimension of high other-focus.

Importantly, this "decentering" of one's focus away from self does more than just shift one's focus to the needs and interests of others—that is, low self-focus is *more than* (and does not necessarily require) high other-focus. After all, someone could have low self-focus without being focused on the interests of others. The converse is true as well—someone could be focused on others while at the same time thinking more of herself than she should. For present purposes, the kind of low self-focus we have in mind involves the reorientation of one's relationship to the outside world, highlighting the importance of keeping things in proper perspective and being mindful of one's place in the larger scheme of things. Accordingly, behavioral manifestations of low self-focus should include (among other things) a lack of desire to self-aggrandize or self-promote and an openness to new and challenging information; a simplicity in self-presentation and/or lifestyle (i.e., modesty, open-mindedness, etc.).

Ultimately, though, humility cannot emerge through low self-focus alone—it requires a shift in one's other-focus as well. This does not mean humility requires us to be moral saints or to reduce ourselves to marginal utility in order to help those in need—although humility is arguably compatible with these supererogatory attitudes and behaviors—but it does require us to be mindful, attentive, considerate, and charitable toward others, especially those who may need help or assistance or who are in harm's way. Accordingly, behavioral manifestations of high other-focus should include, among other things, a greater acceptance of others' beliefs, values, and ideas—even when different from one's own—and an increased desire to help and be of service to others (i.e., tolerance, civic-mindedness, etc.). The humble person is someone who is actively interested in promoting or protecting others' well-being and seeks to make a difference when and where she reasonably can.

On this view, by being invested in the lives of others—rather than merely completely absorbed with satisfying our own selfish interests—we become grounded and embedded in the world. Indeed, looked at properly, humility doesn't reduce the force or scope of one's own needs and interests—rather, it greatly expands them. Others' well-being becomes entangled with our own. Because humility facilitates a realistic appraisal of ourselves, it removes (or reduces) the need to *inflate* or *deflate* our own value or significance, which in turn makes is less necessary to inflate or deflate our estimation of other people's value or significance. And it is this "unencumbered" encountering of others as morally valuable individuals in their own right that facilitates our appreciation for their welfare and our desire to protect and promote their interests (LaBouff et al. 2012).

In this way, we postulate that high other-focus serves as a counterpoint to the existential aftershock that may result from the decentering associated with low self-focus.[12] After all, there is a fine line between existential awareness and existential angst. The latter can leave us feeling isolated, alone, and anxious in a vast, purposeless universe. By immersing herself in the lives and interests of others, the humble person becomes grounded in a way that staves off the existential angst that could otherwise lead to nihilism or egoism rather than well-being. Being humble strikes a middle ground between these two extremes. It is for these and related reasons that we believe that being humble requires *both* low self-focus and high other-focus. The two elements are mutually reinforcing and serve as the twin dimensions of humility.

As part of our research on humility, we developed a new psychometric tool for measuring several facets of humility—namely, religious humility, cosmic humility, environmental humility, other-focus, and the perceived value of humility.[13] Given our account concerning the twin hallmarks of humility, we predicted (and found) that scores on the subscales of the humility scale correlated with a wide variety of positive traits and dispositions—including, (a) a sense of civic responsibility and the desire to meaningfully contribute to their communities (Furco, Muller, and Ammon 1998); (b) a concern for the welfare of people in their lives and appreciation for the importance of community values and traditions (Schwartz and Bilsky 1987); and (c) a commitment to living an honest, principled life (Schlenker, Miller, and Johnson 2009). In addition, all but religious humility were positively correlated with the strength of their humanitarian-egalitarian ideals (Katz and Hass 1988); their commitment to the understanding, appreciation, and tolerance of their fellow humans (Schwartz and Bilsky 1987); the importance of moral virtues and values to their personal identity (Aquino and Reed 2002); the strength of their "individualizing" (harm/care and fairness) moral intuitions (Graham, Haidt, and Nosek 2009);[14] and their desire to be economically charitable toward others—as well as negatively related to their desire for undeserved social status (Wright and Reinhold, in preparation). These are just some of the many benefits of humility.[15]

For present purposes, however, we want to revisit a potential benefit of humility we mentioned a few moments ago when it comes to the relationship between low self-focus and high other-focus—namely, that the latter may help stave off the potential existential anxiety that might result from the decentering of the self that is associated with the former. Indeed, as we

---

12. We would like to thank Regina Rini for pushing us to think more carefully about this issue.
13. For more about our scale, see Wright, Nadelhoffer, Ross, and Sinnott-Armstrong (in preparation).
14. Only religious humility scores were correlated with the "binding" (authority, group membership, purity) intuitions.
15. For more on the positive psychology of humility, see Nadelhoffer and Wright (forthcoming); Wright et al. (2017).

suggested earlier, this may help explain why humility involves *both* low self-focus and high other-focus. On our view, being connected to others helps reground us in the world in the wake of the potentially terrifying realization of our own smallness in relation to the grander scheme of things. If this is correct, then humility may serve as a buffer for existential anxiety.[16] But is this "buffering" hypothesis correct? Up until this point, we have simply speculated that it is. For present purposes, we wanted to put the supposition to the test.

2. FOUR NEW STUDIES

We adopted a two-tiered approach to shedding light on the buffering hypothesis about humility and existential anxiety. First, we wanted to use data from scales that measure humility, free will beliefs, existential anxiety, and existential thinking to explore the correlational relationship between the underlying psychological constructs. Second, we wanted to run some vignette-based studies using primes that were designed to challenge (or deflate) people's beliefs about free will and dualism. Our motivating assumption here was that *decreasing* people's beliefs in free will and dualism would tend to *increase* their existential anxiety. So, our goal was to see whether people who score high in humility are less threatened by these primes than people who score low in humility. We will now discuss our findings, warts and all.

2.1. Study 1

The materials for Study 1 were uploaded to Qualtrics. Participants were 147 people recruited through Amazon's Mechanical Turk (MTurk) service and paid $1 each for completing the survey. Participants had to be at least eighteen years of age and living in the United States. Fifty-six percent of participants reported being male, 44 percent female. Participants reported being represented by the following ethnicities: 87 percent Caucasian, 4 percent African-American, 5 percent Asian, and 4 percent Hispanic. The average age of participants was 35.5 years.

For this first preliminary study, we were simply interested in looking for predicted correlations between humility, existential anxiety, existential thinking, and beliefs about free will, determinism, dualism, and related concepts.

---

16. Kesebir (2014) provides some evidence that humility serves as a buffer against death anxiety. However, the measure that was used for humility was problematic. So, while the buffering hypothesis put forward by Kesebir is interesting, we think more empirical work needs to be done before we can say with any confidence whether the hypothesis is true.

So, each participant provided responses to the following measures (which were presented in random order):

1. The Humility Scale (Wright, Nadelhoffer, Ross, and Sinnot-Armstrong, in preparation)
2. The Free Will Inventory (FWI; Nadelhoffer, Shepard, Nahmias, Sripada, and Ross 2014)
3. The Existential Anxiety Questionnaire (EAQ; Weens, Costa, Behon, and Berman 2004)
4. The Existential Anxiety Scale (EAS; Good and Good 1974)
5. The Scale for Existential Thinking (SET; Allan and Shearer 2012)

As predicted, people's humility was, on average, negatively correlated with their existential anxiety as measured by both the EAQ, $r(147) = -.21, p = .012$, and the EAST, $r(147) = -.41, p < .001$. Also, as predicted, humility was positively correlated with participants' tendency to engage in existential thinking, $r(147) = .35, p < .001$. Moreover, if we split people's humility into its two dimensions—low self-focus and high other-focus—both are correlated to existential anxiety, $rs(147) = -.18$ to $-.34$, ps .029 to $<.001$, and existential thinking, $rs(147) = .26$ to $.36, p < .001$.

Indeed, it would appear that both dimensions of humility are required for the full buffering against existential anxiety. We separated people into one of three categories, based on whether they were above or below the mean split for low self-focus and high other-focus: (1) low humility: −LSF, −HOF, (2) mixed: +LSF, −HOF or −LSF, +HOF, (3) high humility: +LSF, +HOF. We then examined the differences in existential anxiety and existential thinking among these three groups. We ran a multivariate ANOVA (see Figure 15.1) and found that there was indeed a significant group difference for existential anxiety as measured by EAST, $F(2,144) = 9.6, p < .001$—as well as a marginal difference on the EAS, $F(2,144) = 2.6, p = .07$—and for existential thinking, $F(2,144) = 6.6, p = .002$. This difference became even more pronounced if we simply contrasted people with low humility (−LSF, −HOF) and people with high humility (+LSF, +HOF): they were significantly different from one another on both measures of existential anxiety and existential thinking, $t(s) = 2.1$ to 4.6, $ps = .038$ to $.001$.

Why does humility buffer against existential anxiety? One explanation is that humility is positively correlated with free will, $r(147) = .23, p = .006$, and dualism, $r(147) = .51, p < .001$—which are, in turn, negatively correlated with existential anxiety, $rs(147) = -.17$ to $-.38$, ps .045 to $<.001$. So it may be that humility elevates people's beliefs in free will and/or dualism, which in turn protects against feelings of existential anxiety.

To test this hypothesis, we ran partial correlations to see if the relationship between humility and existential anxiety would disappear if we controlled for

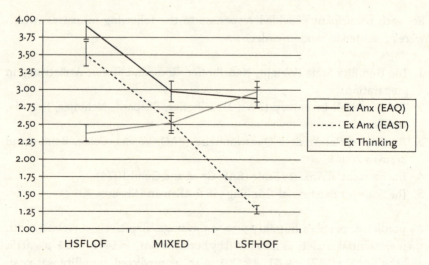

Figure 15.1:
Difference in existential anxiety and thinking across levels of humility.

free will and dualism. In each case, the correlation decreased, becoming only marginally significant for the EAQ, $r(143) = -.16$, $p = .062$ while remaining significant for the EAST, $r(143) = -.31$, $p < .001$ and for existential thinking, $r(143) = .22$, $p = .007$. So, the relationship between humility and existential anxiety cannot be fully explained by stronger beliefs in free will or dualism. This is not surprising. We didn't predict that beliefs about free will and dualism would explain all of the variance in the relationship among humility, existential anxiety, and existential thinking. We simply expected it to play a role—which is what we found in Study 1. But collecting correlational data was just the first step in the process.

2.2. Study 2

The materials for Study 2 were uploaded to Qualtrics. Participants were 213 people recruited through Amazon's Mechanical Turk (MTurk) service and paid $1 each for completing the survey. Participants had to be at least eighteen years of age and living in the United States. Fifty-two percent of participants reported being male, 48 percent female. Participants reported being represented by the following ethnicities: 85 percent Caucasian, 6 percent African-American, 4 percent Asian, and 4 percent Hispanic. The average age of participants was 38.0 years.

In Study 1, we found evidence that was suggestive of our buffering hypothesis when it comes to humility and existential anxiety—that is, the greater people's humility, the lower their existential anxiety. However, in Study 2 we wanted to explore the causal relationship. So we decided to implement a

Figure 15.2:
Mock news banner.

between-subject vignette-based design. Given the angst and anxiety that are often associated with skepticism about free will beliefs—both in the popular press and in academic circles—we thought that challenging free will beliefs might be a useful means of inducing existential anxiety. Based on some of our own previous work on free will beliefs (Schooler et al. 2014), we used three primes in Study 2 that we had used in the past with some success. Each prime was specifically designed to challenge free will or dualism (or both). We did not use a control condition for Study 2 because our goal was to test the buffering hypothesis when it comes to humility and anxiety.[17] So we used these three primes about free will and/or the mind–brain relationship simply as a means of driving down people's beliefs in free will and dualism and thereby driving up their existential anxiety in order to see whether people who were high in humility were less prone to the manipulation (i.e., less existentially anxious). Each participant was presented with one of three primes (see later discussion) and then presented with the same psychometric scales used in Study 1. For Study 2, the format was as follows:

1. Humility Scale (this was given before the primes so that there was no chance for their humility to itself be influenced by the primes)
2. One of three primes
3. FWI, EAQ, EAS, and ETS (once again, presented in random order)
4. Demographics
5. Debriefing

In order to increase the likelihood that participants took themselves to be reading an actual news story rather than a made-up vignette, we included a banner from *New Scientist* with each case (see Figure 15.2)—a strategy that has been used in previous studies on people's judgments about the relationship among free will, responsibility, and advances in neuroscience (see, e.g., Schooler et al. 2014; Shariff et al. 2014). Because some deception is involved with this approach, participants were debriefed afterward and told that while the stories they read were based partly on recent events or stories, these stories were nevertheless modified to suit the ends of our present investigation.

---

17. We did use a control condition in Study 3—which we will discuss later.

Here are the three vignettes we used in Study 2:

The Neuroscience + Anti–Free Will Case

Neuroscientists Discover that Free Will Is an Illusion
By Julia Reed
September 2, 2011

Your decision to read this article was entirely the product of your brain. In a study published this month in the *Journal of Cognitive Neuroscience*,[1] researchers using new brain scanning technology could see exactly which brain processes occurred as people made decisions. "We discovered that our decisions are caused entirely by complex brain processes," says lead researcher Dr. Peter Bernstein of Princeton University's Center for Neuroscience. "This shows that there is no free will. I'll be honest, I find these results very discouraging."

Such discoveries seem to vindicate Dr. Francis Crick—a Nobel Prize winning scientist and co-discoverer of DNA—who once wrote: "Most people take free will for granted, since they feel that they are usually free to act as they please. Although we appear to have free will, in fact, our choices have already been predetermined for us by our brains and we cannot change that. Of course, myths such as free will seem only too plausible. Eventually, we will find the belief in free will every bit as outdated as the belief that the earth is flat."

Professor Bernstein's study used neuroimaging techniques pioneered by Dr. Martha White at Yale University's Center for Advanced Neuroscience. Participants in this experiment were students considering classes to take the next semester. They read descriptions of three psychology courses, considered reasons for and against each, and then pressed one of four buttons indicating their decision to take one of the courses (or none of them). They did all this while lying in a new type of functional Magnetic Resonance Imaging (fMRI) scanner, which measures where and when brain activity occurs, as well as the connections between specific brain processes. The researchers were able to measure how earlier brain processes, such as activity in the temporal striatum, predicted which course participants ultimately decided to take (see figure 1).

Dr. Bernstein explained, "Since their behavior was completely predicted by their brain processes, they clearly did not have free will about which course they selected. Their decisions were nothing more than the inevitable outcome of the processes we observed in their brain." This skeptical view about free will is now the dominant position among the world's leading scientists and philosophers. Dr. White concludes: "These studies confirm that our brains cause our decisions and then we consciously experience the outcome, much like a spectator observing a play. As scientists continue to demystify the mind by uncovering the neural mechanisms that drive our thoughts and behaviors, it has become increasingly clear that there is no role left for free will to play."

[1] Bernstein, P., Yin, R., Smith, J., White, M., and Snyder, H. 2011. "Mapping neural activity during a complex decision-making task." *Journal of Cognitive Neuroscience* 23(5): 1042–1051. | Article |

The Neuroscience + Anti-Dualism Case
Neuroscientists Study Decision-Making in the Brain
By Julia Reed
September 2, 2011

In a study published this month in the *Journal of Cognitive Neuroscience*,[1] researchers using new brain scanning technology could see which brain processes occurred as people made decisions, and they found earlier brain activity that correlated with the decisions people made. "We have discovered some of the complex processes happening in the brain during decision-making," says Dr. Peter Bernstein at the Center for Neuroscience at Princeton University, who led the study.

Such discoveries seem to vindicate Dr. Francis Crick—a Nobel Prize winning scientist and a co-discoverer of DNA—who once wrote: "As we discover how the brain works, we will understand what is going on when people make their decisions. Just as the discovery of DNA explains the mechanism behind something we experience everyday—the similarities between children and their parents—the discoveries of neuroscience will help to explain the mechanisms involved in our everyday experiences of making decisions."

Professor Bernstein's study used neuroimaging techniques pioneered by Dr. Martha White at Yale University's *Center for Advanced Neuroscience*. Participants in this experiment were students considering classes to take the next semester. They read descriptions of three psychology courses, considered reasons for and against each, and then pressed one of four buttons indicating their decision to take one of the courses (or none of them). They did all this while lying in a new type of *functional Magnetic Resonance Imaging (fMRI)* scanner, which measures where and when brain activity occurs, as well as the connections between specific brain processes. The researchers were able to measure how earlier brain processes, such as activity in the temporal striatum, correlated with which course participants decided to take (see figure 1).

Because the results reveal some of the neural states that occur during decision-making, many researchers believe that Dr. Bernstein's team has provided evidence that helps us understand how humans make complex decisions. Dr. White concludes: "Such studies will allow us to explore the neurobiology of even more complex decisions. Our brains have remarkable powers to integrate and process a vast amount of information, and we are just beginning to understand how it does so in human decision making." In light of the gathering data in neuroscience, this view is now the dominant position among the world's leading scientists and philosophers.

[1] Bernstein, P., Yin, R., Smith, J., White, M., and Snyder, H. 2011. "Mapping neural activity during a complex decision-making task." *Journal of Cognitive Neuroscience* 23(5): 1042–1051. | Article |

The Anti-Free Will + Anti-Dualism Prime
Why You Don't Really Have Free Will
By John Reed
January 1st, 2012

Perhaps you've chosen to read this essay after scanning other articles on this website. Or maybe you've decided what to order for breakfast, or what clothes you'll wear today.

You haven't. You may *feel* like you've made choices, but in reality your decision to read this article, and whether to have eggs or pancakes, was determined long before you were aware of it—perhaps even before you woke up today. And your "will" had no part in that decision. So it is with all of our other choices: not one of them results from a free and conscious decision on our part. There is no freedom of choice, no free will. And those New Year's resolutions you made? You had no choice about making them, and you'll have no choice about whether you keep them.

The debate about free will, long the purview of philosophers alone, has been given new life by scientists, especially neuroscientists studying how the brain works. And what they're finding supports the idea that free will is a complete illusion. After all, it is becoming increasingly clear that we are biological creatures, collections of molecules that must obey the laws of physics. All the success of science rests on the regularity of those laws, which determine the behavior of every molecule in the universe. Those molecules, of course, also make up your brain—the organ that does the "choosing." And the neurons and molecules in your brain are the product of both your genes and your environment, an environment including the other people we deal with. Memories, for example, are nothing more than structural and chemical changes in your brain cells. Everything that you think, say, or do, must come down to molecules and physics.

True "free will," then, would require us to somehow step outside of our brain's structure and modify how it works. Science hasn't shown any way we can do this because "we" are simply constructs of our brain. We can't impose a nebulous "will" on the inputs to our brain that can affect its output of decisions and actions, any more than a programmed computer can somehow reach inside itself and change its program. And that's what neurobiology is telling us: Our brains are simply meat computers that, like real computers, are programmed by our genes and experiences to convert an array of inputs into a predetermined output. Recent experiments involving brain scans show that when a subject "decides" to push a button on the left or right side of a computer, the choice can be predicted by brain activity at least *seven seconds* before the subject is consciously aware of having made it. "Decisions" made like that aren't conscious ones. And if our choices are unconscious, with some determined well before the moment we think we've made them, then we don't have free will in any meaningful sense.

The inescapable scientific conclusion is that although we *feel* that we're characters in the play of our lives, rewriting our parts as we go along, in reality we're puppets performing scripted parts written by the laws of physics.

Dr. John Reed is the Walter M. Bernstein Professor of Neuroscience at Princeton University. He is a member of the National Academy of Sciences. His most recent book is titled The Illusion of Free Will: How Our Brains Decide What We Do (MIT Press, 2012).

While the first two primes are loosely based on real articles, the third prime is an excerpt from an editorial biologist Jerry Coyne published in USA Today on New Year's Day in 2012. In previous research, we found this third prime to have a more pronounced and reliable influence on people's beliefs about free will, determinism, dualism, and related concepts (Schooler et al. 2014). Given the rhetorical nature of Coyne's piece and given that he argues not just against free will and dualism but also for outright epiphenomenalism—that is, the view that our conscious deliberations, decisions, and choices don't impact our behavior—it is perhaps unsurprising that this prime tends to be more effective than the other two. We nevertheless wanted a broad range of primes to better put our hypothesis to the test. After all, in order to see whether humility buffers existential anxiety, it was important that the primes we used actually induced existential anxiety. So we selected these three primes under the assumption that they would decrease beliefs in free will and dualism—which would in turn increase existential anxiety. In short, our goal in Study 2 was to explore whether humility continues to buffer against existential anxiety even when people are exposed to primes designed to induce it.

Unfortunately, our priming efforts did not appear to work: we did not find any differences in levels of existential anxiety or existential thinking between the three different prime conditions—or, more importantly, between Study 1, which had no primes (and so can serve as a baseline), and Study 2, $Fs(3,356) < 1.0$, *ns*. So it appears that reading primes failed, at least in this instance, to heighten existential anxiety in the people who read them.

We can still examine whether the two dimensions of humility continue to predict existential anxiety and existential thinking across the different priming conditions—which they did, though the effects were a bit reduced. This time humility was significantly correlated with existential anxiety as measured by the EAST, $r(213) = -.20$, $p = .003$, though not by the EAQ. It was also still positively correlated with people's tendency to engage in existential thinking, $r(213) = .55$, $p < .001$.

We once again separated people into one of three categories based on whether they were above or below the mean split for low self-focus and high other-focus and examined the differences in existential anxiety and existential thinking between these three groups for each condition. It was still the case that level of humility predicted existential anxiety—though in the former case only for the EAST, $t(129) = 2.0$, $p = .044$—and existential thinking, $t(129) = 7.3$, $p < .001$. The other interesting difference between Study 1 and Study 2 emerged when we examined the relationship between existential anxiety and free will/dualism beliefs for each of the three groups of low-humility, mixed, and high-humility people. In Study 1, existential anxiety is more strongly negatively correlated with free will and dualism for people low in humility than for people high in humility. The opposite is true for Study 2—the stronger correlations show up for the people *high* in humility (see Table 15.1). Thus, it appears that the primes were doing *something*—it just isn't clear what.

One worry we had was that by presenting participants with the humility scale before the primes—as we did for Study 2—we were somehow priming people with humility which was in turn having a downstream influence on how they responded to the primes and other psychometric scales. In short, the humility scale might be weakening the effect of the primes—thereby making it hard for us to test our buffering hypothesis. For unless the primes tend to *decrease* people's free will/dualism beliefs and *increase* people's existential anxiety, we can't contrast the responses of people who are high in humility from those who are low in humility. Thus, we decided to move the humility scale to the end of the survey for Study 3 (among other changes).

2.3. Study 3

The materials for Study 3 were uploaded to Qualtrics. Participants were 164 people recruited through Amazon's Mechanical Turk (MTurk) service and paid $1 each for completing the survey. Participants had to be at least eighteen years of age and living in the United States. Fifty-seven percent of participants reported being male, 43 percent female. Participants reported being represented by the following ethnicities: 82 percent Caucasian, 7 percent African-American, 6 percent Asian, and 5 percent Hispanic. The average age of participants in Study 3 was 35.5 years.

Table 15.1. CORRELATIONS BETWEEN FREE WILL, HUMILITY, AND EXISTENTIAL ANXIETY

Humility			Study 1					Study 2				
			FW	DE	DU	FWB	Resp	FW	DE	DU	FWB	Resp
HSFLOF (Low Humility)	EAQ	P	-.508**	.134	-.287	-.351*	-.306*	-.050	.100	-.137	.037	.042
		Si	.000	.379	.056	.018	.041	.711	.454	.307	.785	.757
	EAST	P	-.356*	.180	-.389**	-.289	-.212	-.035	.108	-.320*	-.057	-.073
		Si	.017	.238	.008	.054	.163	.796	.419	.014	.674	.584
		N	45	45	45	45	45	58	58	58	58	58
LSFHOF (High Humility)	EAQ	P	-.297*	.030	-.158	-.189	-.213	-.534**	.333**	-.375**	-.413**	-.334**
		Si	.043	.842	.288	.204	.151	.000	.004	.001	.000	.004
	EAST	P	-.263	-.057	-.285	-.240	-.093	-.398**	.294*	-.295*	-.349**	-.291*
		Si	.074	.701	.052	.104	.534	.000	.012	.011	.002	.012
		N	47	47	47	47	47	73	73	73	73	73

This time around, we wanted to move the humility scale from the front end of the survey to the back end. Consequently, the format of Study 3 was as follows:

1. One of the three primes (including one neutral prime)
2. FWI, EAQ, EAS, ETS (once again, presented in random order)
3. Humility Scale
4. Demographics
5. Debriefing

Moreover, we also decided that we needed to add a neutral prime that would serve as a control condition. While our goal in Study 2 was to move people's free will/dualism beliefs around with each of the three primes, this didn't enable us to see whether the primes were having an overall effect. So, for Study 3, (a) we once again used the Neuroscience + Anti-Free Will Case and the Anti-Free Will + Anti-Dualism Case from Study 2, (b) we eliminated the Neuroscience + Anti-Dualism Case from Study 2, and (c) we added a neutral prime that was about the migratory patterns of whales.

However, even after moving the humility scale to the end of survey (to make sure that it wasn't priming participants), we once again failed to find any differences in levels of existential anxiety or existential thinking between the three different prime conditions or the newly introduced control condition, $Fs(3,144) = .076–1.25$, ns. We also failed to find a difference in people's FWI scores between any of the conditions, $Fs(3,144) = .21–2.02$, ns.

This verifies that the primes were neither influencing attitudes about free will, determinism, or dualism in the people who read them nor were they inducing existential anxiety. While this was disappointing, perhaps it shouldn't have been surprising. In past research, we have often found that people's cherished beliefs about free will and dualism are hard to reliably and consistently move around (and even when these beliefs move around in statistically significant ways, they don't tend to move very much). Unfortunately, because we were unable to get people's intuitions about free will and dualism to budge, we were unable to get their existential anxiety to budge either—which means that we failed in both Study 2 and Study 3 to test our original buffering hypothesis about humility and existential anxiety. So we made one final round of changes to our experimental design in the hopes we could better illuminate our hypothesis.

2.4. Study 4

The materials for Study 4 were uploaded to Qualtrics. Participants were 98 people recruited through Amazon's Mechanical Turk (MTurk) service and paid $1 each for completing the survey. Participants had to be at least eighteen years of age and living in the United States. Fifty-seven percent of participants reported being male, 43 percent female. Participants reported being represented by the following ethnicities: 82 percent Caucasian, 6 percent African-American, 7 percent Asian, and 5 percent Hispanic. The average age for participants in Study 4 was 34.0 years.

For Study 4, we decided that perhaps we should try some of the other primes and psychometric tools that have been used successfully in the past by researchers studying free will beliefs. One of the most widely used paradigms involves providing participants with a brief biographical note about Francis Crick's Nobel-winning career followed by an excerpt about free will and related concepts from Crick's *Astonishing Hypothesis* (Vohs and Schooler 2008). Given that researchers have claimed to have reliable success in pushing around people's free will beliefs with this anti–free will prime, we decided to use this as our main prime in Study 4. Here is the prime in its entirety:

Francis Crick is the British physicist and biochemist who collaborated with James D. Watson in the discovery of the molecular structure of DNA, for which they received the Nobel Prize in 1962. He is the author of *What Mad Pursuit, Life Itself,* and *Of Molecules and Men.* Dr. Crick lectures widely all over the world to both professional and lay audiences, and is a Distinguished Research Professor at The Salk Institute in La Jolla, CA.

Dr. Crick's essay (below) comes from *The Astonishing Hypothesis*.

A Postscript on Free Will

"You," your joys and your sorrows, your memories and your ambitions, your sense of personal identity and free will, are in fact no more than the behavior of a vast assembly of nerve cells and their associated molecules. Who you are is nothing but a pack of neurons.

Most religions hold that some kind of spirit exists that persists after one's bodily death and, to some degree, embodies the essence of that human being. Religions may not have all the same beliefs, but they do have a broad agreement that people have souls.

Yet the common belief of today has a totally different view. It is inclined to believe that the idea of a soul, distinct from the body and not subject to our known scientific laws, is a myth. It is quite understandable how this myth arose without today's scientific knowledge of nature of matter and radiation, and of biological evolution. Such myths, of having a soul, seem only too plausible. For example, four thousand years ago almost everyone believed the earth was flat. Only with modern science has it occurred to us that in fact the earth is round.

> From modern science we now know that all living things, from bacteria to ourselves, are closely related at the biochemical level. We now know that many species of plants and animals have evolved over time. We can watch the basic processes of evolution happening today, both in the field and in our test tubes and therefore, there is no need for the religious concept of a soul to explain the behavior of humans and other animals. In addition to scientists, many educated people also share the belief that the soul is a metaphor and that there is no personal life either before conception or after death. Most people take free will for granted, since they feel that usually they are free to act as they please. Three assumptions can be made about free will. The first assumption is that part of one's brain is concerned with making plans for future actions, without necessarily carrying them out. The second assumption is that one is not conscious of the "computations" done by this part of the brain but only of the "decisions" it makes—that is, its plans, depending of course on its current inputs from other parts of the brain. The third assumption is that the decision to act on one's plan or another is also subject to the same limitations in that one has immediate recall of what is decided, but not of the computations that went into the decision.
>
> So, although we appear to have free will, in fact, our choices have already been predetermined for us and we cannot change that. The actual cause of the decision may be clear cut or it may be determined by chaos, that is, a very small perturbation may make a big difference to the end result. This would give the appearance of the Will being "free" since it would make the outcome essentially unpredictable. Of course, conscious activities may also influence the decision mechanism.
>
> One's self can attempt to explain why it made a certain choice. Sometimes we may reach the correct conclusion. At other times, we will either not know or, more likely, will confabulate, because there is no conscious knowledge of the "reason" for the choice. This implies that there must be a mechanism for confabulation, meaning that given a certain amount of evidence, which may or may not be misleading, part of the brain will jump to the simplest conclusion.

As far as anti-free will primes go, this is a bludgeon. It not only challenges free will but it also challenges dualism, the soul, self-awareness, and even choice itself. For precisely these reasons, it ought to be an effective prime!

Another way in which Study 2 and Study 3 represented a departure from the orthodox approach in the literature on free will beliefs is that we did not use FAD+ (Paulhus and Carey 2011) to measure people's beliefs about free will (and related concepts). Instead, we used the Free Will Inventory (FWI)—which one of us helped develop and validate (Nadelhoffer et al. 2014). But we didn't choose FWI because it just happens to be one of our own. Rather, FWI was designed specifically to improve on a few limitations and shortcomings of FAD+.[18] However, in the past, when we ran direct comparisons between FAD+ and FWI, we admittedly found that the scores on FAD+ were a bit easier to move around with primes (albeit still not always in reliable or predictable ways). At the time, we actually took this to speak *in favor* of the validity of FWI over FAD+. After all, if a good scale is one that reliably picks out *stable* constructs, you wouldn't expect scores on such a scale to be very easy to push around. For present purposes, however, we decided that if we were going to use the Crick prime, perhaps we should pair it with FAD+ rather than FWI—since this has been the most widely used paradigm in the extant literature. In light of these changes, the format for Study 4 was as follows:

1. Crick prime or neutral prime
2. FAD+, EAQ, EAS, and ETS (once again, presented in random order)
3. Humility Scale
4. Demographics
5. Debriefing

This time around, there were two notable differences between the control and experimental condition: people showed a *decrease* in their free will scores in the experimental condition, $t(96) = 2.45$, $p = .016$. But they also showed a marginal *decrease*—rather than the expected increase—in their existential anxiety as measured by EAQ, $t(96) = 1.92$, $p = .057$ (the EAST also trended in the same direction—a reduction—though it wasn't significant). So, on the one hand, the Crick prime successfully decreased people's responses to FAD+ (at least, to the free will beliefs portion—people's scores on the rest of the scale did not move). In this respect, the combination of the two was a success on the surface. After all, since we were trying to use decreased beliefs about free will in order to induce existential anxiety, we needed to accomplish the former in order to test our hypothesis. However, even though we managed to decrease people's beliefs in free will with the Crick essay (as measured by FAD+), this was not accompanied by an increase in existential anxiety as we predicted. Indeed, we found just the opposite—namely, people showed a decrease in existential anxiety after reading the prime. As such, this left us once again unable to test our original hypothesis concerning the relationship between humility and existential anxiety. We nevertheless think this entire exercise was telling (for reasons we will discuss in the final section). But before we talk about the implications of our findings, we first want to briefly highlight some possible limitations and shortcomings of our present studies and what we might do in the future to address them.

18. See Nadelhoffer et al. (2014) for details.

3. CURRENT LIMITATIONS AND FUTURE DIRECTIONS

One obvious limitation of our present studies is that we focused narrowly on the intuitions and beliefs of Americans. So, when we spoke throughout the paper about "folk intuitions," "common sense morality," and the like, we were using those terms as shorthand for the moral beliefs and practices of people living in the United States. We realize this is suboptimal and narrows the application of our present findings. We have recently been working with colleagues in Brazil to collect cross-cultural data about free will beliefs. So this would be a natural place for us to expand our current line of research down the road.

Another limitation of our present studies is that we didn't collect any longitudinal data—we simply primed people with mock news stories and tested whether that had a significant influence on their attitudes and beliefs just a few moments later. Here, too, we acknowledge that this is a shortcoming of our present research and one that we hope to remedy down the road. For instance, in Shariff et al. (2014), they test for shifts in free will beliefs after students have taken an undergraduate course in neuroscience. We think this is a promising approach. However, one downside is that the participants will not only be WEIRD (Western, educated, and from an industrialized, rich, democratic country), they will also be much younger (and less diverse) than is typical of the general population. It is one reason we like MTurk as a tool for recruiting participants—namely, you get a more representative sample than is typically the case in studies involving only undergraduates (even if the MTurk population, too, is less representative than one would like). That said, we nevertheless appreciate that longitudinal data would be both telling and important. Figuring out how best to collect the salient data is a challenge we plan to tackle in the future.

Another limitation of our present studies is that we only used vignettes to try to decrease beliefs in free will and dualism and increase existential anxiety. Other researchers have had success with different approaches. For instance, one commonly used design involves the use of so-called "Velten" statements (Vohs and Schooler 2008). On this paradigm, participants are divided into three conditions—a neutral condition, an anti–free will condition, and a pro–free will condition. In each condition, participants are presented with a series of fifteen statements. Participants are required to look at each statement for an entire minute. However, while this paradigm has been used with some success by other researchers, we have used it in the past with very mixed results. The most obvious problem is that the paradigm appears to be very irritating and frustrating for participants. For instance, in the past when we tried this approach, the dropout rate for our participants was nearly 50 percent by the time they read just half of the statements. It turns out having to stare at and contemplate single sentences for a minute is not a pleasant experience. As such, we don't have much faith in the approach (since it seems to add annoyance as a potential confound). However, while we would prefer not to use this particular paradigm—despite its popularity among some researchers—we appreciate that new primes are needed if we are to (even temporarily) move people's beliefs about free will, dualism, and related concepts.

Just as new primes may be needed, it's also possible that new measures for free will beliefs may be called for as well. While we used both FWI as well as FAD+ in our present studies, there are other scales available.[19] Perhaps we would have more luck with these scales instead. Indeed, some researchers have dispensed with scales altogether—opting instead for simpler and more direct methods of detecting changes in free will beliefs. For instance, in some of our previous research (Schooler et al. 2014), we used a 100-point slider scale with a single statement—namely, "I have free will."[20] Here again, while we have had some success with this measure, there are worries with relying on this as a *sole metric* for free will beliefs. For instance, if you give someone an anti–free will prime and then immediately present them with this type of single-sentence slider, you invite participants to perceive demand characteristics. This is a problem even when scales are being used, of course. But at least with the scales, several related data points are collected and a number of the items of the scale don't obviously hook up to the content of the prime. But, here again, just because we don't think a slider is the best approach, it doesn't mean we aren't open to using or developing better measures for free will beliefs.

The same can be said about the two scales we used for measuring existential anxiety. We used these scales because they had already been developed and validated. But, much like several of the free will scales, we worry about the content validity of some of the items. Indeed, we think that a team of philosophers (who work on existentialism) and psychologists (who work on anxiety) would likely be able to construct a better psychometric tool for measuring existential anxiety. Given that members from our team have already constructed and validated scales for measuring (a) humility, (b) open-mindedness, (c) moral flexibility, and (d) beliefs about free will and related concepts, we are certainly open to working on a new and improved scale for measuring existential anxiety down the road. But, for these preliminary studies, we decided to use tools that were already readily available. Constructing and validating scales is a time-consuming and expensive process. So it is a task we set aside for a later date.

---

19. See, e.g., Rakos et al. (2008); Stroessner and Green (1990); Viney, McIntyre, and Viney (1984); and Viney, Parker-Martin, and Dotten (1988).
20. See, also, Protzko, Ouimente, and Schooler (2016).

4. CONCLUSION

Our primary goal in this chapter was to test our buffering hypothesis about humility—that is, the view that trait humility (the combination of low self-focus with high other-focus) serves as a buffer against existential anxiety. We also predicted that humble individuals would be more likely to engage in existential thinking. While we found some correlational support for these hypotheses in Study 1, the findings from Studies 2–4 either did not provide us with any additional support (or provided only mixed support). In this sense, our studies were not as probative as we had hoped. More work remains to be done before we will have a deeper understanding of the relationship among humility, existential anxiety, and existential thinking.

It appears that our biggest methodological barrier was trying to use decreased free will beliefs as our means of inducing existential anxiety. For although Study 1 here again provided some correlational support for our assumption that people who believe less in free will are likely to have more existential anxiety, we were unable to push around people's beliefs in free will sufficiently in Studies 2–4 to test our primary buffering hypothesis about humility and existential anxiety. In order for our experimental paradigms in Studies 2–4 to shed light on this issue, we needed to be able to accomplish two things: first, we needed to reliably decrease people's free will beliefs using primes that challenge these beliefs. Second, we needed people's existential anxiety to increase in the wake of their decreased belief in free will. Because we were unable to accomplish the former, we were unable to accomplish the later.

Given how fickle the findings on free will beliefs appear to be (more on that later), it is clear that we adopted a wrong-headed approach for testing our buffering hypothesis about humility and existential anxiety. Yet we chose this approach for two important reasons: first, the extant literature gave us reason to think that we should be able to move around people's free will beliefs. Second, existential anxiety and a kind of neuro-angst is often evoked in public discussions about the relationship between free will and the brain sciences.

This means that while Studies 2–4 did not enable us to test our primary hypothesis, our efforts were nonetheless illuminating in two ways. On the one hand, they helped us plot a better way forward. On the other hand, they provided a cautionary tale for those who try to get theoretical and practical mileage out of the growing (but problematic) empirical literature on free will beliefs. Before closing, we would like to say a bit more about each of these lessons in turn.

For starters, it now seems that the best way to test the buffering hypothesis about humility and existential anxiety is to do so directly rather than indirectly. Our main mistake in our preliminary efforts to explore this issue is that we chose free will beliefs as an intermediary between the two—that is, we should not have tried to induce existential humility by inducing disbelief in free will. Rather, we should have developed primes that more directly increase existential humility—for example, we could have used short excerpts from famous existential philosophers ranging from Kierkegaard and Nietzsche to Camus and Sartre. By inducing existential anxiety directly rather than indirectly, we should be in a better position to test our buffering hypothesis. This is a task that we are already undertaking. Hopefully, the findings end up being more promising given our purposes.

Second, our present results provide clear evidence that we must be cautious when it comes to drawing stark conclusions from the extant literature on free will beliefs. As we found in our present empirical efforts, it is very difficult to change people's free will beliefs simply be challenging them. While researchers have reported having success with this approach, the results have been difficult to replicate (both directly and conceptually). Contrary to what one might conclude in light of the published literature, people's beliefs in free will and related concepts are fairly resilient. As such, these beliefs are unsurprisingly hard to reliably move around—for example, it is hard to explain why the Crick prime seems to work with some regularity but the Coyne prime yields more mixed results. Moreover, the extant findings are often presented in misleading ways. For instance, in one of the most recent articles on free will beliefs, the title is "Believing There Is No Free Will Corrupts Intuitive Cooperation" (Protzko et al. 2016). However, if you look at the results between the anti–free will condition and the control condition, you find the following: "When asked to rate on a 1–100 scale on their agreement with the statement 'I have free will,' those in the no free will condition believed significantly less (M = 76.541, SD = 24.227) than if they had read the control passage (M = 86.676, SD = 16.045, p <.001)."

In short, even in the no free will condition, the average response to the slider was 76 out of 100—which means that participants decidedly did not "believe there is no free will." Instead, their pro–free will beliefs were simply *weaker* than the participants in the control condition (where the average response was 86 out of 100). While there is no doubt that moving people's scores from 86 to 76 is statistically significant, it is misleading to frame this is terms of inducing *disbelief* in free will. Indeed, it is not clear that any study that has been conducted thus far has been able to move people's free will beliefs from above the midline to below the midline—but this just suggests that while challenging people's beliefs in free will may weaken these beliefs, it doesn't undermine these beliefs. But you wouldn't think this if you looked at how the results in the extant literature have been framed and discussed.

The recent study by Protzko et al. (2016) also reveals another problematic feature of the literature on free will beliefs—namely, that the results are fickle and hard to replicate. For instance, while Vohs and Schooler (2008) did not report this, it turns out that their findings only hold when people think they are participating in two distinct studies. We found the same thing in some of our earlier work (see, e.g., Schooler et al. 2014). For this reason, Protzko et al. (2016) also told subjects they were participating in two distinct studies. Of course, that's fine as far as it goes. But it provides yet more evidence that moving around people's beliefs about free will is not nearly as simple or straightforward as researchers have made it seem. Surely this partly explains why some of these findings have been difficult to replicate. It seems like you can only get people's beliefs in free will to move using very specific experimental designs, primes, and measures—which raises a serious worry about external validity and cautions against reading too much into the findings.

At a minimum, we believe this should give philosophers, psychologists, and pundits pause for concern. A lot of ink has been spilled in the wake of the growing literature on beliefs in free will and related concepts. But the findings that have been the source of much of the fear-mongering about disbelief in free will in the face of the brain sciences are less stable and reliable than many have assumed. As such, not only have the claims about the death of free will been premature, but so have the claims about the possible ramifications. While people's cherished beliefs about free will can be moved around the margins, none of the extant data suggest that these beliefs can be radically altered—especially not with simple primes. Moving forward, we think researchers need to do a better job of discussing the details of their experimental designs as well as the difficulties they may have had replicating earlier results.

Based on the findings we presented in this chapter, we think more work needs to be done before we'll be in any position to shed light on the possible ramifications of wide-scale skepticism about free will (and the neuro-angst that is supposed to follow on its heels). People's beliefs in free will and related concepts appear to be on pretty firm footing, and the findings to the contrary don't appear to be especially robust, stable, or reliable. As such, in our future work on humility and existential anxiety, we won't explore the possible implications of disbelief in free will. Most people believe very strongly in free will. And, at least for now, it doesn't appear this is going to change anytime soon (not even in response to news stories about the brain sciences and the alleged death of free will).

REFERENCES

Allan, B. A., and C. B. Shearer. 2012. The scale for existential thinking. *International Journal of Transpersonal Studies* 31(1): 21–37.

Aquino, K., and A. I. Reed. 2002. The self-importance of moral identity. *Journal of Personality And Social Psychology* 83(6): 1423–1440. doi:10.1037/0022-3514.83.6.1423

Bargh, J. 2008. Free will is unnatural. In J. Baer, J. Kaufman, and R. Baumeister (Eds.), *Are we free? Psychology and free will*, pp. 128–154. New York: Oxford University Press.

Baumeister, R. F., E. J. Masicampo, and C. N. DeWall. 2009. Prosocial benefits of feeling free: Disbelief in free will increases aggression and reduces helpfulness. *Personality and Social Psychology Bulletin* 35: 260–268.

Cashmore, A. 2010. The Lucretian swerve: The biological basis of human behavior and the criminal justice system. *Proceedings of the National Academy of Sciences of the United States of America* 107: 4499–4504.

Coyne, J. 2012. Why you don't really have free will. *USA Today* (January 1st).

Furco, A., P. Muller, and M. S. Ammon. 1998. *Civic responsibility survey*. Service Learning Research and Development Center, University of California, Berkeley.

Good, L., and K. A. Good. 1974. Preliminary measure of existential anxiety. *Psychological Reports* 34: 73–74.

Graham, J., J. Haidt, and B. A. Nosek. 2009. Liberals and conservatives rely on different sets of moral foundations. *Journal of Personality and Social Psychology* 96: 1029–1046.

Greene, J., and J. Cohen. 2004. For the law neuroscience changes nothing and everything. *Philosophical Transactions of the Royal Society B: Biological Sciences* 359: 1775–1785.

Harris, S. 2012. *Free will*. New York: Free Press.

Haynes, J. D. 2011. Beyond Libet: Long-term prediction of free choices from neuroimaging signals. In W. Sinnott-Armstrong and L. Nadel (Eds.), *Conscious will and responsibility*, pp. 85–96. Oxford: Oxford University Press.

Katz, I., and R. G. Hass. 1988. Racial ambivalence and American value conflict: Correlational and priming studies of dual cognitive structures. *Journal of Personality and Social Psychology* 55(6): 893–905. doi:10.1037/0022-3514.55.6.893

Kesebir, P. 2014. A quiet ego quiets death anxiety: Humility as an existential anxiety buffer. *Journal of Personality and Social Psychology* 106(4): 610–623.

LaBouff, J. P., W. C. Rowatt, M. K. Johnson, J. Tsang, and G. M. Willerton. 2012. Humble persons are more helpful than less humble persons: Evidence from three studies. *The Journal of Positive Psychology* 7: 16–29.

Mele, A. 2009. *Effective intentions*. New York: Oxford University Press.

Mele, A. 2012. Autonomy and neuroscience. In L. Radoilska (Ed.), *Autonomy and mental disorder*, pp. 26–43. Oxford University Press.

Mele, A. 2011. Free will and science. In R. Kane (Ed.), *Oxford handbook of free will*, 2nd ed., pp. 499–514. New York: Oxford University Press.

Mele, A. 2010a. Scientific skepticism about free will. In T. Nadelhoffer, E. Nahmias, and S. Nichols (Eds.), *Moral psychology: Classical and contemporary readings*, pp. 295–305. New York, NY: Wiley-Blackwell.

Mele, A. 2010b. Testing free will. *Neuroethics* 161–172.

Mele, A. 2010c. Conscious deciding and the science of free will. In R. Baumeister, A. Mele, and K. Vohs (Eds.), *Free will and consciousness: How might they work?* pp. 43–65. New York: Oxford University Press.

Mele, A. 2011. Free will and science. In R. Kane (Ed.), *Oxford handbook of free will*, 2nd ed., pp. 499–514. New York: Oxford University Press.

Mele, A. 2012. Autonomy and neuroscience. In L. Radoilska (Ed.), *Autonomy and mental disorder*, pp. 26–43. New York: Oxford University Press.

Mele, A. 2009. *Effective intentions*. New York: Oxford University Press.

Montague, P. R. 2008. Free will. *Current Biology* 18: R584–R585.

Morse, S. 2011. Avoiding irrational neuroLaw exuberance: A plea for neuromodesty. *Mercer Law Review* 62: 837–859.

Nadelhoffer, T. 2011. The threat of shrinking agency and free will disillusionism. In L. Nadel and W. Sinnott-Armstrong (Eds.), *Conscious will and responsibility*, pp. 173–188. New York: Oxford University Press.

Nadelhoffer, T. 2014. Dualism, libertarianism, and scientific skepticism about free will. In W. Sinnott-Armstrong (Ed.), *Moral psychology: Neuroscience, free will, and responsibility* (vol. 4), pp. 209–216. Cambridge, MABoston: MIT Press.

Nadelhoffer, T. 2011. The threat of shrinking agency and free will disillusionism. In L. Nadel and W. Sinnott-Armstrong (Eds.), *Conscious will and responsibility*, pp. 173–188. New York: Oxford University Press.

Nadelhoffer, T., and D. Goya-Tocchetto. 2013. The potential dark side of free will: Some preliminary findings. In G. Carusso (Ed.), *Exploring the illusion of free will and moral responsibility*, pp. 121–140. Lanham, MD: Lexington Books.

Nadelhoffer, T., and T. Matveeva. 2009. Positive illusions, perceived control, and the free will debate. *Mind & Language* 24: 495–522.

Nadelhoffer, T., J. Shepard, E. Nahmias, C. Sripada, and L. Ross. 2014. The free will inventory: Measuring beliefs about agency and responsibility. *Consciousness and Cognition* 25: 27–41.

Nadelhoffer, T., and J. Wright. Forthcoming. The twin hallmarks of humility. In C. Miller and W. Sinnott-Armstrong (Eds.), *Moral psychology: Virtues and vices* (vol. 5). Cambridge, MA: MIT Press.

Nahmias, E. 2011. Why "Willusionism" leads to "bad results": Comments on Baumeister, Crescioni, and Alquist. *Neuroethics* 4: 17–24.

Nahmias, E. 2014. Is free will an illusion? Confronting challenges from the modern mind sciences. In W. Sinnott-Armstrong (Ed.), *Moral psychology* (vol. 4): *Free will and responsibility*, pp. 1–25. New York: Oxford University Press.

Nahmias, E. 2011. Why "Willusionism" leads to "bad results": Comments on Baumeister, Crescioni, and Alquist. *Neuroethics* 4: 17–24.

Open Science Collaboration. 2015. Estimating the reproducibility of psychological science. *Science* 349.: doi:10.1126/science.aac4716

Overbye, D. 2007. Free will: Now you have it, now you don't. *The New York Times* (January 2nd).

Paulhus, D. L., and J. M. Carey. 2011. The FAD-Plus: Measuring lay beliefs regarding free will and related constructs. *Journal of Personality Assessment* 93: 96–104.

Protzko, J., B. Ouimente, and J. Schooler. 2016. Believing there is no free will corrupts intuitive cooperation. *Cognition* 151: 6–9.

Rakos, R., K. Laurene, S. Skala, and S. Slane. 2008. Belief in free will: Measurement and conceptualization innovations. *Behavior and Social Issues* 17: 20–39.

Richards, N. 1988. Is humility a virtue? *American Philosophical Quarterly* 25(3): 253–259.

Rigoni, D., S. Kuhn, G. Sartori, and M. Brass. 2011. Inducing disbelief in free will alters brain correlates of preconscious motor preparation: The brain minds whether we believe in free will or not. *Psychological Science* 22(5): 613–618.

Schlenker, B. R., M. L. Miller, and R. M. Johnson. 2009. Moral identity, integrity, and personal responsibility. In D. Narvaez, D. K. Lapsley, D. Narvaez, D. K. Lapsley (Eds.), *Personality, identity, and character: Explorations in moral psychology*, pp. 316–340. New York: Cambridge University Press.

Schooler, J., T. Nadelhoffer, E. Nahmias, and K. D. Vohs. 2014. Measuring and manipulating beliefs and behaviors associated with free will: The good, the bad, and the ugly. In *Surrounding free will: Philosophy, psychology, neuroscience*, ed. A. Mele, pp.72-94. New York: Oxford University Press.

Schwartz, S. H., and W. Bilsky. 1987. Toward a universal psychological structure of human values. *Journal of Personality and Social Psychology* 53(3): 550–562. doi:10.1037/0022-3514.53.3.550

Shariff, A. F., J. D. Greene, J. C. Karremans, J. Luguri, C. J. Clark, J. W. Schooler, R. F. Baumeister, and K. D. Vohs. 2014. Free will and punishment: A mechanistic view of human nature reduces retribution. *Psychological Science* 25 (8): 1563-70.

Sidgwick, H. 1874/2011. *The method of ethics*, ed. J. Bennett (www.earlymoderntexts.com).
Smilansky, S. 2002. Free will, fundamental dualism, and the centrality of illusion. In R. Kane (Ed.), *The Oxford handbook of free will*, pp. 489–505. Oxford: Oxford University Press.
Smilansky, S. 2001. Free will: From nature to illusion. *Proceedings of the Aristotelian Society* 101: 71–95.
Smilansky, S. 2000. *Free will and illusion*. New York: Oxford University Press.
Smilansky, S. 2001. Free will: From nature to illusion. *Proceedings of the Aristotelian Society* 101: 71–95.
Smilansky, S. 2002. Free will, fundamental dualism, and the centrality of illusion. In R. Kane (Ed.), *The Oxford handbook of free will*, pp. 489–505. Oxford: Oxford University Press.
Stillman, T. F., and R. F. Baumeister. 2010. Guilty, free, and wise: Belief in free will facilitates learning from self-conscious emotions. *Journal of Experimental Social Psychology* 46: 951–960.
Stillman, T. F., R. F. Baumeister, K. D. Vohs, N. M. Lambert, F. D. Fincham, and L. E. Brewer. 2010. Personal philosophy and personnel achievement: Belief in free will predicts better job performance. *Social Psychological and Personality Science* 1: 43–50.
Stroessner, S. and C. Green. 1990. Effects of belief in free will or determinism on attitudes toward punishment and locus of control. *The Journal of Social Psychology* 130(6): 789–799.
Taylor, S. and J. Brown. 1988. Illusion and well-being: A social psychological perspective of mental health. *Psychological Bulletin* 103(2): 193–210.
Viney, W., R. McIntyre, and D. Viney. 1984. Validity of a scale designed to measure beliefs in free will and determinism. *Psychological Reports* 54: 867–872.
Viney, W., P. Parker-Martin, and S. Dotten. 1988. Beliefs in free will and determinism and lack of relation to punishment rational and magnitude. *Journal of General Psychology* 115: 15–23.
Viney, W., R. McIntyre, and D. Viney. 1984. Validity of a scale designed to measure beliefs in free will and determinism. *Psychological Reports* 54: 867–872.
Vohs, K. D., and J. W. Schooler. 2008. The value of believing in free will: Encouraging a belief in determinism increases cheating. *Psychological Science* 19: 49–54.
Weens, C. F., N. M. Costa, C. Dehon, and S. L. Berman. 2004. Paul Tillich's theory of existential anxiety: A preliminary conceptual and empirical analysis. *Anxiety, Stress, and Coping* 17: 383–399.
Wegner, D. 2003. *The illusion of conscious will*. Cambridge, MA: MIT Press.
Wegner, D. 2008. Self is magic. In J. Baer, J. Kaufman, and R. Baumeister (Eds.), *Are we free? Psychology and free will*, pp. 226–247. New York: Oxford University Press.
Wegner, D. 2003. *The illusion of conscious will*. Cambridge, MA: MIT Press.
Wirzba, N. 2008. The touch of humility: An invitation to creatureliness. *Modern Theology* 24(2): 225–244.
Wright, J. C., T. Nadelhoffer, T. Perini, A. Langville, M. Echols, and K. Venezia. 2017. The psychological significance of humility. *Journal of Positice Psychology* 12 (1): 3–12.
Wright, J. C., and E. Reinhold, E. (In preparation). The multi-faceted nature of greed.
Zhao, X., L. Liu, X. Zhang, J. Shi, and Z. Huang. 2014. The effect of belief in free will on prejudice. *PLoS One* 9(3): 1–7.

CHAPTER 16

Purpose, Freedom, and the Laws of Nature

SEAN M. CARROLL

In the popular imagination, existentialism is associated with philosophers sitting in cafes, smoking cigarettes and drinking apricot cocktails (Blakewell 2016). Nothing could be further from the popular image of scientists, performing precise experiments while decked in lab coats. But despite this disparity in stereotypes, there are undeniable connections between existentialism and science, especially biology and neuroscience (Flanagan 2009). Upon reflection, it's not hard to see why. Two primary themes of existentialism are the essentially purposeless nature of the cosmos and the human ability to make free choices. Both of these ideas engage in crucial ways with research from contemporary science.

The connections go beyond biology and neuroscience. An honest grappling with the questions of purpose and freedom in the universe must also involve ideas from physics and cosmology. If a better understanding of the neuroscientific basis for human thought and action calls into question the existence of an authentic self, the same could be said for the idea that we are simply collections of particles and forces obeying the rules of quantum mechanics. If we want to create purpose and meaning at the scale of individual human lives, it behooves us to understand the nature of the larger universe of which we are a part.

In this chapter, I will survey some of the ways in which physics and cosmology have an impact on the foundations of both neuroscience and existentialism. My goals are primarily explanatory and pedagogical, rather than breaking new ground.

1. PATTERNS AND DETERMINISM

The first great edifice of physics was classical Newtonian mechanics. Although today most of Newton's specific principles have been superseded by relativity and quantum mechanics, there is a real sense in which even the most modern physical theories follow the basic classical paradigm. For that reason, it is worth appreciating how much of a break classical physics represents, both with our intuitive picture of the world and with pre-Newtonian attempts to describe it carefully.

The basic idea of classical mechanics is simple. The world consists of a collection of things, located in three-dimensional space and evolving with time. A complete description of the state of each thing is simply a specification of its position and its momentum. There are other qualities that things might have—energy, angular momentum, and so forth—but from the basic quantities of position and momentum, everything else can be derived. Newton's laws provide a unique answer to the question: Given the position and momentum of everything in the universe, how does the universe evolve over time?

It's worth highlighting two central features of this conception that distinguish it from predecessors such as Aristotle's physics (Carroll 2016). First is a principle whose significance is easy to overlook: *conservation of momentum*. This is encapsulated in Newton's First Law of Motion: every object in a state of uniform motion remains in that state of motion unless acted upon by an external force. In other words, the natural tendency of things is to keep moving as they are. This elegant idea stands in stark contrast to Aristotle's view that motion is necessarily the result of some cause; the natural tendency of things is to be still, and motion represents a departure that requires an explanation. It's a simple change of perspective, but one with profound consequences. Aristotle's cosmos was one of teleology and causes; a Newtonian world just keeps moving, without any future goals or external guidance. This change in our conception of the underlying laws of nature doesn't directly lead to the disappearance of "purpose" in human lives, but the two phenomena are clearly related.

This shift is driven home even more dramatically by the second feature of classical mechanics: *conservation of information*. This principle was elucidated first not by Newton himself, but by French mathematician and physicist Pierre-Simon Laplace, in his famous "Laplace's Demon" thought experiment (Laplace 1814). Laplace never mentioned a demon; he imagined a "vast intellect" that knew the exact state of the entire universe—the position and momentum of every piece of matter—and had the calculational ability to use Newton's laws to predict the future (and retrodict the past) from that knowledge. Laplace's thought experiment is not practical, but it highlights an essential aspect of classical physics: the universe does not care about its past or future. Its evolution through time is governed by patterns that relate the state at one moment of time to those immediately earlier and later. The total amount of information contained in any such state is the same and determines the entirety of the evolution.

A Newtonian universe, then, is both deterministic (the state of the universe at one time completely fixes what will be the case at all future times) and reversible (it also completely fixes the past). This "clockwork" aspect of the classical universe contributed to the gradual erosion of belief in cosmic purpose. If we are physical beings, obeying impersonal laws of nature, and those laws work from moment to moment rather than being directed toward a goal, it's not hard to conclude that life itself is purposeless.

It is sometimes claimed that the subsequent development of chaos theory (Strogatz 2016) undermined the idea that the universe is deterministic even at a classical level, but that's misleading. Laplacian determinism says "the exact state of the universe at one time, plus the laws of physics, determine the exact state of the universe at every other time." The lesson of chaos theory is "small deviations between two possible states at one moment in time can develop into large deviations at later times." Those statements are completely compatible with each other. Chaos theory reminds us that, in practice, none of us is ever going to be Laplace's demon; the requirements of precision and computing power are just too large. But that is completely unsurprising. Neither Laplace nor anyone who subsequently understood his claim imagined that he was providing a way to build a realistic prediction machine. The point is merely that the evolution of the universe is determined, not that we can learn what it is determined to be.

2. QUANTUM MECHANICS

The only revolution in physics comparable to—and arguably great than—classical mechanics in importance is the twentieth-century establishment of quantum mechanics, which threatened to completely overthrow the entrenched view of Laplacian determinism.

Revolutions take time to absorb, and decades after its appearance physicists still don't have an accepted view of the foundations of quantum mechanics. What seems clear is this: when we observe a quantum system, the best the theory is able to tell us is the probability of obtaining a given outcome. There is no way, in principle, to reduce that probability to a certainty. In place of classical notions of position and momentum, quantum mechanics posits a "wave function" from which the probabilities of measurement outcomes can be calculated.

Many questions are immediately raised. Does the impossibility of exact prediction mean that the future history of the universe is fundamentally nondeterministic? Does quantum indeterminism suggest a physical basis for the kind of radical freedom that Sartre pointed to as a primary aspect of the human condition? Does the emphasis on "observation" imply a newfound role for human agency or consciousness in the workings of the physical world?

Answers to these questions will depend on one's attitude toward how quantum mechanics should be interpreted, but we can sketch some reasonable possibilities. Even the most basic issue of whether quantum mechanics is deterministic or not is up for debate (Ney and Albert 2013). Certainly, we cannot deterministically predict the outcomes of experiments, but that does not directly imply that the underlying physics is indeterministic. Indeed, there are two famous counterexamples. In "hidden-variable" theories, such as the model developed by Louis de Broglie and David Bohm, there are ontological structures over and above the quantum wave function that do completely determine future outcomes, but their values are impossible to determine until the measurements are made (Bohm 1952). In "many-worlds" theories, pioneered by Hugh Everett, the wave function is all that exists and it evolves deterministically, but that evolution includes branching into separate worlds (Wallace 2012). After branching occurs, self-locating uncertainty arises because there are multiple copies of every person, none of whom initially know which branch they are on. But there are also genuinely indeterministic varieties of quantum theory; "dynamical collapse" models, such as that proposed by Giancarlo Ghirardi, Alberto Rimini, and Tullio Weber (GRW), augment the usual smooth evolution of the wave function by a fundamentally stochastic probability that it will collapse at any moment (Ghirardi, Rimini, and Weber 1985). This GRW approach is amenable to experimental constraint, so future empirical investigation may be able to shed light on the matter once and for all.

What is more clear is that quantum mechanics by itself should *not* affect how we think about human freedom. The quantum formalism does not allow us to predict exact experimental outcomes ahead of time; that does not imply, contrary to the impression one might receive from some of the less reputable corners of popular-science writing, that human volition brings reality into being by the act of observation. Given the wave function of a system, we know the probability of each measurement outcome, and that is all that it is possible to know. Any account according to which human beings can influence those probabilities through their intentions would represent a dramatic incompatibility with quantum mechanics, not simply taking advantage of its features.

Likewise, despite possible implications of words such as "measurement" and "observation," it has become increasingly clear that consciousness plays no role in quantum mechanics. A measurement event happens when an isolated quantum system is brought into contact with a large, macroscopic environment, which becomes entangled with the different possible measurement outcomes. This process of "decoherence" is purely physical and well-understood within the mathematical formalism (Schlosshauer 2007).

All of which is to say: while quantum mechanics represents a profound and puzzling departure from the comfortable rules of classical mechanics, it doesn't actually change the status of our existing ways of thinking about human agency and freedom.

3. THE ARROW OF TIME

The principles of conservation of momentum and information, common to both classical and quantum mechanics, seem so basic and simple that we might wonder why it took so long for them to be discovered. Somehow, these fundamental principles of nature do not manifest themselves straightforwardly in the world of our everyday experience. Ultimately, a single feature of the macroscopic world distinguishes it from the characteristics of the underlying microscopic laws: the arrow of time (Albert 2003; Carroll 2010).

Time's arrow is no more or less than the fact that the past and the future seem like very different things. The past has happened; it's in the books, not something that our present choices can possibly ever affect. The future, by contrast, is unknown; a central existentialist theme is our freedom to make choices (implicitly about the future) and its concomitant responsibility. We have memories and records of the past, but not of the future; we are all born young and then grow old; causes come before effects.

None of these features is inherent in the underlying laws of physics, which treat the past and future on an equal footing. Laplace's demon sees both directions of time with unimpeded clarity. The arrow of time does not arise from the fundamental laws themselves, but from specific arrangement of matter in our actual universe. In particular, from the fact that the universe started out in a very orderly state after the Big Bang, almost fourteen billion years ago and has been growing more disorderly ever since. Physicists characterize this increase in disorder by saying that "entropy tends to increase over time." That's the Second Law of Thermodynamics, developed over the course of the nineteenth century. The question is how the Second Law can be reconciled with deterministic, reversible underlying laws.

That question was largely answered by Austrian physicist Ludwig Boltzmann. His contribution was to notice that, although matter is made of atoms, we don't actually see the individual atoms in an object; we only see certain observable macroscopic characteristics. There are potentially many different arrangements of atoms that would look the same to our eyes. Entropy, according to Boltzmann, is simply (the logarithm of) the number of arrangements of a system that would be macroscopically indistinguishable. It increases over time because the universe started in a very low-entropy state corresponding to a relatively small number of possible configurations, and the state of the universe has naturally been evolving toward higher entropy states ever since simply because there are so many more ways to be high-entropy than to be low-entropy.

This abstract physics discussion relates directly to human-scale questions of how we experience the flow of time through our lives. Consider the basic issue of why we remember the past and not the future. Naïvely, we attribute this distinction to a fundamental ontological difference: the past has already happened, while the future is not yet real. The laws of physics don't reflect any such difference, putting the past and future on an equal footing. Nevertheless, we consider certain kinds of features of the present state of the world—fossils, photographs, historical documents—as (more or less) accurate records of memories of the past. No analogous artifacts serve as reliable records of the future. The imbalance ultimately flows from the low-entropy condition of the early universe.

Without that condition, the existence of a photograph today of you from ten years ago would indicate essentially nothing reliable about the past, any more than it does about the future. Given the existence of such a photo, if the past were high-entropy, it would be overwhelmingly probable that it had assembled itself somewhat randomly from molecules bumping into each other over time, free of any direct connection to the kind of scene the picture portrays. It is only the combination of (1) a photograph in the current universe and (2) a very low-entropy past that lets us conclude that the most likely way for such a image to come to be is through the existence of an actual event that had been photographed. The evolution of entropy explains why we have memories of the past but not of the future.

In modern physics and cosmology, then, the universe is not teleological, pulled toward a future goal. It is, if anything, *ekinological*, pushed from the past, from "εκκίνηση," meaning "start" or "departure." The distinction between past and future arises from special conditions near the beginning of the universe. The story of cosmic evolution is one of gradual winding-down toward equilibrium.

4. EMERGENCE

The arrow of time is an example of a feature that is undoubtedly true about our universe but nowhere to be found in fundamental physics. This situation isn't unusual. There is more than one way to accurately talk about the universe; reality presents itself to us as a series of interconnected levels, each with its own concepts and rules. At the bottom is fundamental or microscopic physics. The many levels "above" this most comprehensive description are *emergent* (Bedau and Humphreys 2008).

Discussions of emergence are notoriously contentious, as different communities (and different commentators within each community) choose to emphasize different varieties of the phenomenon. The basic idea is straightforward. We have some system (a person, an ecosystem, a box of gas) that could, in principle, be described at a "microscopic" or "fundamental" level, in which we would specify literally the precise state of all its constituent atoms and molecules and study its dynamics using the underlying laws of physics. But we can also describe the system at a "macroscopic" or "coarse-grained" level, where we specify it in enormously simpler terms. We could, for example, give the temperature and pressure and velocity of a gas, rather than the position and momentum of every gas molecule. Such a description is said to be emergent if this simplified description suffices to give us a (sufficiently) accurate description of how the system behaves.

This definition corresponds to what is known as "weak" emergence—there is a useful macroscopic vocabulary for describing the system, but, given infinite information and processing power, we could just as well describe it using the microscopic language. A great deal of attention has also been paid to the idea of "strong" emergence, according to which something *truly new* manifests at the macroscopic level, in behavior that cannot even in principle be described by microscopic constituents and fundamental laws. The whole, in the strong-emergentist view, exerts an irreducible causal influence on the individual parts. This would represent a fundamental shift in what we think of as the basic laws of physics.

The problem with strong emergence is that there is no good evidence it ever occurs in nature. There is one plausible candidate: human consciousness (Chalmers 2006). This is not the place to adjudicate the subtle issues surrounding qualia, the Hard Problem, and property dualism. For our present purposes, let us simply note that consciousness is undoubtedly one of the most complex phenomena in the known universe, and it should come as no surprise that the connection between it and the underlying rules of microscopic physics should be difficult to reliably ascertain. Given the tremendous empirical success of those laws in regimes where their consequences are readily apparent, it seems reasonable to imagine that they apply just as well to the atoms in our brains and bodies (including the actions we take and the words we write and speak) as they do in other contexts. Under that view, it would be illegitimate to lean on strong emergence as playing a crucial role when we start thinking about humans as conscious agents.

But weak emergence is enough (Loewer 2012). Even if we grant that the microscopic laws of physics are complete and causally closed on their own terms, it is still possible for emergent higher level theories to invoke entirely distinct sets of ontological concepts—to posit entities and behaviors that are "effectively new" even if they are not "truly new." Weak emergence is more than sufficient to underwrite a reconciliation between impersonal, Laplacian underlying laws and a rich macroscopic world populated by human agents with values, self-awareness, and the ability to make decisions. The fact that ideas such as "purpose" and "choice" are absent in the vocabulary of fundamental physics does not mean they cannot play a role at the human scale (Carroll 2016).

A paradigmatic example of the emergence of seemingly new vocabularies is the existence of cause-and-effect relations. Fundamental physics does not appeal to causes and effects; rather, it describes the patterns obeyed by states of the universe as they evolve from one moment to another. If one asks for the cause of a certain state of the universe at a particular time, the best answer at a microscopic level is simply "the state of the universe at some previous time, plus the laws of physics." Accurate in some strict sense, but neither compatible with our informal everyday notions of causality nor especially helpful.

Emergence, and the existence of an arrow of time, help account for why the macroscopic world seems to be so well-described by chains of cause and effect. Think of memories or records, which are parts of the universe at one time that exert great leverage over past events—had they been slightly different, we could infer a substantial difference in the prior history of the universe. If this photograph showed you wearing a red shirt rather than a blue one, we would have reliably deduced that you actually had been wearing that shirt when the picture was taken. Analogously, we can refer to something as a "cause" if it exerts great leverage over the *future*—a small change in its condition would imply a substantial difference in the subsequent evolution of the universe. It makes sense to say "the flying baseball caused the window to break" because a relatively slight deviation in the ball's trajectory would have left the window unbroken. The move from reversible, deterministic microscopic laws to temporally directed emergent macroscopic laws brings cause-and-effect vocabulary to life.

5. EXISTENTIAL IMPLICATIONS

Existentialism describes human subjects, anxious and on their own, struggling to find meaning in an apparently purposeless existence. How does this story comport with our best understanding of the universe as provided by physics and cosmology?

There is little doubt that modern science has thoroughly undermined any hopes for a higher purpose or meaning inherent in the universe itself. In contrast with an Aristotelian system, governed by teleology and cause-and-effect relations, physics now describes a world characterized by conservation of momentum and information, a world that moves and exists by itself, without any external guidance or goals. The evolution of the universe is undirected and mechanical, proceeding from moment to moment in accordance with unbending rules. We human beings are unimaginably small on the scale of the universe as a whole; life plays no special role in the cosmos. And there is nothing unique or nonphysical about the stuff of which we are made; every person can be thought of as a collection of elementary particles, being and interacting as described by strict mathematical laws.

So the message seems to be mixed. Existentialists face up to the absurdity of the universe but find some solace in our ability to bring meaning to our lives by making free choices. Physics and cosmology seem to reaffirm the absurdity while denying the freedom.

Such a conclusion would be too hasty, however. Freedom and meaning are not useful concepts when we are describing the world in microphysical terms; but that does not imply that they can't be part of our macroscopic human-scale vocabulary.

Indeed, it seems clear that they should be part of that vocabulary. Just as emergence and the arrow of time legitimate talk of causes and effects, they do the same for freedom and meaning (Dennett 1984; Ismael 2016). Just as a "cause" is an aspect of the universe which, had it been slightly different, would have implied a noticeable difference in the future evolution of the macroscopic world, it is sensible to define a "choice" as one of a number of a person's conceivable actions that are compatible with our macroscopic information about that person. Free will can be defined as "the ability to have acted otherwise." Implicit in this definition, however, is a specification of what is kept fixed when we ask whether something different could have happened. If what we keep fixed is the complete microscopic specification of all the particles and forces in a human being, then what happens next is completely determined (at least probabilistically, in a quantum world) by the laws of physics, and "freedom" is not a useful notion. But if what we keep fixed are only macroscopically observable features of the world—those features, in other words, that we actually have access to—then other actions are completely possible, and freedom becomes a sensible concept. This is reflected in the fact that it is impossible to talk coherently about human beings without using a vocabulary of actions defined by choices.

Purpose and meaning have a similar status. Nowhere to be found in the fundamental laws of physics, these ideas are nevertheless perfectly consistent with our best effective description of the macroscopic world. The individual particles of which we are made do not have purposes and do not ascribe meaning to their existence. That doesn't mean that *we* cannot. The universe, vast and impersonal, does not provide us with meaning, out there to be discovered; but by striving for authenticity in our actions we can create meaning for ourselves.

There is one existentialist theme that seems to be refuted, at least in part, by modern physics and cosmology: the emphasis on the fundamental irrationality or mystery of existence. To the contrary, the fantastic success of our investigations into the physical world and the astonishing ability to predict events with exquisite precision speak to the fundamental intelligibility of the universe. It may be countered that this is a matter of mere description, not true explanation, but that seems like a category error. We explain things *within* the world by reference to the context in which they appear, but the world itself simply is. We are able to discover what it is, and, for more than that, it is illegitimate to ask. What we bring to the world remains up to us.

REFERENCES

Albert, D. Z. 2003. *Time and chance*. Boston, MA: Harvard University Press.
Bedau, M. A., and P. Humphreys (Eds.). 2008. *Emergence: Contemporary readings in philosophy and science*. Cambridge, MA: MIT Press.
Blakewell, S. 2016. *At the existentialist café: Freedom, being, and apricot cocktails*. New York, NY: Other Press.
Bohm, D. 1952. A suggested interpretation of the quantum theory in terms of "hidden variables" I. *Physical Review* 85: 166–179.
Carroll, S. M. 2010. *From eternity to here: The quest for the ultimate theory of time*. New York: Dutton.
Carroll, S. M. 2016. *The big picture: On the origins of life, meaning, and the universe itself*. New York: Dutton.
Chalmers, D. J. 2006. Strong and weak emergence. In P. Clayton and P. Davies (Eds.), *The re-emergence of emergence*, pp. 244–256. Oxford: Oxford University Press.
Dennett, D. 1984. *Elbow room: The varieties of free will worth wanting*. Cambridge, MA: MIT Press.
Flanagan, O. 2009. One enchanted being: Neuro-existentialism and meaning. *Zygon: Journal of Science and Religion* 44(1): 41–49.
Ghirardi, G. C., A. Rimini, and T. Weber. 1985. A model for a unified quantum description of macroscopic and microscopic systems. In L. Accardi, et al. (Eds.), *Quantum probability and applications*. New York: Springer.
Ismael, J. T. 2016. *How physics makes us free*. New York: Oxford University Press.
Laplace, P. S. 1814. *A philosophical essay on probabilities*. Reprinted (2015). HardPress Limited.
Loewer, B. 2012. The emergence of time's arrows and special science laws from physics. *Interface Focus* 2: 13–19.
Ney, A., and D. Z. Albert (Eds.). 2013. *The wave function: Essays on the metaphysics of quantum mechanics*. New York: Oxford University Press.
Schlosshauer, M. A. 2007. *Decoherence and the quantum-to-classical transition*. New York: Springer.
Strogatz, S. 2016. *Nonlinear dynamics and chaos: With applications to physics, biology, chemistry, and engineering*, 2nd ed. Westview Press.
Wallace, D. 2012. *The emergent multiverse: Quantum theory according to the Everett interpretation*. New York: Oxford University Press.

PART IV
Neuroscience and the Law

CHAPTER 17

The Neuroscience of Criminality and Our Sense of Justice

An Analysis of Recent Appellate Decisions in Criminal Cases

VALERIE HARDCASTLE

Criminal justice . . . was part of the process that made subordination real. And subordination was real, most notably, for American blacks; also for members of other minority races; and for the poor, the deviant, the unpopular.
—L. M. Friedman (1993: 84)

What is the relationship between the brain, criminality, and just rewards? Versions of this question have received considerable attention as of late (Farahany and Cohen 2009; Gazzaniga and Steven 2005; Glenn and Raine 2014; Greene and Cohen 2004; Hardcastle 2015; Vincent 2013). At the same time, we can trace these discussions back to the late 1700s and the controversial work of the Viennese physicians Franz Joseph Gall and his student Johann Gaspar Spurzheim. They argued that the mind is arranged into separate and distinct "faculties," each of which has a separate and distinct "organ" in the brain (e.g., Gall 1798). Gall went on to argue that one could determine character traits and basic intelligence by reading the information off of the shape of the skull, which roughly traced the brain organs that supported our mental faculties.

Although phrenology per se never became popular in the United States, criminologists of that era certainly discussed the question of whether committing criminal acts were indicative of mental or brain disorders. Isaac Ray, an American psychiatrist and one of the founders of the discipline of forensic psychiatry, hypothesized that moral mania existed in which an "individual without provocation or any other rational motive, apparently in the full possession of his reason, and oftentimes, in spite of his most strenuous efforts to resist, imbrues his hands in the blood of others" (Ray 1838: 197).

But perhaps the event that threw this discussion into sharpest relief was the trial of Charles J. Guiteau for the assassination of President Garfield in 1888, one of the first high-profile cases in the United States that relied on an insanity defense (cf., Fink 1938). Guiteau believed himself responsible for Garfield's victory after he wrote a speech entitled "Garfield vs. Hancock" that he distributed to the Republican National Committee at their summer meeting in 1880. As a reward for his help, he insisted that he be given an ambassadorship to Vienna, or perhaps to Paris. His repeated entreaties were continually rebuffed until, finally, in May 1881, Secretary of State James Blaine personally told Guiteau to leave and never return. On July 2, 1881, Guiteau shot President Garfield twice in the back after lying in wait for him at a train station (and hiring a cab to take him to jail afterward). Eleven weeks later, President Garfield succumbed to infections from the wounds.

Dr. Edward Charles Spitzka, a leading forensic psychiatrist at the time, testified at Guiteau's trial that "Guiteau is not only now insane, but that he was never anything else," and he stated that the condition was due to "a congenital malformation of the brain" (Rosenburg 1995: 278). However, the opinion of the opposing side was that Guiteau was only pretending to be insane for the purposes of the trial (Christianson 2002). Dr. John Gray, superintendent for the New York Utica Asylum, testified for the prosecution that Guiteau acted out of "wounded vanity and disappointment" and not because of any mental defect. He was "depraved" but not deranged. Bound up with this disagreement was the question of how and whether the condition of the mind or brain relates to criminal acts and behavior. The defense insisted that Guiteau was morally insane; the prosecution denied that such a thing existed.

Guiteau insisted on representing himself, and the defense allowed him to ramble on at his trial, hoping that the jury would be able to discern his mental condition for itself. Often testifying in poetic verse, Guiteau explained that he had killed Garfield because God had told him that Garfield was ruining the Republican Party and that he had to die to save the country from the Democrats. He also told the jury that God "the Deity Himself will protect me" after shooting the President (Christianson 2002). Nevertheless, on January 13, 1882, the jury found Guiteau guilty and six months later he was hanged. Housed in the Washington, DC, mental institution St. Elizabeth's Hospital until he was executed, Guiteau clearly was seen at the time as disordered in some fashion. However, it is just as clear that such a view did not stop the judge and jury from also seeing the ultimate sentence as justice served.

As we shall see, contemporary views in this matter have not changed much in the intervening years. A century and a third later, we are still embroiled in conflicts regarding how and whether brain data should affect our sense of just rewards. But one important difference now is that we can look at actual data concerning the brain. We know much better how the brain underlies thought and behavior. One question this chapter addresses is whether (and how) these data and this knowledge influence the outcomes in criminal trials. In particular, this chapter seeks to address whether such evidence is differentially used depending on the race of the defendant. It then speculates on why we see the outcomes we do, and what those outcomes tell us about our criminal justice system, our notions of punishment and responsibility, and changes to our courts of law that are likely coming.

Judges and juries still struggle with the notion from the turn of the last century that "certain moral faculties are presided over by special cell areas, and that injury or disease or these centres may be productive of vicious or criminal impulses" (Lydston 1903: 155). But now these data have appeared in enough criminal cases to allow for some general conclusions regarding how judges and juries are using this material in their decisions and sentencing recommendations. It is one thing to believe that a brain defect can lead to violence; it is quite another to excuse a convicted criminal from punishment due to a faulty brain. This chapter examines the conditions under which we appear willing to do so.

Section 1 of this essay describes the methodology of this study and differentiates it from others. In particular, this study examines only appellate cases from the past five years in which a brain scan was cited as a consideration in the decision. Unlike other recently published analyses (Catley and Claydon 2015; Chandler 2015; de Kogel and Westgeest 2015; Denno 2015), I use "neuroscience" in its narrow meaning, referring to information pertaining to the brain directly and not information derived from behavioral tests that indirectly inform about brain deficits. Section 2 outlines the results of the analysis, focusing on how a defendant's race might be correlated with whether a defendant is able to get a brain scan, whether the scan is admitted into evidence, how the scan is used in the trial, and whether the scan changes the outcome of the hearing. Section 3 provides a deeper comparative analysis of the cases in which imaging data were successful in altering the sentence of defendants and those in which the data were unsuccessful. Section 4 provides some psychosocial reasons for the successes and failures we see with neuroscience's use in particular court cases. Finally, Section 5 points to larger trends in our criminal justice system indicative of more profound changes in how we as a society understand what counts as a just punishment.

1. NEUROSCIENCE IMAGING STUDY METHODOLOGY

Unlike behavioral evidence as it pertains to brains and brain disorders, brain imaging data have only recently been broadly accepted by the courts as admissible evidence in criminal cases. *State v. Kuehn* (2007) upheld the trial court's decision that expert testimony about computed tomography (CT) scans and magnetic resonance imaging (MRI) depicting subdural hematoma, atrophy, or diffuse brain damage met the reliability requirements laid out in *Daubert v. Merrell Dow Pharmaceuticals* (1993), currently the standard in the US federal courts as well as in more than half of its states. (The remainder of the states relies on the older *Frye* standard [*Frye v. United States* 1923], which provides that expert opinion based on a scientific technique is admissible only where the technique is generally accepted as reliable in the relevant scientific community.) Consequently, this study only includes criminal cases with judicial decisions released from January 1, 2007 through December 31, 2012 that reference a brain scan.[1]

Brain imaging tests are complicated and quite expensive. Unlike behavioral exams, which psychologists and psychiatrists can perform on defendants in the prison or jail without much equipment beyond perhaps pen and paper (or laptop), imaging exams usually require the defendant to travel to a facility that maintains the scanning devices. (Electroencephalogram [EEG] recordings are one possible exception, though their use is certainly not routine in prison settings.) Brain scan analysis relies on sophisticated computerized techniques, which in turn necessitates trained technicians and scientists who can maintain the equipment and perform the calculations. While costs for brain scans vary widely depending on the type and location of the facility, they can run into the thousands of dollars. Given the complications and expense of running these scans, who has access, under what conditions, and to what effect become important questions for our criminal justice system. This chapter presents a first start in answering these questions.

Searches for decisions were conducted on the WestLaw database, using the following parameters: "brain," "scan," imag!," "PET," "SPECT," "MR," "CT," "EEG," and "neuro!." At least two people read each decision to determine whether it cited the scan as contributing to the finding in any way. Cases that mentioned that a party had had one or more brain scans but that those data were not later referenced in the decision were excluded from the study. Seventy-six cases met these criteria. We then conducted searches via Google,

---

1. I admit that this time frame is relatively arbitrary. While some types of scans were admitted prior to 2007, the full range of brain scanning technology did not appear in courts until later. I did not want to include the first cases covering the range of scanning technologies in order to give the judicial system time to settle *Daubert* and *Frye* considerations.

local media coverage of the cases, and the written decisions themselves to ascertain the race or ethnicity of the defendants.

We were unable to identify the race or ethnicity of nineteen of the defendants; we categorized the remaining fifty-seven as either: Asian-American, African-American, Hispanic, white, or other. "Other" included resident or nonresident aliens. We also sorted the decisions themselves into three categories: Successful, Unsuccessful, and Neutral. All decision categories are from the defendant's point of view. A Successful case refers to a case in which the sentence was reduced or the defendant won the argument and the case was remanded. (Some remanded cases have yet to be ultimately decided as of this writing.) Unsuccessful cases are those in which the scan was not allowed into the proceedings, the scan was allowed but was specifically and explicitly disregarded, the scan was allowed but it was not significant enough to overcome other arguments (i.e., the defendant did not prevail in the decision), or the scan was of the victim showing damage or injury that could have been caused by the defendant. Neutral cases were those in which the scan was allowed but showed a normal brain, or the scan was attached to other issues not relevant to the degree of responsibility of the defendant.

Figure 17.1 illustrates the broad uses of the scans at trial. The majority of the cases (54 percent) were offered by the defense for the purposes of mitigation, while another 20 percent were used in reference to competency. Sixteen percent were scans of victim's brains, and, in the remaining 10 percent of cases, the discussion of the brain scan was used in other ways (e.g., requesting funds for a scan, suing for a different type of scan).

Twelve of the scans were Neutral (showed no brain damage). Fifteen were Successful, and seven of those resulted in a change of sentence for the defendant (9 percent). For remaining forty-nine cases, the scans were Unsuccessful,

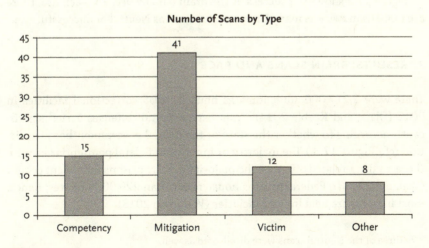

Figure 17.1:
Use of imaging scans in appellate courts (2007–2012).

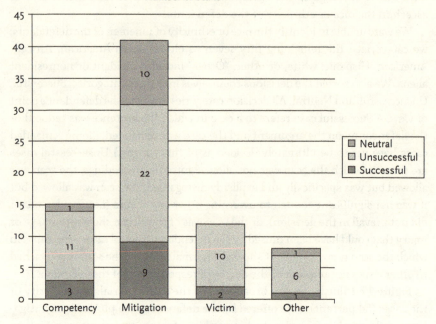

Figure 17.2:
Success of imaging scans in appellate courts by use (2007–2012).

with nine of the scans showing injury to the victim, five of the scans explicitly disregarded in the judicial decision, twenty of the scans disallowed,[2] and eighteen of them admitted, but not enough to overcome aggravating factors. (In twenty-three of the cases total, the brain scans were not admitted as evidence.)

Figure 17.2 shows the success of the brain data by use. For each use, there are more than twice as many Unsuccessful cases as Neutral or Successful ones.

2. RESULTS: BRAIN SCANS AND RACE

There were 2,217,000 adult inmates housed in US correctional facilities in 2013 (Glaze and Kaeble 2014).[3] The distribution of offenses varied significantly according to whether the state or the federal government has jurisdiction (cf., Figure 17.3.) The majority of inmates in federal penitentiaries were there for drug-related offenses; the majority in state prisons for violent crimes (Waldman 2013). Federal prisons house fewer than 220,000 inmates; hence, most inmates are held in state facilities (Waldman 2013).

---

2. Three of the Neutral scans were disallowed as well.
3. These numbers do not include those detained in Indian Country jails, territorial prisons, military prison, immigration detention, juveniles, or those civilly committed.

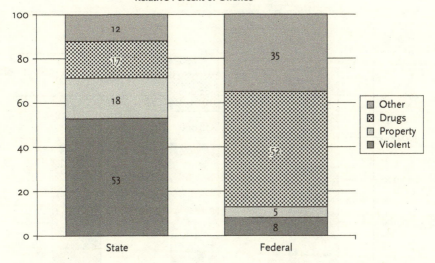

Figure 17.3:
State and federal inmates by offense (2010).
Data from Waldman (2013).

There is significant racial disparity in rates of incarceration: African-Americans are imprisoned at 5.6 times the rate of whites, and Hispanics are imprisoned at 1.8 times the rate of whites (Mauer and King 2007). Focusing just on state prisoners, Figure 17.4 illustrates the number of the race of inmates by type of offense. For violent crimes overall, more African-Americans are imprisoned than whites (284,631 to 228,782), and both groups significantly outnumber Hispanic prisoners (162,489). However, the types of violent crimes for which whites, African-Americans, and Hispanics are convicted follow different patterns. Whites are sentenced for almost twice the number of rapes and other violent sexual assaults than African-Americans or Hispanics (79,282 to 39,975 and 35,863), but African-Americans are convicted of more murders than whites or Hispanics (65,568 to 45,369 and 37,956) (Carson and Golinelli 2013).

I provide this background because most of the cases in this study concern state defendants and some sort of violent crime. Hence, without additional information, and assuming that the availability of scans is constant across race and ethnicity, one would expect that the number of defendants in our set of cases who are African-American would outnumber those who are white. However, as is clear from Figure 17.5, there are fewer identified African-Americans than there are whites. Almost twice as many defendants were white than were Black or African-American (33 to 17). Six were identified as Hispanic or Latino, one as Asian-American, and one as international.

We were not able to identify the race or ethnicity for eighteen defendants, which prevents us from demonstrating the significance of the effect shown.

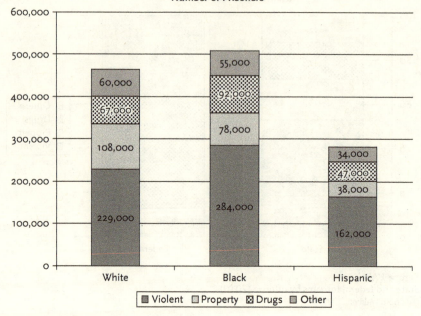

Figure 17.4:
State prisoners by race and offense (2011). Numbers rounded to the nearest 100,000. White includes American Indian, Pacific Islanders, and persons identifying as two or more races. Black excludes persons identifying as Latino or Hispanic.
Data from Carson and Golinelli (2013).

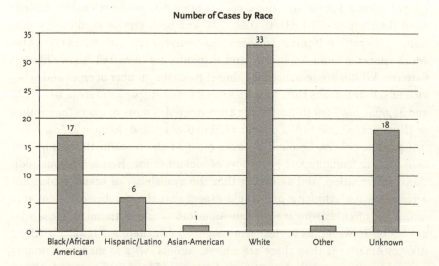

Figure 17.5:
Number of cases by race of defendant.

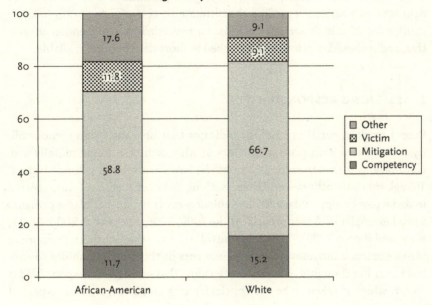

Figure 17.6:
Percentage of brain scans used by race.

Nevertheless, the trend identified is suggestive: it appears that African-American defendants do not have the same access to brain imaging as whites, or even (by percentage of inmates) as Hispanics. However, as can be seen in Figure 17.6, once African-American defendants get a brain scan, it is used in roughly the same way as it is by white defendants, with the majority of the cases centered on mitigation.

On only one measure did African-American defendants appear comparable to white defendants: whether the brain scan was admitted into evidence. Two-thirds of the scans of African-American defendants were admitted into evidence, and almost 60 percent of the white defendants' scans were.[4] However, if the scans were admitted into evidence, they were less likely to be Successful if the defendant were African-American. Out of the ten cases in which brain scans were admitted into evidence (excluding scans of the victims proffered by the prosecutors), only one case with an African-American defendant was Successful (*U.S. v. Williams* 2010, discussed later). This stands in contrast to six Successful scans for white defendants, out of a total of eighteen admitted.

There is an important caveat to this analysis: five of the thirteen scans admitted into evidence were Successful for defendants with an unknown race. That is, a greater percentage of admitted scans for defendants of unknown race were Successful (38.5 percent) than for either identified whites (33.3 percent)

4. These figures exclude the scans of victims.

or African-Americans (10 percent). These data preclude us from drawing any definitive conclusions regarding the ultimate success of using brain imaging studies for African-American defendants; nevertheless, the trend is suggestive, and it should continue to be tracked as more data become available.

3. ASSIGNING RESPONSIBILITY

Recent work in social psychology indicates that how we assign responsibility is correlated with prior judgments of what counts as being morally bad, which are in turn dependent upon other, larger, social and cultural factors. Though there are others—and there is a long and deep history of this research in social psychology—Mark Alicke's culpable control model of blame presents a good exemplar of this approach (Alicke 2000, 2008; Alicke et al. 2008; Alicke, Rose, and Bloom 2008). "Culpable control" refers to the fact that our desire to blame someone intrudes on our assessments of that person's ability to control his or her thoughts or behavior. Deciding that someone is responsible for an act, which is taken to be the conclusion of a judgment, is actually part of our psychological process of assessing blame. If we start with a spontaneous negative reaction, then that can lead to our hypothesizing that the source of the action is blameworthy as well as to an active desire to blame that source. This desire, in turn, skews our interpretations of the available evidence such that it supports our blame hypothesis. We highlight evidence that indicates negligence, recklessness, impure motives, or a faulty character, and we ignore evidence that suggests otherwise. In other words, instead of dispassionately judging whether someone is responsible, we validate our spontaneous reaction of blameworthiness.

This, of course, is just another instance of the confirmation bias, the tendency to search for, interpret, and remember information in such a way that it confirms one's preconceptions (Watson 1960; see also Kahneman 2011; Tversky and Kahneman 1974). This well-known type of systematic error arises in many cases of inductive reasoning. Modern interpretations of this phenomenon suggest that it is a result of informal heuristics that we use to help make quick decisions about our world on the fly (Friedrich 1993; Hergovich, Schott, and Burger 2010; Kunda 1999; Maccoun 1998; Matlin 2004; Nickerson 1998; Oswald and Grosjean 2004). Jon Hanson in particular has argued that these sorts of heuristics often impair policy-makers and our judicial system (Benforado and Hanson 2005, 2008a, 2008b, 2008c; Chen and Hanson 2004a, 2004b; Hanson 2012; Hanson and McCann 2008; Hanson and Yosifin 2003, 2004).

But we actually do more than selectively use information at our disposal. Data suggest that we will even exaggerate a person's actual or potential control over an event to justify our blame judgment, and we will change the threshold for how much control is required for a blame judgment (Alicke et al. 2008; see also Alicke 1994; Berg and Vidmar 1975; Eften 1974; Kalven and Zeisel 1966; Lagnado and Channon 2008; Lerner and Miller 1978; Lerner, Miller, and Holmes 1976; Nemeth and Sosis 1973; Schlenker 1980; Sigall and Ostrove 1975; Snyder, Higgins, and Stuckey 1983; Sosis 1974). Normally, we understand and explain others' behavior along two dimensions: the person's disposition, or personality, and the person's situation, or the surrounding context. If we explain someone's behavior dispositionally, then we are claiming that the main cause of the action under question is a personal characteristic. "He can't help it; that is just the way he is," is a sort of dispositional explanation. In contrast, if we explain some behavior situationally, then we are claiming that the action was a result of the circumstances in which the person found him- or herself. "Any reasonable person would have done this," is a sort of situational explanation.

It turns out that we are strongly inclined to use dispositional explanations for behavior, even when we know that the situation was a very important factor in the response. In fact, we often assume that behavior corresponds to internal dispositions, despite strong evidence to the contrary (Gilbert and Jones 1986; Gilbert and Malone 1995; Jones 1979; Jones and Harris 1967; Jones and Nisbett 1972; Miller 1976; Ross, Lepper, and Hubbard 1975; Snyder, Tanke, and Berscheid 1977; Winter and Uleman 1984). In one classic experiment, for example, students were told to give a pro-Castro presentation to their classmates. Even though the students had no choice in their assignment—and other students were well aware of this fact—the audience ascribed the content of the presentation to the speakers' pro-Castro beliefs (Gilbert and Malone 1995). In short: despite efforts to the contrary, the assignment of responsibility and blame is more about the unconscious response biases of evaluators than it is the rational application of objective standards.

This research paradigm and Alicke's theoretical model dovetail nicely with the trend we identified indicating that brain scans of African-American defendants were less likely to be mitigating when used as evidence in court. If judges and juries were already predisposed to believe that African-Americans are guilty, then, on balance, it would take a greater weight of evidence to overcome any aggravating factors as compared to white or middle-class defendants. And indeed, all-white juries convict African-American defendants significantly more often than white defendants (Anwar, Bayer, and Hjalmaarsson 2012).

Research on implicit bias has repeatedly demonstrated that most Americans associate African-Americans with badness (e.g., Nosek et al. 2002; see also Greenwald and Krieger 2006; Greenwald, McGhee, and Schwartz 1998; and https://implicit.harvard.edu/implicit/aboutus.html). Implicit attitude tests have been shown to affect a wide range of behaviors, including things like classroom performance in school, medical treatment by health care professionals, and even the analysis of memos written by law associates (Darley and Gross 1983; Green et al. 2007; Greenwald et al. 2009; Murray 1996; Reeves 2014). Importantly, recent studies have shown that judges and juries are not only biased in the same ways as the general population in the United States, but that these biases do in fact affect their memories and judgments (Levinson 2007; Rachlinski et al. 2009). Indeed, how to manage implicit bias in the courtroom is becoming a topic of discussion in legal scholarship (Benforado 2010a, 2010b; Benforado, Hanson, and Yosifon 2004; Chen and Hanson 2004a; Jois 2009; Kahan, Hoffman, and Braman 2009; Kang and Lane 2010; Kang et al. 2012; Krieger and Fiske 2006; Seaman 2008; Yosifon 2008). The trends identified here bolster this literature and perspective.

4. BRAINS SCANS AND MITIGATION: REACTIVE VERSUS PREDATORY VIOLENCE

What can we say about the difference between the cases in which proffering a brain scan as evidence in support of the defendant leads either to the sentence being reduced or the case remanded and those that do not? Our analysis indicates that successful deployment of brain scans depends on the type and severity of the crime. In line with Denno's (2015) analysis, we find that the brain imaging data proffered in criminal cases are received and used in exactly the same manner as other sorts of evidence: mitigating evidence is weighed against aggravating evidence; the worse the crime, the less likely brain data indicating reduced behavioral control or capacity for responsibility will ultimately prevail in sentencing.

Restricting our discussion to cases in which the sentence was changed due to proffered brain evidence, we see that there were two cases in which victim scans showed that the defendant did not cause the injuries claimed, one case of fraud, two murders committed during robberies gone bad, and three cases involving minors (one for distributing child pornography and two murders). There was also one case of extended abuse that resulted in the death of a child.

This contrasts with the Unsuccessful cases in which the scan was admitted and used, but it was not enough to overcome other aggravating factors to change the defendant's sentence. Here, as before, there was one case of fraud, then one possession with intent to distribute a significant amount of cocaine, one domestic violence case of a near-fatal stabbing, four predatory murders (two committed during a robbery and one of a police officer), four rape/murders or sexually violent murders, and three familial rampage murders with multiple deaths. There were also two cases involving minors (both unprovoked murders).

The biggest difference between these two groups is the type of violence encountered. The Successful cases with adult defendants that involved violence all were cases of so-called reactive violence. The defendants did not anticipate the murders they ultimately committed; instead, they reacted in the moment and under great duress. Anthony Welch broke into the home of Rufus and Kyoko Johnson, previous neighbors and acquaintances of his, intending to rob them. In the course of the robbery, Mr. Welch beat the Johnsons to death using objects found in the home (*Welch v. State* 2008). Khyle Briscoe went with Shaun Pina to the home of Ben Parovel, a known drug dealer, intending to rob him. A fight broke out among the three, and, in the end, Mr. Pina was shot and killed by Mr. Parovel as Mr. Briscoe was struggling with Mr. Parovel over a gun (*Briscoe v. Scribner* 2009).

The deaths were not planned in the murder trial involving the juvenile either. Kenshawn Maxey, along with his friends Lashuan Levi and Artis Moore, went to the O'Aces Bar and Gill, intending to rob it. Mr. Maxey ended up killing the bartender, Sal Zendano, and Mr. Levi as Mr. Levi was struggling with Mr. Zendano over access to the cash register and yelling to Mr. Maxey to shoot the bartender (*Maxey v. Donat* et al. 2012). In each of these cases, the defendant did not start out with the intention of killing anyone. Rather, as the events unfolded, defendants made (sometimes admittedly poor) decisions that directly resulted in the death of others.

U.S. v. Williams (2010) has a different flavor and perhaps it should be treated separately. Naeem J. Williams beat and abused his stepdaughter over a period of seven months until she died. While one could argue that Mr. Williams did not intend to kill the child, and each instance of harm was an instance of reactive violence, one could also make the case that Mr. Williams had plenty of opportunities over the course of the abuse to affect the ultimate outcome. While cognitive tests indicated that he suffered from borderline intellectual functioning, which perhaps precluded him from forming requisite intent for the crime, it was also the case that Mr. Williams was an active-duty soldier at the time of the death, and the state of Hawaii, which has not tried a capital case since it became a state in 1957, had jurisdiction. It is possible that these latter two factors played a role in overturning the penalty of death, even though they were not cited as reasons in the decision.

In the Unsuccessful cases, however, the defendants anticipated and planned their violent acts, at least to some degree. Brandy Holmes, along with her boyfriend, Robert Coleman, forced their way into Julian and Alice Brandon's home and then shot them both prior to robbing them (*State v. Holmes* 2009).[5] Kevin Mercer shot Sergeant First Class Tracy Davis, an active-duty soldier, in the back of the head after he forced Sergeant Davis to hand over the keys to his car (*State v. Mercer* 2009). Marlon Duane Kiser shot Deputy Sheriff Donald

---

5. Because Rev. Brandon did not die right away, Ms. Holmes and Mr. Colement then stabbed him to death. Mrs. Brandon, whom they believed they had killed, actually survived the attack but died five years later.

Bond multiple times with a high-powered assault rife after Deputy Sheriff Bond interrupted Mr. Kiser trying to set a fruit stand on fire (*State v. Kiser* 2009). At some point after an argument, Jack Lee Townsend retrieved his gun and then shot Mary Ellen Smith, his landlady and roommate, four times in the abdomen, killing her instantly (*Townsend v. McDonald* 2009). Similarly, Clyde J. Rainey, the minor, shot Koupou Saechao twice in the back after attempting to rob him but discovering he had nothing to steal (*Rainey v. Knowles* 2008).[6]

The rape/murders, too, were not crimes of passion or instances of reactive violence. Johnnie Hoskins strangled his victim, Dorothy Berger, after he had tied her up and raped her (*Hoskins v. State* 2009). Richard Leavitt stabbed and mutilated his victim, Danette Jean Elg, after breaking into her home and raping her in her bed (*Leavitt v. Arave* 2011).

The familial rampage murders and the domestic stabbing contain premeditation as well. Brian Nelson beat his ex-girlfriend Sara Tennant to death with a crowbar, then proceeded to beat to death her father, Harold Tennant, his son, Eric Tennant, and Mr. Tennant's girlfriend, Jean Bookwalter, to eliminate any witnesses, and then set their house on fire (*People v. Nelson* 2009). Peter Nong Le shot and killed his brother's girlfriend, Tuyet Le, and her daughter, Jennifer Cu, using two different weapons, after arguing with Ms. Le over her care of his elderly parents and her disrespecting and insulting him (*Nong Le v. Barnes* 2011). Robin Lee Row set her house on fire, burning to death her husband, Randy Row, and her two children, Joshua Cornellier and Tabitha Cornellier (*Row v. Beauclair* 2011). James Phillips stabbed his wife seventeen times in a public venue after she twice obtained orders of protection against him for threatening her with physical harm (*People v. Philips* 2011).

5. IS OUR SENSE OF JUST REWARDS CHANGING?

In many respects, what you see in these cases is exactly what you would expect to find with respect to mitigation considerations: the mitigating factors are to be weighed against aggravating factors, and brain imaging data are no different. One would expect brain scans to be less likely to be mitigating the more violent or predatory the crime. As we have seen, in cases involving reactive, unplanned violence, brain data indicating neural damage or injury can lessen the punishment. However, in cases of premeditated and severe violence, it does not appear that brain data will do much to influence the

---

6. Mr. Rainey's sentence of life without parole (LWOP) was later overturned due to *Miller v. Alabama* (2012), which holds that mandatory LWOP for those under the age of eighteen at the time of their crimes violates the Eighth Amendment. Mr. Rainey was fifteen at the time of his crime.

outcomes. As Stephen Morse has argued (2012), introducing brain data into trials appears to have little effect on how our criminal justice system actually functions.

Over the past half-century, philosophical treatments of punishment have focused primarily and largely on retribution, deterrence, and just deserts instead of rehabilitation, restoration, or therapeutic goals. The reasons for these emphases are many, not the least of which has been the dismal failure of past efforts to rehabilitate convicts. Less discussed is the dualistic assumption in Western legal systems that, by and large, reasons cause behaviors. While medicine, psychiatry, neuropsychology, and neurology all hold that diseases, injuries, and deformities in the brain influence and can even determine a person's thoughts, desires, impulses, and ability to control one's behavior, Western law assumes that all adults are rational beings who act for specific reasons and that, in each instance, an individual could have done otherwise had he or she chosen to. In other words, courts do not care about issues of impulse control, impaired executive functioning, and the like. They focus on rational thought and means–ends reasoning instead. The neuroscience behind criminality might tell us something about the brains of criminals, but it does not tell us anything about a person's reasons for acting. And it is the reasons themselves that are important for legal culpability and the justification of punishment.

And yet we are starting to get small hints of change in our legal system due to a deeper understanding of how brain injuries and disorders can affect patient behavior and control. In many jurisdictions, we now have so-called *diversion courts* in which brain injury or disorder can excuse defendants from punishment for lesser and usually nonviolent crimes (cf., Hardcastle forthcoming). To take a particular example, the American court system is starting to differentiate returning combat vets with traumatic brain injury (TBI) or with other mind/brain disorders from other offenders. Several counties and municipalities have created special courts in which district attorneys can send military personnel and veterans into treatment instead of jail when they commit an offense. Though these courts are relatively new, preliminary data suggest that defendants who go through this system are more likely to become productive and law-abiding citizens than those who are warehoused in our prisons and jails (Marlowe 2010).

Of course, returning vets are not the only group for whom this practice is relevant. For example, just like combat-based TBI, posttraumatic stress arising from civilian trauma exposure strongly increases the risk of involvement in the criminal justice system. And even for a mental disorder as common as attention deficit-hyperactivity disorder (ADHD), rates of criminal convictions for ADHD patients not taking their medication were significantly higher than for those who were. Data such as these have led districts to create diversion courts for mental health issues and addiction as well as for veterans. Since their inception in the late 1990s, these diversion courts have played an increasingly larger role in managing some defendants accused of criminal behavior.

It is clear that we are undergoing no small change in how our legal system understands and assigns responsibility, which also alters how we can connect data regarding the neural underpinnings of behavior with notions of just punishment. In these cases, biological considerations are starting to trump legal questions of culpability and responsibility. Is this the beginning of a greater shift in how our criminal justice system functions relative to evidence from neuroscience?

The traditional verdict of not guilty by reason of insanity requires that defendants either cannot comprehend what they are doing or cannot distinguish right from wrong at the time of the act. Normally, these criteria hold even if we understand what is happening biologically in an individual at the time of the crime. However, diversion courts are based on the notion that their defendants, while not delusional, are still not as rational as "normal" defendants. In these courts, biological explanations for behavior take precedence over reasons-based explanations. As our science of behavior grows more sophisticated, it encroaches more and more on the territory covered by reasons-based behavioral explanations, which means that the scope of our criminal courts will cover less and less over time. If we know our brains cause our behavior, and our brains are the way they are because of their underlying genetics and previous life experiences, it becomes difficult to maintain that punishment as retribution for behavior is a just or even a coherent notion.

The idea of "therapeutic justice," which informs diversion courts, is a misnomer in that it belies conventional notions of justice. "Justice" includes the concepts of retribution and compensation; therapeutic justice does not. Even though diversion treatments result in loss of rights for defendants, they are not punishment in any traditional sense. Like some views of punishment, they are consequentialist, in that one goal is to prevent further lawlessness. But they are not concerned with deterrence, which traditional consequentialist views of punishment include. Indeed, if anything, their aim is the antithesis of deterrence: to encourage more citizens to seek treatment for mental disorders. Both traditional courts and diversion courts have as their goal an orderly society with law-abiding citizens, but the similarities between them end there. Our legal system is perhaps moving beyond a system of punishment to one that leaves time-honored views of justice far behind.

Still, from the point of view of science, there remain troubling inconsistencies. Brain data appear to excuse criminal behavior in one instance, but the very same data seem to have no effect in another. And the difference between the cases has nothing to do with the degree or type of impairments in the defendants. The same brain injuries or disorders across patients give rise to the same problems in impulse control, emotional lability, long-term planning, and executive functioning. Hence, from a purely scientific point of view, the same brain injuries or disorders should have same impact on judging the level of responsibility for an action or behavior. The type of action it is, or how we interpret it morally, should not influence the data's capacity for mitigation.

At some point in the possibly not so distant future, the cognitive dissonance apparent in our criminal justice system between the legal interpretation of responsibility and neuroscience's data will have to be resolved. Nonetheless, this analysis suggests that perhaps Greene and Cohen were right after all when they wrote: "Neuroscience will probably have a transformative effect on the law, despite the fact that existing legal doctrine can, in principle, accommodate whatever neuroscience will tell us. New neuroscience will change the law ... by transforming people's moral intuitions about free will and responsibility" (2004: 1775). Perhaps we are at the beginning of this sea change.

ACKNOWLEDGMENTS

Many thanks are due to Donald Clancy, Frank Faries, Andrew Hakala-Finch, Vincente Raja Galian, Richard Stephenson, and Derek Tucker for their help in collecting and analyzing the data. This project has been generously supported by the Weaver Institute for Law and Psychiatry, a research fellowship at the Medical Humanities at the University of Texas Medical Branch-Galveston, and a senior fellowship in Philosophy and Psychiatry at Ruhr University-Bochum. Special thanks are due to Jason Glenn for his lessons on the history of the US criminal justice system and Albert Newen for pointing me toward the literature on blame assignment.

REFERENCES

Alicke, M. D. 1994. Evidential and extra-evidential evaluations of social conduct. *Journal of Social Behavior and Personality* 9: 591–615.
Alicke, M. D. 2000. Culpable control and the psychology of blame. *Psychological Bulletin* 12: 56–574.
Alicke, M. D. 2008. Blaming badly. *Journal of Cognition and Culture* 8 (1-2): 179–186.
Alicke, M. D., J. Buckingham, E. Zell, and T. Davis. 2008. Culpable control and counterfactual reasoning in the psychology of blame. *Personality and Social Psychology Bulletin* 3: 371–1381.
Alicke, M. D., D. Rose, and D. Bloom. 2008. Causation, norm violation, and culpable control. *Journal of Philosophy* 10: 70–696.
Anwar, S., P. Bayer, and R. Hjalmaarsson. 2012. The impact of jury race in criminal trials. *The Quarterly Journal of Economics* 127: 1017–1055.
Benforado, A. 2010a. Frames of injustice: The bias we overlook. *Indiana Law Journal* 85: 1332–1378.
Benforado, A. 2010b. The body of the mind: Embodied cognition, law, and justice. *St. Louis University Law Journal* 54: 1185–1216.

Benforado, A., and J. Hanson. 2005. The costs of dispositionism: The premature demise of situationist law and economics. *Maryland Law Review* 64: 24–84.

Benforado, A., and J. Hanson. 2008a. The great attributional divide: How divergent views of human behavior are shaping legal policy. *Emory Law Journal* 57: 311–338.

Benforado, A., and J. Hanson. 2008b. Naïve cynicism: Maintaining false perceptions in policy debates. *Emory Law Journal* 57: 499–574.

Benforado, A., and J. Hanson. 2008c. Legal academic backlash: The response of legal theorists to situationist insights. *Emory Law Journal* 57: 1087–1146.

Benforado, A., J. Hanson, and D. Yosifon. 2004. Broken scales: Obesity and justice in America. *Emory Law Journal* 53: 1645–1806.

Berg, K. S., and N. Vidmar. 1975. Authoritarianism and recall of evidence about criminal behavior. *Journal of Research in Personality* 9: 147–157.

Briscoe v. Scribner, No. CIV S-04-2175 FCD GGH (E.D. Cal. 2009).

Carson, E. A., and D. Golinelli. 2013. Prisoners in 2012–advance counts. *Bureau of Justice Statistics Bulletin NJC 242467*.

Catley, P., and L. Claydon. 2015. The use of neuroscientific evidence in the courtroom by those accused of criminal offenses in England and Wales. *Journal of Law and the Biosciences* 2: 510–549.

Chandler, J. A. 2015. The use of neuroscientific evidence in Canadian criminal proceedings. *Journal of Law and the Biosciences* 2: 550–579.

Chen, R., and J. Hanson. 2004a. Categorically biased: The influence of knowledge structures on law and legal theory. *Southern California Law Review* 77: 1103–1254.

Chen, R., and J. Hanson. 2004b. The illusion of law: The legitimating schemas of modern policy and corporate law. *Michigan Law Review* 103: 1–149.

Christianson, S. 2002. Charles Guiteau trial: 1881. In E. W. Knappman, S. Christianson, and L. O. Paddock (Eds.), *Great American trials, 2nd edition*, pp.187–192. Farmington Hills, MI: Gale Group/Thompson Learning.

Darley, J. M., and P. H. Gross. 1983. A hypothesis-confirming bias in labeling effects. *Journal of Personality and Social Psychology* 44: 20–33.

Daubert v. Merrell Dow Pharmaceuticals, Inc. (509 U.S. 579 1993).

de Kogel, C. H., and E. J. M. C. Westgeest. 2015. Neuroscientific and behavioral genetic information in criminal cases in the Netherlands. *Journal of Law and the Biosciences* 2 (3): 580–605.

Denno, D. W. 2015. The myth of the double-edged sword: An empirical study of neuroscience evidence in criminal cases. *Boston College Law Review* 5: 493–551.

Eftan M. G. 1974. The effect of physical appearance on the judgment of guilt, interpersonal attraction, and severity of recommended punishment in a simulated jury task. *Journal of Research in Personality* 8: 45–54.

Farahany, N. A., and J. E. Cohen. 2009. Genetics, neuroscience, and criminal responsibility. In N. Farahany (Ed.), *The impact of behavioral sciences on criminal law*, pp. 183–240. New York: Oxford University Press.

Fink, A. E. 1938. *Causes of crime: Biological theories in the United States, 1800–1915*. Philadelphia: University of Pennsylvania Press.

Friedman, L. M. 1993. *Crime and punishment in American history*. New York: Basic Books.

Friedrich, J. 1993. Primary error detection and minimization (PEDMIN) strategies in social cognition: a reinterpretation of confirmation bias phenomena. *Psychological Review* 100: 298–319.

Frye v. United States, 293 F. 1013 (D.C. Cir. 1923).

Gall, F. J. 1798. Letter from Dr. F. J. Gall, to Joseph Fr[eiherr] von Retzer, upon the functions of the brain, in man and animals. *Der Neue Teutsche Merku* 3: 311–332.

Gazzaniga, M. S., and M. S. Steven. 2005. Neuroscience and the law. *Scientific American Mind* 1: 2–49.

Gilbert, D. T., and E. E. Jones. 1986. Perceiver-induced constraint: Interpretation of self generated reality. *Journal of Personality and Social Psychology* 50: 269–280.

Gilbert, D. T., and P. S. Malone. 1995. The correspondence bias. *Psychological Bulletin* 117: 21–38.

Glaze, L. E., and D. Kaeble. 2014. Correctional populations in the United States, 2013. *Bureau of Justice Statistics Bulletin NJC 248479*.

Glenn, A. L., and A. Raine. 2014. Neurocriminology: Implications for the punishment, prediction, and prevention of criminal behavior. *Nature Reviews Neuroscience* 1: 4–63.

Green, A. R., D. R. Carney, D. J. Pallin, L. H. Ngo, K. L. Raymond, L. I. Iezzoni, and M. R. Banaji. 2007. Implicit bias among physicians: Prediction of thrombolysis decisions for black and white patients. *Journal of General Internal Medicine* 22: 1231–1238.

Greene, J., and J. Cohen. 2004. For the law, neuroscience changes nothing and everything. *Philosophical Transactions of the Royal Society London B* 35: 1775–1785.

Greenwald, A. G., and L. H. Krieger. 2006. Implicit bias: Scientific foundations. *California Law Review* 94: 945–968.

Greenwald, A. G., D. E. McGhee, and J. K. L. Schwartz. 1998. Measuring individual differences in implicit cognition: The implicit association test. *Journal of Personality and Social Psychology* 74: 1464–1480.

Greenwald, A. G., A. T. Poehlman, E. L. Uhlmann, and M. R. Banaji. 2009. Understanding and using the implicit association test: III. meta-analysis of predictive validity. *Journal of Personality and Social Psychology* 97: 17–41.

Hanson, J. (Ed.). 2012. *Ideology, psychology, and law*. New York: Oxford University Press.

Hanson, J., and M. McCann. 2008. Situationist torts. *Loyola Law Review* 41: 1345–1454.

Hanson, J., and D. Yosifin. 2003. The situation: An introduction to the situational character, critical realism, power economics, and deep capture. *University of Pennsylvania Law Review* 152: 129–346.

Hanson, J., and D. Yosifon. 2004. The situational character: A critical realist perspective on the human animal. *Georgetown Law* 93: 1–179.

Hardcastle, V. G. 2015. Would a neuroscience of violence aid in understanding moral or legal responsibility? *Cognitive Systems Research* 3: 4–53.

Hergovich, A., R. Schott, and C. Burger. 2010. Biased evaluation of abstracts depending on topic and conclusion: Further evidence of a confirmation bias within scientific psychology. *Current Psychology* 29: 188–209.

Hoskins v. State, 965 So.2d 1 (Fla. 2007).

Jois, G. U. 2009. Stare decisis is cognitive error. *Brooklyn Law Review* 75: 63–141.

Jones, E. E. 1979. The rocky road from acts to disposition. *American Psychologist* 34: 107–117.

Jones, E. E., and V. A. Harris. 1967. The attribution of attitudes. *Journal of Experimental Social Psychology* 3: 1–24.

Jones, E. E., and R. E. Nisbett. 1972. The actor and the observer: Divergent perceptions of the causes of behavior. In E. E. Jones, D. E. Kanouse, H. H. Kelley, R. E. Nisbett, S. Valins, and B. Weiner (Eds.), *Attribution: Perceiving the causes of behavior*, pp. 79–94. New York: General Learning Press.

Kahan, D. M., D. A. Hoffman, and D. Braman. 2009. Whose eyes are you going to believe? *Scott v. Harris* and the perils of cognitive illiberalism. *Harvard Law Review* 122: 837–906.

Kahneman, D. 2011. *Thinking fast and slow*. New York: Farrar, Strauss, and Giroux.

Kalven, Jr., H., and H. Zeisel. 1966. *The American jury*. Chicago, IL: University of Chicago Press.

Kang, J., and K. Lane. 2010. Seeing through colorblindness: Implicit bias and the law. *UCLA Law Review* 58: 465–520.

Kang, J., M. Bennett, D. Carbado, P. Casey, N. Dasgupta, D. Faigman, R. Godsil, A. G. Greenwald, J. Levinson, and J. Mnookin. 2012. Implicit bias in the courtroom. *UCLA Law Review* 59: 1124–1186.

Krieger, L. H., and S. T. Fiske. 2006. Behavioral realism in employment discrimination law: Implicit bias and disparate treatment. *California Law Review* 94: 997–1062.

Kunda, Z. 1999. *Social cognition: Making sense of people*. Cambridge, MA: MIT Press.

Lagnado, D. A., and S. Channon. 2008. Judgments of cause and blame: The effects of intentionality and foreseeability. *Cognition* 108: 754–770.

Leavitt v. Arave, 646 F.3d 608 (9th Cir. 2011).

Lerner, M. J., and D. T. Miller. 1978. Just world research and the attribution process: Looking back and ahead. *Psychological Bulletin* 85: 1030–1051.

Lerner, M. J., D. T. Miller, and J. G. Holmes. 1976. Deserving and the emergence of forms of justice. In L. Berkowitz and E. Walster (Eds.), *Advances in experimental social psychology*, vol. 9, pp. 133–162. New York: Academic Press.

Levinson, J. D. 2007. Forgotten racial equality: Implicit bias, decision making, and misremembering. *Duke Law Journal* 57: 345–424.

Lydston, G. F. 1903. *The diseases of society*. Philadelphia, PA: Lippincott.

Maccoun, R. J. 1998. Bias in the interpretation and use of research results. *Annual Review of Psychology* 49: 259–287.

Marlowe, D. B. 2010. Research update on adult drug courts. *National Association of Drug Court Professionals*. http://www.nadcp.org/sites/default/files/nadcp/Research%20Update%20on%20Adult%20Drug%20Courts%20-%20NADCP_1.pdf.

Matlin, M. W. 2004. Pollyanna principle. In R. F. Pohl (Ed.), *Cognitive illusions: A handbook on fallacies and biases in thinking, judgment and memory*, pp. 255–273. New York: Psychology Press.

Mauer, M., and R. S. King. 2007. Uneven justice: State rates of incarceration by race and ethnicity. *The sentencing project*. http://www.sentencingproject.org/doc/publications/rd_stateratesofincbyraceandethnicitypdf.

Maxey v. Donat, 2012 WL 295632, No. 3:09–cv–00012–ECR–VPC (D. Nevada 2012).

Miller v. Alabama, 567 U.S. [183 L.Ed.2d 407, 132 S.Ct. 2455 (2012).

Miller, A. G. 1976. Constraint and target effects in the attribution of attitudes. *Journal of Experimental Social Psychology* 12: 325–339.

Morse, S. J. 2012. Neuroimaging evidence in law: a plea for modesty and relevance. In J. R. Simpson (Ed.), *Neuroimaging in forensic psychiatry: From the clinic to the courtroom*, pp. 341–358. Chichester: Wiley-Blackwell.

Murray, C. M. 1996. Estimating achievement performance: A confirmation bias. *Journal of Black Psychology* 22: 67–85.

Nemeth, C., R. H. and Sosis. 1973. A simulated jury: Characteristics of the defendant and the jurors. *Journal of Social Psychology* 90: 221–229.

Nickerson, R. S. 1998. Confirmation bias: A ubiquitous phenomenon in many guises. *Review of General Psychology* 2: 175–200.

Nong Le v. Barnes, No. SACV 12-393-DSF (CW) (C.D. Cal. 2012).

Nosek, B. A., M. R. Banaji, and A. Greenwald. 2002. Harvesting implicit group attitudes and beliefs from a demonstration web site. *Group Dynamics: Theory, Research, and Practice* 6: 101–115.

Oswald, M. E., and S. Grosjean. 2004. Confirmation bias. In R. F. Pohl (Ed.), *Cognitive illusions: A handbook on fallacies and biases in thinking, judgment and memory*, pp. 79–96. New York: Psychology Press.

People v. Nelson, 235 Ill.2d 386 (Ill. 2009).

People v. Philips, 16 N.Y.3d 510 (N.Y. 2011).

Rachlinski, J. J., S. Johnson, A. J. Wistrich, and C. Guthrie. 2009. Does unconscious racial bias affect trial judges. *Notre Dame Law Review* 84: 1195–1246.

Rainey v. Knowles, 2008 WL 4104285, Case No. 08-17222 (N.D. Cal. 2008).

Ray, I. 1838. *A treatise on the medical jurisprudence of insanity*. Boston: Little, Brown.

Reeves, A. N. 2014. Written in black: Exploring confirmation bias in racialized perceptions of writing skills. *Yellow Paper Series*: Nextions. http://www.nextions.com/wp-content/files_mf/14468226472014040114WritteninBlackandWhiteYPS.pdf.

Rosenburg, C. E. 1995. *The trail of the assassin Guiteau: Psychiatry in the gilded age*. Chicago, IL: University of Chicago Press.

Ross, L., M. R. Lepper, and M. Hubbard. 1975. Perseverance in self-perception and social perception: Biased attributional processes in the debriefing paradigm. *Journal of Personality and Social Psychology* 32: 880892.

Row v. Beauclair, 1:98-cv-00240, No. 545 (D.Idaho Aug. 29, 2011).

Schlenker, B. R. 1980. *Impression management: The self-concept, social identity, and interpersonal relations*. Boston: Brooks/Cole.

Seaman, J. 2008. Hate speech and identity politics: A situationalist proposal. *Florida State University Law Review* 36: 99–123.

Sigall, H., and N. Ostrove. 1975. Beautiful but dangerous: Effects of offender attractiveness and nature of the crime on juridic judgment. *Journal of Personality and Social Psychology* 31: 410–414.

Snyder, C. R., R. L. Higgins, and R. J. Stuckey. 1983. *Excuses: Masquerades in search of grace*. Hoboken, NJ: Wiley.

Snyder, M., E. D. Tanke, and E. Berscheid. 1977. Social perception and interpersonal behavior: On the self-fulfilling nature of social stereotypes. *Journal of Personality and Social Psychology* 35: 656–666.

Sosis, R. H. 1974. Internal-external control and the perception of responsibility of another for an accident. *Journal of Personality and Social Psychology* 30: 393–399.

State v. Holmes, 5 So.3d 42 (La. 2009).

State v. Kiser, 284 S.W.3d 227 (Ten. 2009).

State v. Kuehn, 728 N.W.2d 589 (Neb. 2007).

State v. Mercer, 381 S.C. 149 (S.C. 2009).

Tversky, A., and D. Kahneman. 1974. Judgment under uncertainty: Heuristics and biases. *Science* 185: 1124–1131.

U.S. v. Williams, 731 F. Supp. 2d 1012 (D. Haw. 2010).

Vincent, N. A. (Ed.). 2013. *Neuroscience and legal responsibility*. New York: Oxford University Press.

Waldman, P. 2013. Six charts that explain why our prison system is so insane. *The American Prospect*. http://prospect.org/article/six-charts-explain-why-our-prison-system-so-insane.

Watson, P. C. 1960. On the failure to eliminate hypotheses in a conceptual task. *Quarterly Journal of Experimental Psychology* 12: 129–140.
Welch v. State, 992 So.2d 206 (Fla. 2008).
Winter, L., and J. S. Uleman. 1984. When are social judgments made. Evidence for the spontaneousness of trait inferences. *Journal of Personality and Social Psychology* 47: 237–252.
Yosifon, D. 2008. Legal theoretic inadequacy and obesity epidemic analysis. *George Mason Law Review* 15: 681–740.

CHAPTER 18

The Neuroscientific Non-Challenge to Meaning, Morals, and Purpose

STEPHEN J. MORSE

1. INTRODUCTION

As millennia of philosophizing attest, there are challenging questions about the existence, source, and content of meaning, morals, and purpose in human life, but present and foreseeable neuroscience will neither obliterate nor resolve them. Neuroscience, for all its astonishing recent discoveries, raises no new challenges in these domains. It poses no unique threat to our life hopes or to our ability to decide how to live and how to live together. The supposed challenges were best summed up by an editorial warning in *The Economist*: "Genetics may yet threaten privacy, kill autonomy, make society homogeneous and gut the concept of human nature. But neuroscience could do all of these things first" (The Economist 2002).

The primary quarries of those who think that neuroscience poses a challenge to meaning, morals, and purpose are the related concepts of responsibility and desert, especially as they play a role in criminal law. After all, responsibility and desert are intrinsic features of present moral and criminal legal concepts, practices, and institutions, including the imposition of punishment. Most of those who challenge responsibility and desert think that these concepts are philosophically questionable and lead to primitive, prescientific practices, such as overly harsh punishments. Responsibility and desert are also intrinsic to civil law, but virtually none of these challengers considers how their views would affect desert theories in contracts, torts, and property law, for example. Although critics have a duty to embed their criticisms of criminal law in a wider understanding of the implications of the criticisms, this chapter will nonetheless engage with the dominant critique by limiting itself to the potential effect of neuroscience on criminal law and moral responsibility more generally.

As is well known, the primary challenges neuroscience allegedly presents to responsibility, desert, and retributive justifications of punishment are the threat from determinism and the specter of the person as simply a "victim of neuronal circumstances" (VNC) (Greene and Cohen 2006) or "just a pack of neurons" (PON) (Crick 1994). Allegedly, no one is responsible for any of his behavior, and no one deserves a proportionate response to his behavior either because determinism is true and inconsistent with responsibility or because mental states are epiphenomenal and we are therefore not the sort of creatures that can be guided by reason. I have argued repeatedly that no such soul-bleaching outcome as *The Economist* dreads is remotely justified by neuroscience at present (e.g., Morse 2011a) or by any other science for that matter. This chapter reiterates some of these arguments and responds to the challenge of the new determinism, termed "hard incompatibilism" (HI) by its proponents.

The chapter begins by reviewing the law's psychology, concept of personhood, and criteria for criminal responsibility. It then turns to the two primary challenges, determinism and VNC/PON, suggesting that neither is new to neuroscience and neither at present justifies revolutionary abandonment of moral and legal concepts and practices that have been evolving for centuries in both common law and civil law countries. I then turn to HI. First, I suggest some concerns internal to the approach. Then, because the metaphysical premises for responsibility or jettisoning it cannot be decisively resolved, I suggest that the real issue is the type of world we want to live in and that the hard incompatibilist vision is not normatively desirable, even if it is somehow achievable.

2. PSYCHOLOGY, PERSONHOOD, AND RESPONSIBILITY

This section offers a "goodness of fit" interpretation of current Anglo-American criminal law. It does not suggest or imply that the law is optimal "as is," but it provides a framework for thinking about the challenges presented by neuroscience, behavioral genetics, or any other material or social science that purports to be deterministic in some broad sense.

Law presupposes the "folk psychological" view of the person and behavior. This psychological theory, which has many variants, causally explains behavior in part by mental states such as desires, beliefs, intentions, willings, and plans (Ravenscroft 2010). Biological, sociological, and other psychological variables also play a role, but folk psychology considers mental states fundamental to a full explanation of human action. Lawyers, philosophers, and scientists argue about the definitions of mental states and theories of action, but that does not undermine the general claim that mental states are fundamental. The arguments and evidence disputants use to convince others itself presupposes the folk psychological view of the person. Brains don't convince each other; people do. The law's concept of the responsible person is simply an agent who can be responsive to reasons.

For example, the folk psychological explanation for why you are reading this chapter is, roughly, that you desire to understand the relation of the new sciences to agency and responsibility, you believe that reading the chapter will help fulfill that desire, and thus you formed the intention to read it. This is a "practical" explanation rather than a deductive syllogism.

Brief reflection should indicate that the law's psychology must be a folk psychological theory, a view of the person as the type of creature who can act for and respond to reasons. Law is primarily action-guiding and is not able to guide people directly and indirectly unless people are capable of using rules as premises in their reasoning about how they should behave. Unless people could be guided by law, it would be useless (and perhaps incoherent) as an action-guiding system of rules.[1] Legal rules are action-guiding primarily because these rules provide an agent with good moral or prudential reasons for forbearance or action. Human behavior can be modified by means other than influencing deliberation, and human beings do not always deliberate before they act. Nonetheless, the law presupposes folk psychology, even when we most habitually follow the legal rules. Unless people are capable of understanding and then using legal rules to guide their conduct, the law is powerless to affect human behavior. The law must treat persons generally as intentional, reason-responsive creatures and not simply as mechanistic forces of nature.

The legal view of the person does not hold that people must always reason or consistently behave rationally according to some preordained, normative notion of rationality. Rather, the law's view is that people are capable of minimal rationality according to predominantly conventional, socially constructed

---

1. See Sher (2006: 123), stating that although philosophers disagree about the requirements and justifications of what morality requires, there is widespread agreement that "the primary task of morality is to guide action," as well as Shapiro (2000: 131–132) and Searle (2002: 22, 25). This view assumes that law is sufficiently knowable to guide conduct, but a contrary assumption is largely incoherent. As Shapiro writes:

> Legal skepticism is an absurd doctrine. It is absurd because the law cannot be the sort of thing that is unknowable. If a system of norms were unknowable, then that system would not be a legal system. One important reason why the law must be knowable is that its function is to guide conduct. (Shapiro 2000: 131)

I do not assume that legal rules are always clear and thus capable of precise action guidance. If most rules in a legal system were not sufficiently clear most of the time, however, the system could not function. Furthermore, the principle of legality dictates that criminal law rules should be especially clear.

standards. The type of rationality the law requires is the ordinary person's commonsense view of rationality, not the technical, often optimal notion that might be acceptable within the disciplines of economics, philosophy, psychology, computer science, and the like. Rationality is a congeries of abilities, including *inter alia* getting the facts straight, having a relatively coherent preference-ordering, understanding what variables are relevant to action, and the ability to understand how to achieve the goals one has (instrumental rationality). How these abilities should be interpreted and how much of them are necessary for responsibility may be debated, but the debate is about rationality, a core folk psychological concept.

Virtually everything for which agents deserve to be praised, blamed, rewarded, or punished is the product of mental causation and, in principle, is responsive to reasons, including incentives. Machines may cause harm, but they cannot do wrong, and they cannot violate expectations about how people ought to live together. Machines do not deserve praise, blame, reward, punishment, concern, or respect because they exist or as a consequence of the results they cause. Only people, intentional agents with the potential to act, can do wrong and violate expectations of what they owe each other.

Many scientists and some philosophers of mind and action might consider folk psychology to be a primitive or prescientific view of human behavior. For the foreseeable future, however, the law will be based on the folk psychological model of the person and agency described. Until and unless scientific discoveries convince us that our view of ourselves is radically wrong, a possibility that is addressed later, the basic explanatory apparatus of folk psychology will remain central. It is vital that we not lose sight of this model lest we fall into confusion when various claims based on the new sciences are made.

Folk psychology does not presuppose the truth of contracausal free will, it is consistent with the truth of determinism, it does not hold that we have minds that are independent of our bodies (although it, and ordinary speech, sound that way), and it presupposes no particular moral or political view. It does not claim that all mental states are conscious or that people go through a conscious decision-making process each time that they act. It allows for "thoughtless," automatic, and habitual actions and for nonconscious intentions. It does presuppose that human action will at least be rationalizable by mental state explanations or that it will be responsive to reasons under the right conditions. The definition of folk psychology being used does not depend on any particular bit of folk wisdom about how people are motivated, feel, or act. Any of these bits, such as that people intend the natural and probable consequences of their actions, may be wrong. The definition insists only that human action is *in part* causally explained by mental states.

Legal responsibility concepts involve acting agents and not social structures, underlying psychological variables, brains, or nervous systems. The latter types of variables may shed light on whether the folk psychological responsibility criteria are met, but they must always be translated into the law's folk psychological criteria. For example, demonstrating that an addict has a genetic vulnerability or a neurotransmitter defect tells the law nothing per se about whether an addict is responsible. Such scientific evidence must be probative of the law's criteria, and demonstrating this requires an argument about how it is probative.

Consider criminal responsibility as exemplary of the law's folk psychology. The criminal law's criteria for responsibility are acts and mental states. Thus, the criminal law is a folk psychological institution (Sifferd 2006). If an agent meets the criteria for criminal responsibility, the law considers the ascription of blame and the imposition of negative sanctions fair. Blame and punishment, the intentional infliction of some degree of pain on an agent are harmful and therefore require justification in a fair legal system. In Anglo-American jurisprudence, the primary justifications are deontological desert and the consequential goal of achieving social safety. It is, of course, the former that is under assault from the critics.

Let us consider what the criteria are for allegedly deserved punishment. First, the agent must perform a prohibited intentional act (or omission) in a state of reasonably integrated consciousness (the so-called *act requirement*, usually confusingly termed the "voluntary act"). Second, virtually all serious crimes require that the person had a further mental state, the *mens rea*, regarding the prohibited harm. Lawyers term these definitional criteria for prima facie culpability the "elements" of the crime. They are the criteria that the prosecution must prove beyond a reasonable doubt. For example, one definition of murder is the intentional killing of another human being. To be prima facie guilty of murder, the person must have intentionally performed some act that kills, such as shooting or knifing, and it must have been his intent to kill when he shot or knifed. If the agent does not act at all because his bodily movement is not intentional—for example, a reflex or spasmodic movement—then there is no violation of the prohibition against intentional killing. There is also no violation in cases in which the further mental state required by the definition is lacking. For example, if the defendant's intentional killing action kills only because the defendant was careless, then the defendant may be guilty of some homicide crime but not of intentional homicide.

Criminal responsibility is not necessarily complete if the defendant's behavior satisfies the definition of the crime. The criminal law provides for so-called affirmative defenses that negate responsibility even if the prima facie case has been proved. Affirmative defenses are either justifications or excuses. The former obtain if behavior otherwise unlawful is right or at least permissible under the specific circumstances. For example, intentionally killing someone who is wrongfully trying to kill you, acting in self-defense, is certainly legally permissible and many think it is right. Excuses exist when the defendant has done wrong but is not responsible for his behavior. Using generic descriptive language, the excusing conditions are lack of reasonable capacity for rationality and lack of reasonable capacity for self-control (although the latter is more controversial than the former [Morse 2016]). The so-called cognitive and control tests for legal insanity are examples of these excusing conditions. Both justifications and excuses consider the agent's reasons for action, which is a completely folk psychological concept. Note that these excusing conditions are expressed as capacities. If an agent possessed a legally relevant capacity but simply did not exercise it at the time of committing the crime or was responsible for undermining his capacity, no defense will be allowed. Finally, the defendant will be excused if he was acting under duress, coercion, or compulsion. The degree of incapacity or coercion required for an excuse is a normative question that can have different legal responses depending on a culture's moral conceptions and material circumstances.

In short, all law as action-guiding depends on the folk psychological view of the responsible agent as a person who can be properly responsive to the reasons the law provides. If an agent violates the prima facie criteria for a criminal offense, is not justified, and is, roughly speaking, capable of rationality and self-control under the circumstances, the criminal law (and ordinary morality) will consider the agent responsible and deserving of blame and punishment. The agent is also a good candidate for blame and punishment on consequential grounds, such as incapacitating a dangerous person or promoting general deterrence. This is a very familiar picture of responsibility that has been evolving for centuries in Anglo-American and continental legal systems. The question, to which this chapter now turns, is whether neuroscience provides any reason to abandon responsibility, desert, and punishment and to adopt a system of social control based solely on social safety.

3. THE USUAL SUSPECT CHALLENGES: DETERMINISM AND VNC/PON

This section first provides background for the discussion to follow and then addresses the two common challenges. The chapter's next section discusses in more detail the newly emerging form of the determinist challenge, hard incompatibilism, which may seem to avoid the defects of the challenges this section considers.

3.1. Background

Challenges to doctrines, practices, and institutions (which I will refer to generically as the "system") can be internal or external. Although this distinction may not always be clear-cut, it provides a useful framework. In the former case, the basic coherence and acceptability of the system is assumed and criticisms are meant to increase the accuracy, efficiency, and justice of the system. For example, criticisms of the criminal justice system that argue that its responsibility criteria are too narrow or too broad accept the coherence of responsibility and a system to address it that is based on desert and consequential concerns. An external challenge denies that the system is coherent *vel non*. For example, a criticism based on the claim that no one is genuinely responsible for wrongdoing is an external criticism because it assumes that a partial or wholly responsibility-based system is incoherent because it rests on a fundamental mistake. The determinist and VNC/PON challenges are external.

As proponents of these challenges fully recognize, they provide no basis whatsoever for internal reform of a responsibility-based criminal justice system. Determinism is not selective or partial. If it grounds a moral or legal practice, it applies to all who come within the practice, and it cannot make the distinctions concerning guilt and desert that are at the heart of criminal justice. Relatedly, determinism is not the equivalent of a folk psychological compulsion excuse. If it were, then everyone would be compelled, and all would be excused. This, too, would fail to make the distinctions our criminal justice system makes. The same is true of the VNC/PON challenge. It denies the possibility of responsibility, applies to all, and would entail abandoning the moral responsibility distinctions our system now makes.

As is apparent, the consequences of accepting these critiques would be nothing short of radical and completely unmoored from standard views of responsibility. This is no reason not to adopt such changes if they are justified, but it seems clear that the burden of proof should clearly be placed on the proponents of radical change. They seek to abandon a system that has evolved for centuries and that is in accord with commonsense and with moral, political, and legal theories that are widely endorsed and that seems to work, albeit imperfectly (see MacIntyre [2007: ch. 13] for a defense of this stance). These challenges cannot borrow internal reformist changes because they are total critiques. If no one is really morally responsible, and, consequently, no one deserves any blame or punishment, then only a radically new system of social control is justified. What reason would there be to replace the time-tested system based on an unresolvable metaphysical argument (incompatibilist) or an unproven scientific (VCN/PON) claim?

Proponents of the challenges often claim that responsibility and desert theorists should bear the burden of persuasion because they are justifying harsh treatment. But this counterclaim begs the question. It assumes that all punishment is unjustifiably harsh because no one deserves any punishment at all, but that is precisely what desert theorists deny. More important, until the brave new world the radical challenges propose is fully described, it is not clear that it is either workable or that it will not be even more inconsistent with human flourishing than the current system based on responsibility and desert.

3.2. The Determinist Challenge

In one form or another, the challenge from determinism to "free will" and responsibility has been mounted for millennia. Neuroscience poses no new challenge in this respect. No science can prove the truth of determinism (although some seem to push intuitions harder in that direction), and the answers to the determinist threat are the same to neuroscience as they have been to any of its also deterministic predecessors such as psychodynamic psychology or genetics. Neuroscience poses no new assault on meaning, morals, and purpose if these aspects of life are at risk from the truth of determinism.

The primary reason people care about the issue is because it allegedly underwrites conceptions of responsibility, agency, and dignity that are crucial to our image of ourselves and to our political and legal practices and institutions. If responsibility is undermined, meaning, morals, and purpose may indeed be undermined. The alleged incompatibility of determinism and free will and responsibility is therefore a foundational metaphysical and moral issue. Determinism is not a continuum concept that applies to various individuals in various degrees. To the best of our knowledge, there is no partial or selective determinism. If the universe is deterministic or something quite like it, responsibility is possible or it is not. If human beings are fully subject to the causal laws of the universe, as a thoroughly physicalist, naturalist worldview holds, then many philosophers claim that "ultimate" responsibility is impossible (e.g., Pereboom 2001; Strawson 1989). On the other hand, plausible "compatibilist" theories suggest that, for many different reasons, responsibility is possible in a deterministic universe even if no human being has the godlike contracausal, ultimate freedom that incompatibilists require (Moore 2016; Vihvelin 2013; Wallace 1994). Compatibilists hold that human beings possess whatever degree of freedom or control is necessary for responsibility.

Compatibilism is the dominant view among philosophers of responsibility, and it most accords with common sense. Much about who we are and what we do is not a product of our rational action, including our genetic endowment, early environment, and the opportunities that present themselves to us. Luck clearly plays an immense role in human life (Frank 2016). Nonetheless, the compatibilist claims that we retain sufficient capacity to be guided by reason and to choose otherwise when we act or omit (Moore 2016). This is our ordinary view of ourselves and our agency. When any theoretical notion contradicts common sense, the burden of persuasion to refute common sense must be very high, and no metaphysics that denies the possibility of responsibility exceeds that threshold.

There seems no resolution to the incompatibilism/compatibilism debate in sight, but our moral and legal practices do not treat everyone or no one as responsible. Determinism cannot be guiding our practices. If one wants to excuse people because they are genetically and neurally determined or determined for any other reason, one is committed to negating the possibility of responsibility for everyone.

Our criminal responsibility criteria and practices have nothing to do with determinism or with the necessity of having so-called free will (Morse 2007). The metaphysical libertarian capacity to cause one's own behavior uncaused by anything other than oneself—the strongest conception of free will—is neither a criterion for any criminal law doctrine nor foundational for criminal responsibility. Criminal responsibility involves evaluation of intentional, conscious, and potentially rational human action. And few participants in the debate about determinism and free will and responsibility argue that we are not conscious, intentional, potentially rational creatures when we act. The truth of determinism does not entail that actions and nonactions are indistinguishable and that there is no distinction between rational and nonrational actions or compelled and uncompelled actions. Our current responsibility concepts and practices use criteria consistent with and independent of the truth of determinism.

In short, the hard determinist incompatibilists (including the hard incompatibilists) are proposing a radical revision of practices and institutions that have evolved for centuries on the basis of an unresolvable metaphysical claim and in the face of common sense. The case simply is not proved. But, in my view, because this viewpoint makes no obvious internal mistakes, it must be taken seriously, at least in so far as proponents offer a different vision of social ordering rather than just a pure metaphysical argument. Section 4 of this chapter returns to that vision. Rather than attempt to resolve the unresolvable, it assumes that a compatibilist also makes no internal mistakes and turns to a comparative analysis of the world as we know it and the world that hard incompatibilist is offering us. But first let us consider the most radical claim that neuroscience arguably presents: VNC/PON.

4. THE TRULY RADICAL CHALLENGE TO MEANING, MORALS, AND PURPOSE

This section addresses the hyperreductive claim that the new sciences, and especially neuroscience, will cause a paradigm shift in our view of ourselves as agents who can direct our own lives and can be responsible by demonstrating that we are "merely victims of neuronal circumstances" or a "pack of neurons" (or some similar claim that denies human agency). This claim holds that we are not the kinds of intentional creatures we think we are. If our mental states play no role in our behavior and are simply epiphenomenal, then traditional notions of agency and responsibility based on mental states and on actions guided by mental states would be imperiled. More broadly, neurons, or even a big pack of them like the connectome, do not have meaning, morals, and purpose. They are just biophysical mechanism. They are not agents. Meaning, morals, and purpose are products of people, not mechanisms. But is the rich explanatory apparatus of agency and intentionality simply a post hoc rationalization that the brains of hapless *Homo sapiens* construct to explain what their brains have already done? Will the criminal justice system as we know it wither away as an outmoded relic of a prescientific and cruel age? If so, criminal law is not the only area of law in peril. What will be the fate of contracts, for example, when a biological machine that was formerly called a person claims that it should not be bound because it did not make a contract? The contract is also simply the outcome of various "neuronal circumstances."

Before continuing, we must understand three things. The compatibilist metaphysics discussed earlier does not save agency, responsibility, morals, meaning, and purpose if the radical claim is true. If determinism is true, two states of the world concerning agency are possible: agency and all that it entails exists or it does not. Compatibilism assumes that agency is true because it holds that agents can be responsible in a determinist universe. It thus essentially begs the question against the radical claim. If the radical claim is true, then compatibilism is false because no responsibility is possible if we are not agents. It is an incoherent notion to have genuine responsibility without agency. Second, those forms of hard determinism that accept agency but not responsibility are equally false if the radical claim is true. Finally, this challenge is not new to neuroscience. This type of reductionist speculation long precedes the contemporary neuroscientific era fueled by noninvasive imaging. Neuroscience appears to at last provide the ability to prove the claim by discovering the underlying neural mechanisms that are doing all the work that mental states allegedly do, but the claim is not novel to it. The question remains, however: Is the radical claim true?

Given how little we know about the brain–mind and brain–mind–action connections—we do not know how the brain enables the mind and action (Adolphs 2015)—to claim that we should radically change our conceptions of ourselves and our legal doctrines and practices based on neuroscience is a form of "neuroarrogance." It flies in the face of common sense and ordinary experience to claim that our mental states play no explanatory role in human behavior; and thus the burden of persuasion is firmly on the proponents of the radical view, who have an enormous hurdle to surmount. Although I predict that we will see far more numerous attempts to use the new sciences to challenge traditional legal and common sense concepts, I have elsewhere argued that, for conceptual and scientific reasons, there is no reason at present to believe that we are not agents (Morse 2008; 2011*a*: 543–554).

In particular, I can report based on earlier and more recent research that the "Libet industry" appears to be bankrupt. This was a series of overclaims about the alleged moral and legal implications of neuroscientist Benjamin Libet's findings, which were the primary empirical neuroscientific support for the radical claim. This work found that there is electrical activity (a readiness potential) in the supplemental motor area of the brain prior to the subject's awareness of the urge to move his body and before movements occurred. This research and the findings of other similar investigations (e.g., Soon et al. 2008) led to the assertion that our brain mechanistically explains behavior and that mental states play no explanatory role. Recent conceptual and empirical work has exploded these claims (Mele 2009, 2014; Moore 2012; Nachev and Hacker 2015; Schurger et al. 2012; Schurger and Uithol 2015). Moreover, even hard incompatibilists (see Section 5) agree that the conceptual and empirical criticisms of the implications of the Libetian program are decisive (Pereboom and Caruso, Chapter 11 this volume). In short, I doubt that the Libet industry will emerge from whatever chapter of the bankruptcy code applies in such cases. It is possible that we are not agents, but the current science does not remotely demonstrate that this is true. Just for completeness, I should add that similar radical views from psychology, such as "the illusion of conscious will" (Wegner 2002) and the "automaticity juggernaut" (Kihlstrom 2008), suffer from the same defects (Morse 2011a). The burden of persuasion is still firmly on the proponents of the radical view.

Most important, contrary to its proponents' claims, the radical view entails no positive agenda. If the truth of pure mechanism is a premise in deciding what to do, no particular moral, legal, or political conclusions follow from it.[2] This includes the pure consequentialism that Greene and Cohen incorrectly think follows. The radical view provides no guide as to how one should live or how one should respond to the truth of reductive mechanism. Normativity depends on reason, and thus the radical view is normatively inert. Reasons are mental states. If reasons do not matter, then we have no reason to adopt any particular morals, politics, or legal rules or to do anything at all.

Suppose we are convinced by the mechanistic view that we are not intentional, rational agents after all. (Of course, what does it mean to be "convinced" if mental states are epiphenomenal? Convinced usually means being persuaded by evidence and argument, but a mechanism is not persuaded; it is simply physically transformed. But enough.) If it is really "true" that we do not have mental states or, slightly more plausibly, that our mental states are epiphenomenal and play no role in the causation of our actions, what should we do now? If it is true, we know that it is an illusion to think that our deliberations and intentions have any causal efficacy in the world. We also know,

---

2. This line of thought was first suggested by Professor Mitchell Berman in the context of a discussion of determinism and normativity (Berman 2008: 271 n. 34).

however, that we experience sensations—such as pleasure and pain—and we care about what happens to us and to the world. We cannot just sit quietly and wait for our brains to activate, for determinism to happen. We must, and will, deliberate and act. And if we do not act in accord with the "truth" that the radical view suggests, we cannot be blamed. Our brains made us do it.

Even if we still thought that the radical view was correct and standard notions of agency and all that it entails were therefore impossible, we might still believe that the law would not necessarily have to give up the concept of incentives. Indeed, Greene and Cohen concede that we would have to keep punishing people for practical purposes (Greene and Cohen 2006). The word "punishment" in their account is a solecism because, in criminal justice, it has a constitutive moral meaning associated with guilt and desert. Greene and Cohen would be better off talking about positive and negative reinforcers or the like. Such an account would be consistent with "black box" accounts of economic incentives that simply depend on the relation between inputs and outputs without considering the mind as a mediator between the two. For those who believe that a thoroughly naturalized account of human behavior entails complete consequentialism, this conclusion might be welcomed.

On the other hand, this view seems to entail the same internal contradiction just explored. What is the nature of the agent that is discovering the laws governing how incentives shape behavior? Could understanding and providing incentives via social norms and legal rules simply be epiphenomenal interpretations of what the brain has already done? How do we decide which behaviors to reinforce positively or negatively? What role does reason—a property of thoughts and agents, not a property of brains—play in this decision?

As the eminent philosopher of mind and action, Jerry Fodor, reassured us:

> [W]e have . . . no decisive reason to doubt that very many commonsense belief/desire explanations are—literally—true.
>
> Which is just as well, because if commonsense intentional psychology really were to collapse, that would be, beyond comparison, the greatest intellectual catastrophe in the history of our species; if we're that wrong about the mind, then that's the wrongest we've ever been about anything. The collapse of the supernatural, for example, didn't compare; theism never came close to being as intimately involved in our thought and our practice . . . as belief/desire explanation is. Nothing except, perhaps, our commonsense physics—our intuitive commitment to a world of observer-independent, middle-sized objects—comes as near our cognitive core as intentional explanation does. We'll be in deep, deep trouble if we have to give it up.
>
> I'm dubious . . . that we *can* give it up; that our intellects are so constituted that doing without it (. . . *really* doing without it; not just loose philosophical talk) is a biologically viable option. But be of good cheer; everything is going to be all right. (Fodor 1987: xii)

Everything is going to be all right. Given what we know and have reason to do, the allegedly disappearing person remains fully visible and necessarily continues to act for good reasons, including the reasons currently to reject the radical view. We may be a pack of neurons, but that's not all we are. We are not Pinocchios, and our brains are not Geppettos pulling the strings. And this is a very good thing. Ultimately, I believe that the radical view's vision of the person, of interpersonal relations, and of society bleaches the soul. In the concrete and practical world we live in, we must be guided by our values and a vision of the good life. I do not want to live in the radical's world that is stripped of genuine agency, desert, autonomy, and dignity. In short, in a world that is stripped of meaning, morals, and purpose. For all its imperfections, the law's vision of the person, agency, and responsibility is more respectful and humane.

5. THE HARD INCOMPATIBILIST VISION AND ITS DISCONTENTS

In recent years, a group of theorists who adopt a position called "hard incompatibilism" (HI) have recognized that if scholars care about the practical implications of a philosophical position, then they have an obligation to propose the details of psychology, politics, and law that would follow (Pereboom and Caruso, Chapter 11 this volume).[3] HI does not deny the causal theory of action and that human beings are capable of being reasons-responsive. It accepts the possibility of meaning, morals, and purpose because we are agents, but it denies that anyone is responsible for any conduct, and that backward-looking blame and punishment is never justified. Value is preserved, including the distinction between right and wrong, on at least axiological grounds. The reactive attitudes such as anger, indignation, and resentment that P. F. Strawson (1962) famously thought were appropriate when a responsible agent violated a justified prohibition are never justified according to HI. We may feel them, but we should not. Instead, HI proposes that we should feel emotions such as sorrow and regret for harms that have been caused by the harmdoer's actions. We can try to guide such people to do right in the future by attempting to persuade them to see the good reasons not to behave in such injurious ways, but responsibility plays no appropriate role in the social ordering. The role of law and social ordering practices to address harmful behavior is entirely forward-looking. The moral goals of the HI approach are protection, moral formation, and reconciliation, goals that justify forward-looking blame but never deserved punishment.

3. Although Professors Pereboom and Caruso have many contributions to this position, I focus on their chapter in this volume because it is exemplary.

Even though HI claims that no one is responsible, it properly concedes that dangerous people who commit crimes exist and must be controlled for the good of society at large. It recognizes the common criticisms of a purely consequential social control system, such as the potential for intervening more harshly than an individual's dangerous conduct might suggest in order to further general deterrence. To avoid these problems, it adopts a public health–like system of quarantine or of lesser but still intrusive measures to control dangerous agents that are justified by societal self-defense. HI also borrows the proportionality principle from individual self-defense. That is, unnecessary intervention to achieve social safety is unjustified because it would be akin to using disproportionate force when one has a right to defend oneself. Because no one is responsible and deserves to suffer, these interventions should never be painful unless inflicting pain is ultimately unavoidable. Even then, the pain should be minimized consistent with the preventive goal of imposing it. Thus, pure quarantine would have to be under conditions of relatively comfortable confinement. Moreover, when a dangerous agent is being controlled, society has a duty to try to engage the agent's reason or otherwise to offer him a "fix" so that he no longer needs to be controlled. Finally, society has a duty to create the social conditions that decrease the probability of harmdoing generally.

One possible response in a world without responsibility is to treat people as good and bad bacteria, a story that treats potential human harmdoers no differently from any other organic (or inorganic) harmdoer (Morse 1999). It is a pure prediction/prevention scheme. Human animals must simply be managed to increase beneficent action and to decrease harmful action, and any means calculated to achieve this end would be justified because bacteria have no rights or even interests. HI attempts to avoid this dystopian vision by arguing that we are reasons-responsive and thus agents even though we are not responsible.

The HI vision has been criticized by others, but I wish to add to these cautions. HI seems benevolent and humane, but what would be the costs? My major response is that HI tries to have it both ways; that is, to preserve most of the desirable attitudes and practices associated with responsible agency, such as a right to liberty, autonomy, and dignity, but to do so without a robust conception of agency. It is agency on the cheap, much as some compatibilists, such as Dennett and Pinker, try to have compatibilism on the cheap. They suggest that there must be responsibility and punishment but without adopting a robust conception of responsibility (Moore 2012). I shall suggest that HI provides only a pale simulacrum of agency and does not fully comprehend the implications of its view. Although I am a mixed theorist about punishment who believes that we do have basic desert and that retribution is at least a necessary precondition for fair blame and punishment, the argument of this section does not claim that we have basic desert. Rather, I try to stay internal to HI and to explore its implications assuming that we lack basic desert.

Let us begin by noting that neuroscience plays no necessary role in the HI program and by exposing an error about retributivism that seems to, in part, motivate HI's criticism of retributivism. The HI proposal is perfectly defensible independently of any of the alleged truths of neuroscience (or psychology). Neuroscience does not prove the truth of either determinism or the lack of control that is a premise in the HI claim that persons also lack basic desert. HI may try to use neuroscience findings, such as for predicting future harmful behavior or for reducing the potential for future harmdoing, to suggest that its proposal is practicable. HI often overclaims (see Morse 2006, 2013b on "brain overclaim syndrome") for the science (e.g., see Poldrack 2013, providing a much less optimistic view of neuroprediction), but neuroscience is not necessary to the argument. Any successful prediction or behavioral change method, for example, would provide the type of practical assistance crucial to the fair workings of the HI scheme.

Pereboom and Caruso (Chapter 11 this volume), for example, tie excessive and harsh punishments to retributivism, but most of the excessive and harsh criminal justice practices they and I and many other decry have been justified on consequential grounds such as deterrence and incapacitation. It is the retributivists who have had the resources to criticize such programs. HI claims that it, too, has the resources to avoid harsh and intrusive interventions, but does it? Now let us turn to the central difficulties.

It is crucial to understand that nothing is "up to" the person, according to HI, because no one has sufficient "control" to justify what HI terms "basic desert." Whether a person acts well or poorly is not up to them. Whether they feel joy or sorrow in response to their beneficence or harmdoing is not up to them. They cannot be morally praised or blamed for what they have already done or felt (although the action or emotional response may be judged good or bad itself), and they deserve nothing thereby. Whether they respond to moral address and concern is not up to them. Whether they are moved by the right sorts of moral and personal reasons is not up to them. Whether an intervention that addresses or bypasses their reasons is effective is not up to them. They may be reasons-responsive in theory, but reasons in HI's vision are simply mechanisms for changing behavior and are no different in principle from any other mechanism for doing so. Those who believe in responsibility, in contrast, believe that the capacity for reasons-responsiveness or something like it occupies a privileged position because such a capacity indicates that much of our behavior is up to us and we deserve appropriate responses to our action. To treat reasons as simply a mechanism, as HI implicitly does, suggests that we are simply creatures to be manipulated in the right ways to do the right thing rather than being genuinely autonomous agents. We are suspiciously close to the good bacteria/bad bacteria story, although bacteria do not have access to reasons as a mechanism (or so we think).

Although whether we respond to reasons and what reasons we respond to is not up to us, according to HI, we nonetheless have a right to liberty, autonomy, and dignity. But why? Let us compare the current legal regime concerning responsibility and preventive state action, remembering that prevention of danger is the crux of the radical HI proposal. Anglo-American law in this respect is governed by what I have termed desert-disease jurisprudence (Morse 1999, 2002). The state may clutch an individual to promote social safety only if the agent deserves to be clutched because he has committed a crime (including attempts, of course) or if the agent is dangerous but is not responsible for his dangerous conduct. Agents who are dangerous but responsible must be left alone until they commit a crime, no matter how dangerous they are and no matter how certain they are to offend if they are not controlled. The best explanation for this jurisprudence that limits infringements of liberty as described is the respect the law grants responsible agents because they are responsible agents. We can act preventively in the absence of harmful action in the case of younger children, some people with mental disorder or intellectual disability, and some people with dementia precisely because they are not capable of rationality and thus are not responsible for themselves. But we cannot intervene with responsible agents unless they commit crimes because we respect their capacity to act well even if we are quite sure they will not. It is responsibility that underpins the full panoply of rights that protect liberty, autonomy, and dignity.

If nothing is up to us, what is the source of our liberty, dignity, and autonomy? It is understandable that we would have a sense of good (not harmful; beneficent) and bad (harmful; not beneficent) in the HI society, but why, for example, do we have a right to be at liberty in the absence of actual harmdoing? What is the meaning of autonomy when nothing is up to us? Why does the fact that reasons are sometimes successful mechanisms confer dignity on us if those reasons themselves are not up to us in any important way? In assessing reason-bypassing interventions, especially of a more radical type such as deep brain stimulation (DBS, which a completely experimental, unproved intervention for behavioral abnormalities), Pereboom and Caruso (Chapter 11 this volume) suggest that to respect autonomy, decisions should be left up to the subject as far as possible? But why respect a decision that is not really up to the subject? In short, the agency HI concedes is a simulacrum of what we usually mean by agency.

Consider a response of the agent who has acted harmfully and feels no sorrow or regret. He claims that doing it was not up to him and neither was lack of what would be the appropriate emotions. He is now addressed by the agents of social control with moral concern and guidance, with the goal of future protection, reconciliation, and moral formation. HI tells the agent that he is not being blamed for what he did, but he should be altered so as properly to be guided in the future to do what he ought to do. Is the HI proposal internally consistent in this regard? Mitchell Berman doubts it in the context of determinist concerns. Here is his formulation:

> [D]eterminism presents a challenge not only for any retributivist response to Jn(P) [the demand for the need to justify punishment] but—more fundamentally—even to Jn(P) itself. That is because, in my view, determinism threatens not only judgments of responsibility, blameworthiness, or desert but all normative judgments on the ground that the forward-looking judgment "ought to" implies the cogency of the corresponding backward-looking judgment "ought to have." If so, then to the extent that determinism threatens the latter, it threatens the former too. (2008: 271, n. 34)

I believe that the same critique applies to HI. If the agent is not responsible for what he has done in the past, he is also certainly not responsible for what he will do in the future. What does it mean to say that he ought to do better? Whether he continues to behave badly or is manipulated into behaving well by reasons or by any other mechanism, it is not up to him, and he deserves neither praise nor blame for his future conduct. We can say that it would be a good thing if he behaved well and a bad thing if he behaved badly, but whether he acts that way or not is not up to him. The "ought" in HI, like the concepts of agency, dignity, and autonomy, is pallid. If, in response, it is claimed that the agent can be guided by reason, then we might ask whether he is guided because he directs himself to be or simply because he is. If the answer is the former, then this sounds like compatibilist agency, and why is this not sufficient for desert? If it is the latter, the guidance is ephemeral and certainly not up to the person.

Before turning to the systemic and practical implications of HI, let us consider how our reactions to wrongdoing would be affected by the genuine, intellectually internalized adoption of HI. We know that we have the psychological capacity to behave consistently with the truth of desert and the justifiability of blame and punishment. That is the world we live in now. Over time, HI hopes that HI-appropriate emotions and responses will replace those fueled by unjustified acceptance of desert. I will assume that we could remake our psychology consistent with the prescription of HI, although I do think it is an open question whether this is possible. The question is whether one would want to remake ourselves psychologically in the way HI desires. Recall that I am assuming that HI makes no mistakes and that the underlying metaphysical dispute between it and a desert regime is unresolvable. The question, then, is to which regime we lend our support.

Let us start with an everyday case: an act of interpersonal disloyalty and betrayal, such as the infidelity of one's allegedly committed partner. Assume for the sake of simplicity that, according to traditional, pre-HI standards, there is no mitigating or excusing condition present. According to HI, anger, resentment, and indignation would be unwarranted, and no blame would be deserved for the betrayer's behavior. HI concedes that understandable but unwarranted emotional reactions such as anger will occur, but the betrayed partner cannot be blamed for them, even if they have fully embraced the truth of HI. This would be so even if they apparently tried but failed to suppress the unjustified emotions and their equally unjustified expression, such as blaming the betrayer. If the betrayed partner morally addresses the betrayer, and the latter seems to have seen the error of his ways, what should the betrayed partner do if the betrayer lapses again? Anger and blame are not warranted. Just more sorrow and regret? Is this possible? Even if it is, is it really a preferable way to live?

The next example is Bernard Madoff, whose extraordinary fraud wreaked havoc in the lives of countless innocent people and ruined many. The fraud was fully intentional and knowing. To the best of our knowledge, Madoff is not a psychopath and suffers from no major mental disorder, and he had no other potential traditional mitigating or excusing condition. US Federal District Court Judge Denny Chin imposed a sentence of 150 years on Madoff, whose crime he described as extraordinarily evil (*United States v. Madoff* 2009). As Judge Chin recognized, the ferocious length of the sentence was symbolic and unnecessary because Madoff's life expectancy at that time was thirteen years, and he presented no future danger to the community. Nevertheless, Judge Chin emphasized that retribution, deterrence, and justice for the victims demanded a harsh sentence. Now, how would HI respond? As usual, sorrow and regret would be justified, but not anger and blame. Madoff was not responsible. But what should society do with Madoff? He fully understood that he had done grievous wrongs, he seemed genuinely regretful, and he presents no further danger to society. On the public health quarantine model for social self-defense, nothing needs to be done. And general deterrence is barred in this model. If one tries to assimilate general deterrence to social self-defense, then all the familiar problems with consequentialism arise. Which world do you prefer to live in? Judge Chin's—that is, ours—or the world HI proposes.

Consider a more extreme, exceptional final example, Timothy McVeigh. McVeigh, a decorated and honorably discharged veteran who had seen combat service, was an extreme critic of the US government who believed that it was systematically destroying our precious liberties. He was particularly enraged by the siege at Waco. After years of embracing and espousing such attitudes, he decided to act. On April 19, 1995, McVeigh methodically planned and executed the bombing of the Alfred P. Murrah Federal Building in Oklahoma City. The blast killed 168 people, including 19 children in a day care center in the building, and injured 684 others. Although on a few occasions McVeigh said he regretted the loss of life, he also said that it was necessary. There was apparently no genuine moral regret; it was just the cost of doing business. And his last words about the attack were that he wished he had leveled the entire building, which would, of course, have caused far more loss of life. To the best of our knowledge, McVeigh was not a psychopath and had no major mental disorder.

HI would claim that the only appropriate emotions and responses to the carnage would be sorrow and regret and moral address with a view to altering McVeigh's reasons, but not anger, blame, and punishment. Suppose McVeigh had a "conversion" not long after being apprehended. Even without any intervention, he came to understand the enormous monstrosity of his deeds. Of course, guilt, remorse, and asking for forgiveness would be unwarranted because he was not responsible for his mass murder, but assume that his expressions of regret were genuine and we believed that he would not do it again. I assume that HI would demand that he be released, perhaps with some minimal supervision to ensure that his conversion was genuine and permanent. If no such conversion occurs, then he will presumably be confined in a most comfortable place that allows him to pursue his projects while he is subject to moral address and attempts to change his reasons. If he cannot be changed or remains predictively dangerous for any reason, his comfortable confinement will continue. Even if such a set of reactions to McVeigh's outrage is possible, which I consider an open question, do you want to adopt such attitudes and all that they entail?

It is worth pausing to note that the shift in emotions and attitudes HI prescribes would be radical. Experimental work indicates that people are intrinsically retributivist (e.g., Fehr and Gachter 2002). This does not mean that retributive attitudes and practices either should be fostered or cannot be altered. Assuming it is possible to change ourselves as HI desires, however, we would be new people. The history of attempts to remake human beings to achieve a radical new vision of society is one of failure and, indeed, horror. Furthermore, not only criminal justice would be affected. Deontological theories abound in many central legal doctrines, including torts, contracts, and property (Moore 2007). Much of the legal system would have to be redesigned. HI cannot cherry pick criminal justice. Caution is warranted.

Now let us consider the systemic and practical implications of HI, some of which have been alluded to. There are many criticisms that could be leveled, some of which HI has tried to address, but space constraints compel me to focus on perhaps the most important: How much preventive intervention of what type will HI permit? How intrusive will this regime be? HI tries to avoid the potentially intrusive and objectionable interventions that purely consequential justifications would permit by using the societal analog to individual self-defense. It thereby hopes to avoid the familiar problems consequentialism generates, but I fear that these hopes will not be realized. The self-defense theory HI deploys will not prevent massive preemptive intervention, with all the problems this will entail.

Self-defense (and defense of others) has an individual and collective (international law) body of doctrine that has developed for centuries in Western legal tradition. Would either limit intervention only to those cases in which harmdoing has actually occurred or is imminent? Countless trees have been felled, rivers of ink have been spilled, and innumerable pixels have been illuminated about the justification for self-defense and its criteria. HI is a radical proposal whose practical program heavily relies on self-defense, so we are entitled to a fully developed account of self-defense that defends a position on the important questions, but none has been provided. Whatever account is finally provided will almost certainly be theoretically and practically controversial, which is again a reason for caution. In what follows, however, I shall begin to sketch some of the problems that will have to be addressed.

In both individual and self-defense, the potential victim must be wrongfully threatened, the threat must be immediate or imminent, the belief that the wrongful threat is imminent must be reasonable, and the response of the defender must be proportionate to the wrongful harm threatened. In international law, the justification of an anticipatory strike when attack is not temporally imminent is controversial, but many think it is entirely defensible (e.g., Van de hole 2003) because waiting for imminent harm may unjustifiably increase the risk to the victim. Think of the Israeli strike on Iraq's Osiris nuclear plant as an example. Thus, preemptive action based on predicted danger is arguably justified, and there is no indisputable counterargument. In individual self-defense, the imminence of the harm or a reasonable belief that it is imminent is required. It is easy to imagine cases in which the threatened harm is predictively quite certain but not technically imminent (e.g., *State v. Norman* 1986; *State v. Schroeder* 1978) and the influential Model Penal Code has tried to soften this requirement a bit (American Law Institute 1985, Sec. 3.04(1)). Nonetheless, no genuinely preemptive strikes are permitted.

If the potential victim forms a reasonable but mistaken belief that he is in danger and harms the perceived attacker, no liability attaches. There is a rich theoretical debate about whether the reasonably mistaken potential victim is justified or excused, but it is uncontroversial that there is no criminal liability. Reasonableness is, of course, a normative standard, not a bright-line rule, and it is capable of adapting to shifting notions of what interests should be protected. The question to be addressed here is what will be considered a reasonable belief in the need to take preemptive action in the HI scheme.

The more interesting question is why imminence is required. One important reason is that there is typically no opportunity to involve law enforcement rather than to use self-help, and, therefore, "private justice" is permitted. This is a fine rationale, but it does not always apply. Even if attack is certain and no resort to law enforcement is practicable, individual self-defense is not justified unless harm is imminent. Imminence is also justified on epistemic grounds. Until harm is imminent, the individual, fallible potential victim might find it difficult to be sure self-defense was required. This, too, is a good rationale, but it, too, does not always apply.

The most profound reason for the imminence requirement is respect for responsible agency. Self-defense doctrine is always balancing the rights of potential victims and perceived attackers. If the latter are responsible agents, we believe that they may desist simply because good reason convinces them not to act wrongly. Thus, they cannot be defended against unless it seems clear that desistance will not occur. Respect for responsibility also explains why imminence is not required for various forms of preventive involuntary commitment. Such forms of preemptive action are part of disease jurisprudence in which the usual grounds to maximize liberty are abridged because the agent is not responsible for his or her dangerousness. In short, even though the perceived attacker can always desist, and we can never be certain until the "last act" is performed, the last act is not necessary for justifiable self-defense. If the perceived attacker comes sufficiently close to the last act, his interest in liberty yields to the safety of the individual defender or society. Some preemption is thus allowed, but not much, as a result of respect for agency.

HI does not treat all cases of dangerous propensity as examples of a behavioral disorder (or any other kind of disorder), but neither the disordered dangerous agent nor the nondisordered dangerous agent is responsible and neither can be blamed. The only justification for differential treatment of those with and without disorder—and generally for the treatment of any potentially dangerous agent—would have to be purely consequential. Do the authorized decision-makers reasonably believe that the agent will be dangerous in the future, and what needs to be and can be reasonably done to prevent the danger? Having already committed a dangerous action should simply be data to support the decision. Some people who have done great harm, such as Madoff, are unlikely to do it again, and some people who have not yet done a dangerous action might be highly predictively likely to do so without intervention. Any prediction will likely have false positives, especially if predicting low base-rate behavior, even if our predictive tools become increasingly sensitive.

The questions for HI are what belief about the likelihood of an impending danger should be considered reasonable, and whether harm should have to be imminent? Why not adopt a "screen and intervene" system that systematically screens all members of society and takes whatever interventions are necessary to prevent predicted danger? HI cannot, of course, borrow responsible agency from traditional self-defense doctrine as a limit on anticipatory intervention. Nor can it use responsible agency as a foundation for the right to liberty. It can use only consequential justifications to decide when anticipatory intervention is justifiable. Compare, for example, the HI criticisms of Saul Smilansky's (2011) well-known, clever "funishment" argument, which says that those incarcerated for public safety reasons in a world without desert should be compensated for their undeserved deprivation of liberty. The response is entirely consequential (Pereboom and Caruso, Chapter 11 this volume). In effect, the use of self-defense to avoid the problems with consequentialism will not succeed because self-defense will inevitably collapse into consequentialism in the absence of responsible agency.

On consequential grounds, whether and to what degree anticipatory intervention is justifiable will be theoretically controversial. How many unnecessary anticipatory interventions would be justified to prevent one horrific crime? One cannot simply say that unacceptable false-positive rates will limit anticipatory intervention because it begs the questions of how many false positives will be acceptable for what goals. In the involuntary civil commitment system, for example, we know the false-positive rate is very high for low base-rate, serious harmful behavior because severely mentally disordered people do not commit serious crimes as a result of mental disorder alone (Monahan et al. 2001). The practice continues nonetheless because lack of responsible agency abridges traditional liberty interests. Notice, too, that as our prediction and prevention tools became more sensitive, the balance would shift toward ever more anticipatory intervention and thus more liberty intrusion. Even if there was social consensus on the precise interests to be weighed, doing that weighing is well-nigh impossible. HI offers no real limiting principle for the application of anticipatory intervention and thus for massive intrusions on liberty.

Finally, consider how the HI system will be administered. If all members of a society believe that no one is ever responsible for anything they do or will do in the future, what will be the effect on the balance of liberty versus safety? Will we value liberty as much? Will there be a shift toward a stronger preference for safety as the system proceeds? In the Anglo-American legal system, there are constant attempts to "fill the gaps" in desert-disease jurisprudence either by responding harshly to responsible agents or by increasing the category of those considered not responsible (Morse 2011*b*, 2013*a*). Recidivist sentencing enhancement is an example of the former; involuntary civil commitment of so-called mentally abnormal predators is an example of the latter. All of these "gap-fillers" are justified consequentially on public safety grounds, and virtually all have been condemned by informed commentators of every theoretical stripe as abusive of basic rights. Yet the political process produced them because the public demands them, and the state agents that apply them are complicit. Desert-disease jurisprudence, which entirely depends on the distinction between responsible and nonresponsible agency, at least sets imperfect limits on state overreach. What similar constraints will be available in HI? It will be a massive system if it is done properly, even if most interventions might be reasonably minimal. Will there be sufficient tools and experts? Unless those charged with administering the system have objective tools and no discretion, what will be done to prevent excesses and abuses, especially if the administrators do not think they are responsible for what they do? Even if there are sufficient experts and tools, there will be no limits on state intervention except a controversial, manipulable, and ultimately unknowable social welfare calculus. It will be a brave new world, indeed.

A few last points deserve mention. If our society were to increase equality and mental health services, a move compatible with both desert and HI regimes, there would still be a crime problem and differential rates of criminality among various demographic groups for many different reasons. The HI system would thus have disparate impact on some members of society who are already less advantaged, much as our current criminal justice system now does. Furthermore, a desert-based regime has all the resources necessary to soften what seem to be disproportionately harsh and inutile criminal justice doctrines, procedures, and practices and to adopt what HI considers more enlightened social welfare policies. I have recommended desert-based softening myself (e.g., Morse 2003). Recall that virtually all the worst excesses in criminal justice were the product of consequential justifications and that desert had the resources to criticize them. We can have a fairer, kinder criminal justice system without abandoning desert.

Whether to adopt HI or to retain our regime of desert and responsibility is a political question. We are now ready to vote. Before you do, however, please recognize—as we saw previously—that the new neuroscience plays no proper justificatory role in your decision because it raises no new issues in the debate between HI and desert. And be careful what you wish for.

6. CONCLUSION

The existence, source, and content of meaning, morals, and purpose in human existence are controversial, but neuroscience poses no new threat to these desirable human goods, and it surely does not resolve the controversies. In particular, neuroscience raises no new determinist worries, nor does it threaten our familiar sense of ourselves as agents who can act for and be guided by reasons. A new form of incompatibilism, "hard incompatibilism," that denies the possibility of responsibility, is benignly motivated, but its argument, too, does not depend at all on neuroscience. Despite its admirable motivations and attempts to avoid the pitfalls of consequentialism, it nonetheless offers a pallid conception of agency and a radical proposal for reengineering social ordering that is more frightening than our current, imperfect regime. As C. L. Lewis recognized long ago (1953), a system that treats people as responsible agents is ultimately more humane and respectful.

REFERENCES

Adolphs, R. 2015. The unsolved problems of neuroscience. *Trends in Cognitive Sciences* 19: 173.

American Law Institute. 1985. *Model penal code and commentaries*. Washington, DC: The American Law Institute.

Berman, M. 2008. Punishment and justification. *Ethics* 118: 258.

Crick, F. 1994. *The astonishing hypothesis: The scientific search for the soul*. New York: Scribner.

The Economist. 2002 May 25. The ethics of brain science: Open your mind. http://perma.cc/3DKJ-9GAZ.

Fehr, E., and S. Gachter. 2002. Altruistic punishment in humans. *Nature* 415: 137.

Fodor, J. A. 1987. *Psychosemantics: The problem of meaning in the philosophy of mind*. Cambridge, MA: MIT Press.

Frank, R. H. 2016. *Success and luck: Good fortune and the myth of meritocracy*. Princeton, NJ: Princeton University Press.

Greene, J., and J. Cohen. 2006. For the law, neuroscience changes nothing and everything. In S. Zeki and O. Goodenough (Eds.), *Law and the brain*, pp. 207–226. Oxford University Press.

Kihlstrom, J. F. 2008. The automaticity juggernaut—or are we automatons after all? In J. Baer, J. C. Kaufman, and R. Baumeister (Eds.), *Are we free? Psychology and free will*, pp. 155–180. Oxford University Press.

Lewis, C. S. 1953. The humanitarian theory of punishment. *Res Judicatae* 6: 224.

MacIntyre, A. 2007. *After virtue, 3d edition*. South Bend, IN: University of Notre Dame Press.

Mele, A. R. 2009. *Effective intentions: The power of conscious will*. New York: Oxford University Press.

Mele, A. R. 2014. *Free: Why science hasn't disproved free will*. New York: Oxford University Press.

Monahan, J., et al. 2001. *Rethinking risk assessment: The MacArthur study of mental disorder and violence*. New York: Oxford University Press.

Moore, M. S. 2007. Four reflections on law and morality. *William and Mary Law Review* 48: 1523.

Moore, M. S. 2012. Responsible choices, desert-based legal institutions, and the challenges of contemporary neuroscience. *Social Philosophy and Policy* 29: 233.

Moore, M. S. 2016. Stephen Morse and the fundamental psycho-legal error. *Criminal Law and Philosophy* 10 (1): 45–89.

Morse, S. J. 1999. Neither desert nor disease. *Legal Theory* 5: 265.

Morse, S. J. 2002. Uncontrollable urges and irrational people. *Virginia Law Review* 88: 1025.

Morse, S. J. 2003. Diminished rationality, diminished responsibility. *Ohio State Journal of Criminal Law* 1: 289.

Morse, S. J. 2006. Brain overclaim syndrome and criminal responsibility: A diagnostic note. *Ohio State Journal of Criminal Law* 3: 397.

Morse, S. J. 2007. The non-problem of free will in forensic psychiatry and psychology. *Behavioral Sciences and the Law* 25: 203.

Morse, S. J. 2008. Determinism and the death of folk psychology: Two challenges to responsibility from neuroscience. *Minnesota Journal of Law, Science and Technology* 9: 1.

Morse, S. J. 2011a. Lost in translation? An essay on law and neuroscience. In M. Freeman (Ed.), *Law and Neuroscience* 13: 529.
Morse, S. J. 2011b. Protecting liberty and autonomy: Desert/disease jurisprudence. *San Diego Law Review* 48: 1077.
Morse, S. J. 2013a. Preventive detention of psychopaths and dangerous offenders. In K. A. Keihl and W. A. Sinnott-Armstrong (Eds.), *Handbook on psychopathy and law*, pp. 321–345. Oxford University Press.
Morse, S. J. 2013b. Brain overclaim redux. *Law and Inequality* 31: 509.
Morse, S. J. 2016. Moore on the mind. In K. K. Ferzan and S. J. Morse (Eds.), *Legal, moral and metaphysical truths: The philosophy of Michael S. Moore*, pp.233-250. New York: Oxford University Press.
Nachev, P., and P. Hacker. 2015. The neural antecedents to voluntary action: Response to commentaries. *Cognitive Neuroscience* 6: 180.
Pereboom D. 2001. *Living without free will*. Cambridge, MA: Cambridge University Press.
Poldrack, R. 2013. How well can we predict future criminal acts from fMRI data? http://perma.cc/X5TP-LGZ8.
Ravenscroft. I. 2010. Folk psychology as a theory. *Stanford Encyclopedia of Philosophy* http://plato.stanford.edu/entries/folkpsych-theory/.
Schurger, A., and S. Uithol. 2015. Nowhere and everywhere: The causal origin of voluntary action. *Review of Philosophy and Psychiatry* 1 http://perma.cc/7DJL-5BWZ.
Schurger, A., et al. 2012. An accumulator model for spontaneous neural activity prior to self-initiated movement. *Proceedings of the National Academy of Sciences* 109: E2904.
Searle, J. R. 2002. End of the revolution. *New York Review of Books* 49: 33.
Shapiro, S. J. 2000. Law, morality, and the guidance of conduct. *Legal Theory* 6: 127.
Sher, G. 2006. *In praise of blame*. New York: Oxford University Press.
Sifferd, K. 2006. In defense of the use of commonsense psychology in the criminal law. *Law and Philosophy* 25: 571.
Smilansky, S. 2011. Hard determinism and punishment: A practical *Reductio*. *Law and Philosophy* 30: 35.
Soon, C. S., et al. 2008. Unconscious determinants of free decisions in the human brain. *Nature Neurosicence* 11: 543.
State v. Norman. 1989. 324 N.C. 253.
State v. Schroeder. 1978. 261 N.W. 2d 759.
Strawson, G. 1989. Consciousness, free will and the unimportance of determinism. *Inquiry* 32: 3.
Strawson, P. F. 1962/2008. Freedom and resentment. In P. F. Strawson (Ed.), *Freedom and resentment and other essays*. New York: Routledge.
United States v. Bernard L. Madoff. 2009. United States District Court (S.D.N.Y.).t https://www.justice.gov/sites/default/files/usao-sdny/legacy/2012/04/16/062909sentencing.pdf
Van de hole, L. 2003. Anticipatory self-defence under international law. *American University International Law Review* 19: 69.
Vihvelin, K. 2013. *Causes, laws and free will: Why determinism doesn't matter*. New York: Oxford University Press.
Wallace, R. J. 1994. *Responsibility and the moral sentiments*. Boston, MA: Harvard University Press.
Wegner, D. 2002. *The illusion of conscious will*. Cambridge, MA: MIT Press.

INDEX

action(s)
  cause(s) of, 7, 195–196, 264, 321, 343
  control of, 146–159, 258, 266
  free, 163, 196, 263, 266
  freedom of, 182, 184, 197
  goal(s) for, 71–80, 83
  human, 185, 198, 334, 336, 341
  immoral, 55, 200, 202–203, 236, 244
  preemptive, 352–353
  rational, 340–341
  reasons for, 255, 338
  responsibility for, 183, 188, 194–195, 198, 205, 208, 223, 226, 232–233, 240, 270, 327
  See also freedom; (moral) responsibility
act requirement, 337
actualia, 165–170. *See also* possibilia
adaptability, 16, 147, 154, 159
adaptive value, 71, 73, 75
affection, 41–44, 46–47, 49–51. *See also* love; *storge*
affordances, 122
  action-affordances, 137
  social, 139–140
agape, 41–42. *See also* love; *storge*
agency, 7, 9, 345
  and autonomy, 16, 21
  consciousness in, 147–149, 255
  explanation of, 146, 153
  free, 151, 252
  human, 236–241, 243–244, 246–247, 257, 301–302, 341
  moral, 50n17, 147, 164, 243, 270
  normative aspect of, 152
  and responsibility, 270, 335, 340, 342

  responsible, 235, 252, 346, 353–354
  threat of shrinking, 198
  See also autonomy; free will; (moral) responsibility
Ainslie, George, 117
Alicke, Mark, 20, 320–321
alienation, 103
anxious, 131
altruism, 26–29, 79, 81
amygdala, 72, 98, 155, 213, 242
  right, 135
anarchic hand, 117
angst, 16, 19, 147, 263, 265, 279
  existential, 19, 147, 153, 199, 253–254, 270, 275
  See also anxiety; neuro-angst
answerability, 194
  moral, 18
  See also moral responsibility; responsibility
anxiety, 1–3, 5, 8, 10–11, 123, 127, 130, 132, 135, 140, 151, 154–155, 276n16, 279
  existential, 7, 19, 272, 275–279, 283–286, 289–294
  See also angst
Arendt, Hannah, 93–94, 103. *See also* totalitarianism
Aristotle (Aristotelian), 2, 11, 13, 182–183
  physics, 299
  system, 305
arrow of time, 20, 231, 302–303, 305–306
attention, 39, 47–48, 57, 139, 242, 246, 257n5, 273
  volitional (voluntary), 163, 180–181

Attention Deficit Hyperactivity Disorder (ADHD), 55, 212, 325. *See also* attention
attunement, 132–134, 137, 139, 141, 211
   emotional or affective, 133
authenticity, 45, 112, 126–142, 306
   existential, 126–131, 134–136, 141
   relational, 15, 140, 134–142
   See also autonomy; Heidegger; Sartre
automaticity, 147, 197–198, 343
autonomy, 12, 16, 21, 45, 48, 129, 134, 139–140, 212, 333, 345–349. *See also* authenticity

Balaguer, Mark, 183–188
Bargh, John, 147
basal ganglia, 34, 72–73, 150
basic desert, 18, 193–195, 198–203, 205, 207, 211, 216–218, 236, 239–240, 346–347. *See also* moral responsibility; punishment
Befindlichkeit. *See* disposition
behavior(s), 12, 14, 25–27, 30–34, 39, 42, 51–52, 71, 73–75, 80–82, 87, 89, 94, 96, 100–101, 115, 117, 119, 121, 123, 134, 138, 159, 197–201, 214
   active or passive, 71
   control, 16, 146–159, 180
   cooperative, 34, 132
   criminal, 17–19, 194–199, 203–204, 206–212
   emotional, 72, 78
   flexible, 16, 147, 150, 159–160
   habitual, 92, 135, 214
   modules that drive, 121–122
   moral, 12–13, 25, 27, 55
   social, 12, 31–34
   See also action(s); agency; freedom; free will
behavioristic therapy(-ies), 211–212
being-as-object, 130
being-for-myself, 129
being-for-others, 16, 126–127, 129–130
being-in-the-world, 128, 131–132, 137, 140
being-toward-death, 127, 135
being-with (*Mitsein*), 16, 126–128, 128n2, 130, 133, 136–137, 140–141

benevolent sexism. *See* sexism
Benjamin, Walter, 131
Berman, Mitchell, 203, 343n2, 349
Bernstein, Peter, 280–283
binding, 74, 151
biological predisposition(s), 241–242
blame, 20, 152, 182, 203, 207, 211, 216, 244, 320–321, 336, 350–351
   and anger, 350
   backward-looking, 209
   forward-looking, 345
   and/or praise, 238, 240, 270, 349
   and punishment, 55, 266, 337–339, 345–346, 349
   See also punishment
blameworthiness, 200–201, 320, 347
Block, Ned, 169
Bohm, David, 301
Bok, Hilary, 194
Boltzmann, Ludwig, 302
brain
   activity(-ies), 45, 98–99, 153, 155–156, 179, 195, 228–229, 231, 258, 280–281, 283
   damage, 314–315
   disorder(s), 311, 314, 325
   event(s), 9, 17, 163, 189, 257n5
   imaging, 32, 96–97, 101, 155, 314, 319–320, 322, 324
   injury(-ies), 116–117, 227, 325–327
   process(es), 15–16, 100, 126, 151, 158, 195, 259, 280–281
   scan(ning), 20, 228, 256, 280, 283, 313–315, 319, 321–322, 324
   See also neural event(s); neuroscience
Breivik, Anders, 238–239
Bunge, Mario, 232
Buridan's ass, 184
bypassing intuition(s), 257–259

Camus, Albert, 1, 4, 90–91, 102–103, 293
Capgras syndrome, 116
Caruso, Gregg, 21, 56, 164n2, 206, 214, 235–236, 347, 348
causal chain(s), 164n2, 169, 196, 232
   informational, 170–173, 177, 179
   neural, 170
   physical, 170–174, 177
causal closure, 164–168
causal exclusion problem, 153, 156

causation
- criterial, 17, 162–163, 172–177, 179–188
- downward, 169, 174, 177, 299–231
- informational, 162–163, 168n3, 169, 171, 176
- interventionist conception/theory of, 17, 254, 260, 264
- mental (*see* mental causation)
- physical, 17, 162, 164, 168–169, 186
- upward, 230

chaos theory, 300

charity, 41–42, 48, 52n19, 120. *See also* love

Chisholm, Roderick, 7

choice(s), 7, 14–17, 80, 87, 111, 112–113, 118–120, 122–123, 132, 146–147, 157–158, 163, 178–182, 184, 187–189, 211, 213, 224, 226, 230–231, 233, 235–237, 240–246, 251–252, 255, 257n5, 258, 263, 266, 280, 282–283, 288, 298, 302, 305–306
- conscious, 16, 147, 196, 198
- free, 112, 180, 226, 235–237, 298, 306
- moral, 28, 186
- precocial, 26
- unfree, 263
- *See also* action(s); free will; moral responsibility

choice blindness, 120

cognitive behavioral therapy (CBT), 155–156

Cohen, Jonathan, 254, 256–258, 261, 263, 343–344

communication, 90, 133–134, 139
- NMDA channel of, 181

compatibilism, 8n4, 178, 196, 199, 340–342, 346. *See also* incompatibilism

compulsion(s), 150–151, 338–339

consciousness, 8–10, 12, 45, 74, 76, 91, 113–114, 117–119, 121–122, 130, 146–148, 150, 152, 157, 169, 179, 181, 185, 195–196, 224, 231, 252, 253n4, 255, 257, 263n10, 301, 304, 337
- deflationary view of, 198
- higher order syntactic thought (HOST) theory of, 78
- neural basis of, 181

consciousness-raising, 103–104

consequentialism, 205, 343–344, 350–351, 354–355

conservative(s), 62–64

constraint(s), 149–151, 154, 158, 172–173, 179, 184, 205, 226–227, 354
- cultural, 102
- neural, 147

cooperation, 33–34, 79, 118

Copernicus, Nicolaus, 6, 253

Corrado, Michael, 194, 207, 210, 213–214, 216

correction, 207

cortex, 32, 150
- anterior insular, 72
- cerebral, 150, 157
- cingulate, 72, 99, 135
- frontal, 3
- medial parietal, 139
- motor, 271
- orbitofrontal (orbital frontal), 72, 74, 76, 95
- prefrontal, 151, 155, 256–257
  - dorsomedial, 58, 139
  - ventromedial, 213
- pre-motor, 256
- visual, 76–77

criminality, 20, 92, 215, 311, 325, 355. *See also* desert; justice

Darwin, Charles, 1, 2, 4–6, 8

Dasein (human existence), 127–136
- an ontological dimension of, 130
- *See also* Heidegger

Dawkins, Richard, 118

death, 7, 112, 127, 130, 135, 140, 276n16

de Beauvoir, Simone, 1, 4, 91

de Broglie, Louis, 301

decision(s), 7, 19–20, 29, 35, 64, 72, 74, 96, 114–115, 121–122, 148, 153, 158, 164, 171, 174, 178, 182–184, 188, 195–197, 199, 212, 228–229, 232, 235, 239, 251–252, 251n1, 254–264, 280–283, 288, 304, 313–316, 320, 323, 344, 348, 353, 355
- future volitional, 182
- moral, 28, 96, 140

Index [361]

decision(s) (cont.)
   self-forming, 183, 188
   torn, 183–188
   See also action(s); choice(s)
decision-making, 25, 28, 117–119, 147–151, 185, 187, 213, 216, 233, 241, 254–260, 280–282, 336
   moral, 28, 139, 237–238, 240–241, 243–244, 246–247, 252
deep brain stimulation (DBS), 212–213, 348
default network (in the brain), 138–139
dehumanization, 98–99
Dennett, Daniel, 8, 226, 233, 238, 239, 253n3, 346
Descartes, René, 6–7, 259n7
determinism, 7, 17, 20–21, 163, 164n2, 166, 168–171, 174, 178–179, 182, 186, 198, 200–201, 224, 225–226, 232, 253n3, 254, 260n8, 263–265, 276, 283, 286, 300, 334, 338–340, 343n2, 344, 347, 349
   hard, 198–199, 217, 334, 342
   See also free will; indeterminism; moral responsibility
deterrence, 207, 325–326, 338, 346–347, 350
   theories, 204, 206
   See also punishment
de Waal, Frans, 26
dignity, 211, 340, 345, 346, 34, 349. See also agency; autonomy
disgust, 61–62, 64, 94, 99, 236
   sexual, 58
disposition, 133, 154. See also Heidegger
diversion courts, 325–326
Dostoevsky, Feodor, 1, 3
Doyle, John, 229–231
Dreyfus, Hubert, 131, 136n5
dualism, 9, 16, 255, 259n7, 276–277, 283, 286, 288
   free will and/or, 277–279, 283–284, 286, 290–291
   See also free will
dualist(ic), 9
   conception of mind and agency, 253
   intuitions, 253n4, 257, 260n8
   reaction, 252
   theory, 257

Eccles, Sir John, 224–225
Eigentlighkeit. See authenticity
eliminativism, 103
emergence, 20, 232, 303–306
emotion(s), 12, 14, 28, 30, 34–35, 41, 43, 68–71, 73–74, 77, 79, 83, 134, 150, 154–155, 200, 209, 244, 345, 348–351
   and moral judgments, 57, 59, 61–62, 95–97, 100
   self-conscious, 271
   See also disgust; fear; love; morality
emotional closure, 236, 240, 243–244
empathy, 62, 104, 158, 184
empirical methods, 87, 100–101, 104
Enlightenment, 3, 11
entropy, 302–303
eudaimonia, 2, 11. See also Aristotle
Everett, Hugh, 301
exclusion argument (EA), 17, 162–163
   causal, 262
   See also Kim
existentialism, 1–5, 14–15, 20, 87–88, 97, 102–103, 126, 136, 236, 298, 309
   hard-incompatibilist, 17
   realistic, 114, 123
   Sartre's, 112–113, 146
   and science, 20
   third-wave, 4–5
   See also Heidegger; neuroexistentialism; Sartre
extended mind hypothesis (EMH), 138

face blindness, (prosopagnosia), 116
Fanon, Frantz, 4, 92–93, 100–101, 103–104
fear, 5, 27, 34, 69–70, 98, 116, 199
   existential, 135
   as an indicator of danger, 62
   system, 154–155
   See also emotion(s)
Feinberg, Todd, 154
flourishing, 2, 11–12. See also eudaimonia
Fodor, Jerry, 344
folk psychology, 334–337
   dualistic, 253n4
for-itself, 113
4E (embodied, embedded, enactive, and extended), 126, 136–141
4M (mind, meaning, morals, and modality), 16, 126, 140

freedom, 90–92, 94, 102, 104, 113,
    130–131, 134, 233, 235, 237,
    239, 241, 246
  of action, 182, 184, 197
  compatibilist, 237, 240
  to choose, 118, 182–183
  existential/radical/absolute/ultimate,
    7, 14, 87, 98, 101, 114, 123, 129,
    301, 340
  feeling of, 237
  versus free will, 239–240
  human, 135, 163n1, 244, 301
  and human agency, 243, 302
  purpose and, 298
  See also action; choice; compatibilism;
    free will; moral responsibility
free will, 7, 22, 146, 162, 224, 235, 271
  and behavioral control, 148–152
  beliefs, 19, 271n8–n10, 272, 276, 279,
    287, 254
  death of, 19–20, 270, 272, 294
  and determinist challenge, 340–341
  and dualism, 276–291
  first-order/second-order, 17, 163
  versus freedom, 239–240
  libertarian, 7, 16, 149, 169, 171,
    178–189
  and moral responsibility, 18, 55,
    170–171, 186, 193, 238
  skepticism, 193
  weak/strong, 178, 181
  See also action; choice; compatabilism;
    determinism; freedom;
    incompatibilism; indeterminism;
    moral responsibility;
Freud, Sigmund, 6, 114
Frueh, Christopher, 208–209
Fuchs, Thomas, 153

Galileo, Galilei, 4–6, 253–254
Gall, Franz Joseph, 311
genotype, 74, 81–83. *See also* phenotype
genuine order effect(s), 60
Ghirardi, Giancarlo, 301
gratitude, 97, 201–202, 265
Gray, John, 312
Greene, Joshua, 64, 96, 254, 256–258,
    261, 263, 327, 343–344
guilt, 18, 79, 97, 201–202, 216, 236–239,
    243, 247, 262, 265, 351
  and desert, 339, 344

  moral, 240, 243–244, 247
Guiteau, Charles J., 312

Hanson, Jon, 320
hard incompatibilism (HI). *See*
    incompatibilism
hard problem, 2, 9–10, 304. *See also*
    really hard problem
Harris, Sam, 239, 258
Haynes, John Dylan, 193, 195, 228
Heidegger, Martin, 93, 126–136, 141–142.
    See also Dasein; Sartre
Henning, Kris, 208–209
high other-focus, 274–277, 284, 292. *See
    also* low self-focus
hippocampus, 76
Hobbes, Thomas, 83
Homo neanderthalis, 28–29
Homo sapiens, 5, 27, 29, 342
human existence, 15, 111, 126, 128, 134,
    136, 139, 355. *See also* Dasein;
    4E; Heidegger
humanistic image. *See* images of
    humanistic
Hume, David, 3, 13, 28, 178–179, 273
humility, 19, 51, 272–279, 283–286,
    289, 291–294
Husak, Douglas, 203
Husserl, Edmund, 88, 136n5

Iacoboni, Marco, 136–140
ideology, 94
illusionism, 271
images of persons, 1, 10
  humanistic, 1, 2, 5–6, 10
  scientific, 2, 5, 9
inauthenticity. *See* authenticity
incapacitation, 17–18, 194, 203–207,
    213–218, 347
incompatibilism, 178, 182, 264, 341, 355
  hard, 21, 198–199, 217, 334, 338,
    345, 355
  leeway, 198
  source, 198
  See also compatibilism
indeterminism, 17, 162, 164n2,
    165–171, 174, 177–179,
    182–184, 186, 188–189, 300
  quantum domain, 165
  See also determinism; quantum
    mechanics

Index [363]

individualism, 135
  methodological, 15, 126, 135–136, 136n6, 139
individuality, 15, 126, 130, 140
  of action, 128
individualization, 131
  primordial, 127, 132
information, 114–115, 118–119, 121–122, 133, 135, 147, 150–152, 154, 162–164, 169–174, 177–179, 181, 185–186, 189, 233, 242, 244, 255, 258–259, 274, 282, 300, 304, 306, 313, 317, 320
  conservation of, 299, 302, 305
  mental, 17, 76–77, 84, 162
  processing, 97
  systems, 229
informational criteria, 163, 169–174, 177, 179
informational parameters, 170, 174, 176, 178–180
informational reparameterization, 170–172, 174–177
intention, 94, 148–149, 153, 171, 184, 200, 262, 264, 301, 323, 334–336, 343
  conscious, 149, 195–197, 228
  distal, 149, 157
  proximal, 149, 196
intentionality, 138, 342
interpersonal relationship(s). *See* personal relationship(s)
interpreter, 18, 223, 227–228, 231
intersubjectivity, 127
  phenomenological conceptions of, 16, 126
  primary, 139, 141
  secondary, 139
intervention(s), 171, 175–176, 209, 242, 260–262, 346–348, 351–355
  behavioral and/or neurobiological, 241
interventionism, 175, 262
  casual, 264n11

justice, 35, 83, 313, 326, 339, 350
  criminal, 19, 20, 206–207, 213, 237, 244–245, 311, 313–314, 325–327, 339, 342, 344, 347, 351, 355
  and gender, 4
  and injustice(s), 102, 104, 26, 237

  private, 352
  racial, 4
  restorative, 209–210, 243, 244, 246–247
  without retribution, 236
  sense of, 20
  social, 101, 206, 207
  therapeutic, 326
  See also criminality; retributivism

Kane, Robert, 182–188
Kierkegaard, Søren, 1, 3, 132, 193
Kim, Jaegwon, 17, 162–165, 167–169, 171, 186
kindness, 13, 21, 41–42, 48–49, 51–52, 52n19, 99, 202. *See also* love
Kleiman, Mark, 217

Laplace, Pierre-Simon, 299. *See also* determinism
law(s), 12, 21, 28, 35, 80, 82–83, 95, 165, 190, 214, 225, 302, 304, 327, 334–338, 342, 345, 348
  causal, 340
  civil, 21, 333–334
  common, 21, 334
  criminal, 21, 333–334, 337–338, 341, 342
  macroscopic/microscopic, 302–306
  of nature (natural), 19–20, 198, 252, 254, 264, 299–300
  quantum mechanical, 165–166
  rights and, 80
  of society, 79
law-giver(s), 27
law's (folk) psychology, 334–337
Lemos, John, 18, 194, 203, 215
Levy, Neil, 167–168, 168n4, 186–187, 196–197, 200, 217
libertarianism, 171, 196, 199. *See also* determinism; free will; indeterminism
Libet, Benjamin, 17, 134, 148–149, 163, 193, 195–197, 228, 231, 343. *See* choice; free will
love, 12–13, 31, 38
  affectionate, 44, 50–51
  Christian, 41
  as kindness, 13, 41–42, 48–49, 51–52
  and morality, 13, 38–46, 51

romantic, 40–42, 45–46, 48–49
  See also kindness; morality
low self-focus, 273–277, 284, 292. *See also* high other-focus
luck, 163, 186–188, 340
Ludovico method, 208
Luria, Alexander, 157

MacKay, Donald, 224–225
Madoff, Bernard, 350
mammalian reward system, 34
manipulation argument, 264
Marxism, 102–103
Marx, Werner, 141
materialism, 153–154, 156
McKenna, Michael, 235
McVeigh, Timothy, 350–351
meaning, 2–12, 14, 16–17, 20, 68, 76–78, 84, 90, 112–113, 126, 137, 139–140, 146–147, 157, 159, 193–194, 298, 305–306, 333, 340, 342, 344–345, 355. *See also* freedom; morality; purpose
Mele, Alfred, 149, 196
mental act(ivities), 100, 178, 258
mental capacities, 158–159
mental causation, 17, 56, 162–169, 171, 173, 180, 186, 263n10
mental disorder(s), 237, 239, 245, 311, 325–326, 348, 350–351, 354
mental function(s), 16, 147, 150, 158, 162–164, 177
mental health, 206, 245–246, 325, 355
mental information, 17
mental layer(s), 230
mental operations, 140
mental process(es), 4, 19, 148–150, 152–154, 197, 230, 252, 258, 259
mental properties, 153–154, 156
mental representation(s), 15, 126, 138
mental state(s), 8, 16, 96, 120, 146–147, 149–151, 153–156, 158, 165–166, 169, 212, 229–233, 258, 260n8, 334–337, 341–343
Merleau-Ponty, Maurice, 16, 88–89, 91, 93, 97, 100–101, 103, 126, 136–138
mind-brain
  interaction, 156, 158
  interface, 231–233

layer(s), 230
relation(ships), 5, 16, 146–147, 153–154, 230, 279
system, 232
Mitbefindlichkeit (emotional contagion or affective resonance), 133. *See also* Heidegger
mitigation, 237, 315–316, 319, 324, 327
Mitsein (being-with; being with others), 127. *See also* Heidegger
Module(s), 115–119, 121–123, 227–229
moral education theory(-ies), 204, 206
morality, 6–9, 12–13, 17, 26, 89
  love and, 13, 38–46, 51
  neuroscience and, 55–57, 64–65
  platform for (of), 31, 39
  and responsibility, 193, 200
moral judgment(s), 6–9, 12–13, 17, 26. *See also* reliability/unreliability
moral motivation, 25, 27–28
moral rationalism, 95
moral responsibility, 8, 12, 18, 55–56, 168n4, 170–171, 186–188, 193–194, 198–201, 207, 210, 214, 216–217, 232, 235–238, 240, 242–243, 247, 259, 265, 334, 339. *See also* basic desert; free will; luck
moral sedimentation, 14, 88–95, 97, 99–104. *See also* sedimentation
Morris, Hubert, 208
Morse, Stephen, 20–21, 215, 325
Morsella, Ezequiel, 147

Nahmias, Eddy, 4n3, 19, 196–197, 257n15, 258, 265
naturalism, 11, 254
naturalistic fallacy, 80
natural selection, 6, 29, 69, 73, 79, 227
neocortex, 30
neo-Darwinism, 4, 11
neural activity(-ies), 19, 148, 174, 178–179, 254–256, 257n5, 259, 261, 264
neural base(s) (or basis), 14, 68, 117, 162, 170, 171, 174, 179, 181
neural causal chains, 170, 173
neural circuit(s), 99, 101, 154–155, 157, 179, 181–182
neural code, 17, 162–164, 170–171, 174, 177, 180

neural event(s), 149, 157, 159, 179–180, 198, 228
neural firing(s), 173–174, 177, 179
neural form, 262
neural function(s), 16, 147, 150, 155–156, 158
neural mechanism(s), 16, 147, 263n10, 281, 342
neural network(s), 146, 150, 154–155, 227
neural pathways, 148, 152, 154, 262
neural process(es), 4, 19, 58, 126, 137, 147, 149, 153, 156, 171, 252, 254, 258, 261, 262
neural representation(s), 14, 84, 139
neural spike(s), 170, 177, 186
neural state(s), 263n10, 281
neural system(s), 56, 117
neural variable(s), 262, 263n10
neuro-angst, 292, 294. *See also* angst; anxiety
neurobiological process(es), 148, 256
neurobiological underpinning, 78, 146, 158
neurobiology, 4, 27, 35, 69, 159, 282–283
neuroexistentialism, 1–2, 5, 8, 10–12, 18, 270n1. *See also* existentialism; naturalism
neurofeedback (NFB), 155–156, 212–213
neuron(s), 29–30, 32–33, 76–77, 138, 146, 151, 165, 170, 172–177, 179–180, 236, 256–257, 282, 287, 334, 341–342, 345. *See also* VNC/PON
neuronaturalism, 4, 4n3, 19, 252–258, 260–261, 263n10, 265
neuroscience, 4–6, 8, 10–12, 13–14, 16–17, 19–20, 35, 83, 87, 95, 99–101, 104, 116, 126, 135, 137–138, 140, 146–148, 152, 159, 162, 163, 193, 195, 197, 225, 233, 254–255, 259, 270, 279, 281, 282, 290, 298, 313, 325–327, 333–334, 338, 340–342, 347, 355
    affective, 39, 41, 44n12, 48
    existential, 134, 136, 136n7, 140–141
    moral, 96
    and morality, 54–59, 63–65

social, 96–97, 136
Newton, Sir Issac, 299–300
Nichols, Shaun, 202
Nietzsche, Friedrich, 1, 3, 14, 87, 90–91, 97, 101–102, 132, 273, 293

objectification, 100–101, 130
obsession(s), 151
obsessive-compulsive disorder (OCD), 151, 213
Other, 129–130, 132–133. *See also* Sartre
oxytocin, 12–13, 31–33, 40–42, 51–52

Papineau, David, 165, 168
parameter(s), 170, 174–176, 178, 189
    informational, 170–172, 174, 176, 178–180, 189
    physical, 176
Pattee, Howard, 230–231
Pearl, Judea, 162, 176
Pereboom, Derk, 21, 56, 186, 207–208, 347–348
personal relationship(s), 17, 193–194, 199–202
phenotype, 14, 75, 78, 81–84, 230. *See also* genotype
physicalism, 4n3, 9, 171, 252n2
Pickard, Hannah, 244
possibilia, 165–170. *See also* actualia
praiseworthiness, 201. *See also* blame; blameworthiness
preproximal-intention group (PPG), 196
primary otherness, 141
principle of least infringement, 206
prosopagnosia. *See* face blindness
psychopath, 213, 239, 350–351
public health ethics, 206
punisher(s), 69, 71–72, 79–80
    reward(s) and/or, 69–73
    See also reward(s)
punishment, 17, 83, 90, 210, 215–217, 239–240, 242, 313, 324, 326, 333, 344, 346–347
    blame and, 55, 266, 337–339, 345–346, 349, 351
    capital, 96
    criminal, 12, 203–204, 207
    desert and, 265
    divine, 27
    just, 20, 313, 326

justification(s) for, 203–204, 236, 325, 334, 349
  non-desert-based approach to, 245
  retributive, 18, 203, 236–237, 240, 243, 246–247, 265, 326
  reward and, 70–74, 77
  See also basic desert; blame; deterrence theories; moral responsibility; reward(s); retributivism
purpose, 83, 305, 333
  for actions, 73
  biological sense of, 68–69, 71
  cosmic, 300
  and freedom, 20, 298
  human, 2
  in human lives, 299
  meaning and, 6, 9, 11–12, 14, 20, 68, 76, 112, 298, 305–306, 340–342, 345, 355
  See also meaning

qualia, 163, 180, 304
quantum domain randomness, 170–171. *See also* indeterminism
quantum mechanics, 20, 165, 168, 180, 198, 298–302
quantum physics, 169
quarantine, 18, 21, 194, 203, 205–206, 214, 346, 350
Quirk, Douglas, 212

racism, 93, 98, 102–104, 207. *See also* sexism
randomness, 17, 163, 170–171, 178–180, 186, 188, 240. *See also* determinism; indeterminism; quantum mechanics
rationalism, 95–96. *See also* sentimentalism
rationality, 3, 29, 43n11, 335–336, 338, 348
Ray, Issac, 311–312
reactive attitude(s), 201, 265, 345. *See also* Strawson
readiness potential(s) (RP), 148–149, 163, 195, 228, 263n10, 343
really hard problem, 2, 9–10. *See also* hard problem
reasons-responsiveness, 198, 207, 209, 211–212, 347

rehabilitation, 17, 193–194, 203, 205, 211, 213, 216, 218, 240–241, 244–247, 325
reinforcement, 70, 75, 78, 138
  contingency, 70, 79
reinforce(s), 69–72, 74, 77–79
reliability/unreliability
  of moral judgments, 59–63
reliability requirements, 314
representation(s), 14, 15, 71, 121–122, 137, 180–181
  content and meaning in, 76–78, 84
  mental, 126, 138
  See also neural representation(s)
reproduction, 51–52, 68, 75, 81
res cogitans, 7–8
respect, 201, 336, 348, 353
response
  affective, 48, 116, 133, 156
  emotional, 79, 95, 97, 116, 155–156, 202, 347
  narrow-profile, 202
  wide-profile, 202
responsibility, 18, 21, 90, 132, 141, 146, 152, 163, 181–183, 200, 208, 211, 223, 225–226, 232–233, 240, 244, 255, 266, 270, 279, 302, 315, 322, 326–327, 334, 342, 345–349, 353, 355
  assignment of, 321, 326
  causal, 246
  civic, 275
  criminal, 21, 334, 337, 341
  and desert, 333, 339, 340, 355
  legal, 240, 242, 255, 258, 265, 270, 336
  moral (*see* moral responsibility)
  personal, 232–233
  take-charge, 237–238, 241, 243–244, 246
  ultimate, 183, 187–188, 340
  See also basic desert; desert; moral responsibility
retributivism, 193, 203–204, 207, 347. *See also* punishment
Reuter, Shane, 258
reward(s), 34, 69, 74–75, 77, 79–81, 117–118, 151, 180, 243, 311–313, 336
  and punisher(s), 69–73
  See also punisher(s); punishment

Ricoeur, Paul, 141
Rimini, Alberto, 301
role model(s), 133, 141

Sartre, Jean Paul, 1, 4, 7, 88, 90–93, 102–103, 112–114, 118–119, 121, 123, 127, 129–130, 132n3, 134, 134n4, 136, 136n5, 140, 251n1, 293, 301. *See also* angst; existentialism
Scanlon, Thomas, 194
Schoeman, Ferdinand, 214–215
second-order belief, 62–64
sedimentation, 14–15, 87. *See also* moral sedimentation
self-causation, 17, 163, 177–178, 180. *See also* causation
self-consciousness, 224. *See also* consciousness
self-defense, 204–205, 207, 216, 337, 346, 350–354
self-forming act(s) (action(s)), 163, 183, 187
self-forming decision(s), 183, 188
self-forming resolution(s), 185, 187–188
Sellars, Wilfred, 5
sentimentalism, 95–98. *See also* rationalism
sexism, 120, 107
   benevolent, 99
   and racism, 102, 104
   See also racism
shared activities, 127, 131–132, 134
Shepard, Jason, 258
situationism, 197–198
Smilansky, Saul, 18, 216–217, 353
social contract(s), 14, 81–84, 116
social environment, 16, 126, 137–138, 157–158
Spence, Sean, 148, 150, 154, 156–157
Sperry, Roger, 223–226, 229, 232
Spitzka, Edward Charles, 312
split-brain, 223–224, 227–228
Spurzheim, Johann Gaspar, 311
storge. *See* affection; *agape*; love
Strawson, P. F., 201, 253n3, 345
subjectivity, 9–10, 16, 146–147
syntactic linking. *See* binding
System1/System, 2, 147–8

Taylor, Charles, 3
terror management theory, 135
thalamus, 72, 150–151
tit-for-tat, 80–81
totalitarianism, 93–94, 103. *See also* Arendt
transparent bottleneck, 254–255
traumatic brain injury (TBI), 325

Uneigentlichkeit (inauthenticity), 126. *See also Eigentlighkeit* (authenticity)
unreliability. *See* reliability
utilitarianism, 205. *See also* consequentialism

van Inwagen, Peter, 187
vasopressin, 31–32, 40–41, 51–52
violence, 93, 208, 212, 215, 313, 322–324
VNC/PON, 21, 334, 338–339, 341
Vorausspringen ("jumping ahead"), 132

Waller, Bruce, 202, 237–238, 243
Weber, Tullio, 301
Wegner, Daniel, 193
Woodward, John, 162, 176, 262
working memory, 58, 96, 172, 180–181, 189

Zazetsky, Lyova, 157–158
Zeman, Adam, 158